APPLICATIONS OF COMPUTER ALGEBRA

APPLICATIONS OF COMPUTER ALGEBRA

edited by

Richard Pavelle
Symbolics, Inc.

KLUWER ACADEMIC PUBLISHERS
Boston/Dordrecht/Lancaster

Distributors for North America:
Kluwer Academic Publishers
190 Old Derby Street
Hingham, MA 02043, U.S.A.

Distributors for all other countries:
Kluwer Academic Publishers Group
Distribution Centre
P.O. Box 322
3300 AH Dordrecht
The Netherlands

Consulting Editor: Jonathan Allen

Library of Congress Cataloging in Publication Data
Main entry under title:

Applications of computer algebra.

Papers based on the proceedings of a symposium on
symbolic algebraic manipulation in scientific computation
presented by the ACS Division of Computers in Chemistry
at the 188th meeting of the American Chemical Society, Aug. 26–31, 1984.
Includes index.
1. Algebra—Data processing—Congresses.   I. Pavelle,
Richard.   II. American Chemical Society.   Division of Computers in Chemistry.
QA155.7.E4A67   1985   512'.0028'5   85-9760

ISBN-13: 978-1-4684-6890-8   e-ISBN-13: 978-1-4684-6888-5
DOI: 10.1007/978-1-4684-6888-5

Second printing, 1986

CONTENTS

v

CONTRIBUTORS

Professor Carl M. Bender
Department of Physics
Washington University
St. Louis, MO 63130

Dr. John T. Bendler
General Electric Corporate Research
and Development
Schenectady, NY 12301

Professor R. Stephen Berry
Department of Chemistry
The James Franck Institute
5735 South Ellis Avenue
University of Chicago
Chicago, IL 60637

Dr. Robert H. Berman
Research Laboratory of Electronics
Massachusetts Institute of Technology
Cambridge, MA 02139

Dr. Walter Bloss
GTE Laboratories, Inc.
40 Sylvan Road
Waltham, MA 02154

Richard L. Brenner
Symbolics, Inc.
11 Cambridge Center
Cambridge, MA 02142

Dr. Gene Cooperman
GTE Laboratories, Inc.
Fundamental Research Laboratory
40 Sylvan Road
Waltham, MA 02154

Dr. William J. Frawley
GTE Laboratories, Inc.
Fundamental Research Laboratory
40 Sylvan Road
Waltham, MA 02154

Dr. Lionel Friedman
GTE Laboratories, Inc.
Fundamental Research Laboratory
40 Sylvan Road
Waltham, MA 02154

Mohammad Golnaraghi
Theoretical and Applied Mechanics
Cornell University
Ithaca, NY 14853

C. Gomez
INRIA
Domaine de Voluceau
Roquencourt B.P. 105
78153 LE CHESNAY Cedex
FRANCE

Professor W. H. Hui
Department of Applied Mathematics
University of Waterloo
Waterloo, Ontario, N2L 3G1
CANADA

Dr. M. A. Hussain
General Electric Corporate Research
and Development
Schenectady, NY 12301

Dr. W. L. Keith
Theoretical and Applied Mechanics
Cornell University
Ithaca, NY 14853

Professor Jeffrey L. Krause
Department of Chemistry
The James Franck Institute
5735 South Ellis Avenue
University of Chicago
Chicago, IL 60637

Professor Donald R. McLaughlin
Department of Chemistry
University of New Mexico
Albuquerque, NM 87131

Professor F. C. Moon
Theoretical and Applied Mechanics
Cornell University
Ithaca, NY 14853

Professor B. Noble
Mathematics Research Center
University of Wisconsin
610 Walnut Street
Madison, Wisconsin 53705

Dr. Andrew M. Odlyzko
AT&T Bell Laboratories
Room 2C-370
Murray Hill, NJ 07974

Dr. Richard Pavelle
Symbolics, Inc.
11 Cambridge Center
Cambridge, MA 02142

J. P. Quadrat
INRIA
Domaine de Voluceau
Roquencourt B.P. 105
78153 LE CHESNAY Cedex
FRANCE

Professor Thomas E. Raidy
Department of Chemistry
University of South Carolina
Columbia, SC 29208

Professor R. H. Rand
Theoretical and Applied Mechanics
Cornell University
Ithaca, NY 14853

Dr. Patrick Roache
Ecodynamics Research Associates, Inc.
Post Office Box 8172
Albuquerque, NM 87198

Dr. Michael F. Shlesinger
Physics Division
Office of Naval Research
Arlington, VA 22217

Professor Stanly Steinberg
Department of Mathematics and Statistics
University of New Mexico
Albuquerque, NM 87131

A. Sulem
INRIA
Domaine de Voluceau
Roquencourt B.P. 105
78153 LE CHESNAY Cedex
FRANCE

Professor G. Tenti
Department of Applied Mathematics
University of Waterloo
Waterloo, Ontario, N2L 3G1
CANADA

Professor Carl Trindle
University of Virginia
Department of Chemistry
Charlottesville, VA 22901

Professor Paul S. Wang
Department of Mathematical Sciences
Kent State University
Kent, OH 44242

Professor Stanley J. Watowich
Department of Chemistry
The James Franck Institute
5735 South Ellis Avenue
University of Chicago
Chicago, IL 60637

PREFACE

Today, certain computer software systems exist which surpass the computational ability of researchers when their mathematical techniques are applied to many areas of science and engineering. These computer systems can perform a large portion of the calculations seen in mathematical analysis. Despite this massive power, thousands of people use these systems as a routine resource for everyday calculations. These software programs are commonly called "Computer Algebra" systems. They have names such as MACSYMA, MAPLE, muMATH, REDUCE and SMP. They are receiving credit as a computational aid with increasing regularity in articles in the scientific and engineering literature.

When most people think about computers and scientific research these days, they imagine a machine grinding away, processing numbers arithmetically. It is not generally realized that, for a number of years, computers have been performing non-numeric computations. This means, for example, that one inputs an equation and obtains a closed form analytic answer. It is these Computer Algebra systems, their capabilities, and applications which are the subject of the papers in this volume.

On August 26–31, 1984, the American Chemical Society held their 188th national meeting in Philadelphia. On August 26–27, 1984, the ACS Division of Computers in Chemistry held a symposium on Symbolic Algebraic Manipulation in Scientific Computation. This was the first symposium ever organized on applications of Computer Algebra. The symposium was broken into four sessions. The first session gave an introduction to Computer Algebra and explained the uses of these systems as opposed to numeric computation systems. Also included were discussions of the interface between algebraic and numeric systems as well as the application of algebraic systems to the field of mathematics. The second session mainly dealt with applications of Computer Algebra to the field of chemistry and chemical education. The chemistry community is the most recent major scientific group to discover the benefits of using Computer Algebra systems. The third session dealt with the engineering applications of Computer Algebra systems to problems in spectral analysis, robotics, finite element methods, and optimal control. The fourth session dealt primarily with applications of Computer Algebra to computations in physics and applied mathematics.

This volume provides a broad introduction to the capabilities of Computer Algebra systems and gives many examples of applications to real problems in engineering and the sciences. It is my hope that this information will create new users of Computer Algebra systems by showing what one might expect to gain by using them and what one will lose by not using them.

Richard Pavelle
Cambridge, MA

1

MACSYMA:
CAPABILITIES AND APPLICATIONS TO
PROBLEMS IN ENGINEERING
AND THE SCIENCES

Abstract

MACSYMA™ is a large, interactive computer system designed to assist engineers, scientists, and mathematicians in solving mathematical problems. A user supplies symbolic inputs and MACSYMA yields symbolic, numeric or graphic results. This paper provides an introduction to MACSYMA and provides the motivation for using the system. Many examples are given of MACSYMA's capabilities with actual computer input and output. Also presented are several applications where MACSYMA has been employed to deal with problems in engineering and the sciences.

Richard Pavelle
Symbolics, Inc.
MACSYMA Group
11 Cambridge Center
Cambridge, MA 02142

1

1. INTRODUCTION

1.1 MACSYMA: A Personal Perspective

My purpose in this paper is to provide a broad introduction to the capabilities of MACSYMA and to give some examples of its applications to real problems in engineering and the sciences. It is my hope that this information will create new users of Computer Algebra systems by showing what one might expect to gain by using them and what one will lose by not using them.

MACSYMA output is used and CPU times are often given. In some cases I have modified the output slightly to make it more presentable. The CPU times correspond to a Symbolics 3600 and to the MACSYMA Consortium machine (MIT-MC) which is a Digital Equipment KL10. These are about equal in speed and about twice as fast as a Digital Equipment VAX 11/780 for MACSYMA computations. When CPU times are not given one may assume the calculation requires less than 10 CPU seconds.

I have been involved with the MACSYMA project since 1973. Between 1972 and 1973, as a postdoctoral research fellow at the University of Waterloo, I worked on a calculation (by hand of course) that took me about 3 months to complete. I did not submit it for publication because I was not sure the result was correct. I had been working in General Relativity and tensor analysis for several years and massive hand calculations were the normal means of getting a result. Indeed, even today, many mistakenly believe this is the only way to do analysis. In 1973 I joined a small firm outside of Boston to work on a DARPA contract on gravitation. The terms of the contract required us to build component and indicial tensorial capabilities into MACSYMA. At that time MACSYMA was in its early development, and I remember my first encounter with it vividly. I had heard about MACSYMA's algebraic capabilities but, not trusting computers, I thought it was trickery. I recall my amazement when, at that first encounter, one of the MIT hackers had MACSYMA compute the components of the Einstein tensor for a spherically symmetric metric in about 15 minutes of real time. It had a sign error that we later corrected, but it was fantastic to see a machine perform a complicated algebraic calculation that most experts would require several hours to accomplish. I was sold and since then I have been promoting MACSYMA. I also entered into an agreement with myself to never again spend more than 15 minutes on a hand calculation. This agreement is still in effect. Some time later I repeated my earlier hand calculation on MACSYMA, verified my computation in about 2 minutes, and published the paper [Pa1].

Computer Algebra systems can actually boost a person's ability to perform analysis and get useful or publishable results. I shall illustrate this with a personal experience. In 1974 Yang [Ya1] proposed a new gauge theory of gravitation that was meant to incorporate Einstein's General Relativity as a special case. The paper was published in Physical Review Letters, one of the premier journals in the sciences. Yang did not give any solutions to his equations and this seemed to me like a perfect opportunity to test MACSYMA. I knew that Yang's equations were identical, under certain conditions, to others proposed by Kilmister and Newman several years before and that they had studied them carefully [Ki1]. Nevertheless, even if Yang had some solutions and Kilmister and Newman had others, I believed that with MACSYMA I could quickly discover something that might have been missed. After two days of analysis with MACSYMA, I did find solutions that enabled me to criticize the theory on physical grounds [Pa2]. As a consequence I was fortunate to publish three papers in Physical Review Letters about Yang's theory [Pa3,Pa4], and I believe the criticisms I offered are generally accepted to be valid. One of the papers [Pa4] contains an exact solution of a partial differential equation of 4th degree in two variables containing 404 terms. This is the largest and most complex differential equation I have seen for which exact solutions have been found. Does the Guinness Book of World Records accept solutions of PDEs?

I have not used other Computer Algebra systems extensively because MACSYMA has provided all the tools I have needed. I shall try to give the reader ideas about the capabilities of Computer Algebra systems as a whole. However, one should bear in mind that not all of the capabilities of MACSYMA that I present reside on other systems.

Of the general-purpose systems [Pa5,va1,Yu1], MACSYMA offers the most features. It has been licensed to about 400 sites worldwide since 1982. REDUCE is the most widely distributed system. It has been licensed, over the last 20 years, to about 1000 sites worldwide, and its clones (a polite term for unlicensed programs) may be on several thousand additional machines. I do not know about the distribution of the other general systems. However, MAPLE (inexpensive) and SMP (very expensive) will likely have a distribution inversely proportional to their cost. muMATH has had a very large distribution; several thousand copies have been sold to the micro market. The distribution of the special-purpose systems, those that have a very restricted range of capabilities, has been limited.

Computer Algebra systems are very far from saturating the user base. I estimate that perhaps only 20% of the people who need these systems are even aware of their existence and less than 1/4 of these actually use them. I gave a MACSYMA talk at MIT about three years to two engineering departments. Only half of the 50 attendees had even heard of MACSYMA. It is clear that the field of Computer Algebra needs a good deal of public relations.

1.2 What is MACSYMA

The development of MACSYMA began at MIT in the late 60s, and its history has been described elsewhere [Mo1]. A few facts worth repeating are that a great deal of effort and expense went into MACSYMA. There are estimates that 100 man-years of developing and debugging have gone into the program. While this is a large number, let us consider the even larger number of man-years using and testing MACSYMA. At MIT, between 1972 and 1982, we had about 1000 MACSYMA users. If we had 50 serious users using MACSYMA for 50% of their time, 250 casual users at 10% and 700 infrequent users at 2% then the total is over 600 man-years. MACSYMA has been at 50 sites for 4 years and at 400 sites today. Well, we can conclude that at least 1000 man-years have been spent in using MACSYMA. MACSYMA is now very large and consists of about 3000 lisp subroutines or about 300,000 lines of compiled lisp code joined together in one giant package for performing symbolic mathematics.

These huge numbers show why it is difficult to construct new general purpose Computer Algebra systems. A new system must be very large if it is to have many capabilities. Given the technology of the 1980's, it will still require a substantial fraction of the time it took to build MACSYMA, a minimum of 1/5th say, to construct, test and debug a system with MACSYMA's capabilities. One would be left with a system perhaps 1/4 to 1/2 the size. Perhaps the most reasonable approach is to build upon the current knowledge base rather than starting a new system from the ground up.

While this paper is directed towards MACSYMA, the development of MACSYMA and other Computer Algebra systems has really been the result of an international effort. There are many systems, world-wide, of various sizes and designs which have been developed over the past fifteen to twenty years [val]. Research related to the development of these systems has lead to many new results in mathematics and the construction of new algorithms. These results in turn helped the development of MACSYMA as well as other systems.

1.3 Why MACSYMA is Useful or Necessary

Here are some of the more important reasons for using MACSYMA:

1. The answers one obtains are exact and can often be checked by independent procedures. For example, one can compute an indefinite integral and check the answer by differentiating; the differentiation algorithm is independent of the integration algorithm. Since exact answers are given, the statistical error analysis associated with numerical computation is unnecessary. One obtains answers that are reliable to a high level of confidence.

2. The user can generate FORTRAN expressions that allow numeric computers to run faster and more efficiently. This saves CPU cycles and makes computing more economical. The user can generate FORTRAN expressions from MACSYMA expressions. The FORTRAN capability is an extremely important feature combining symbolic and numeric capabilities. The trend is clear, and in a few years we will have powerful, inexpensive desktop or notebook computers that merge the symbolic, numeric and graphic capabilities in a scientific workstation.

3. The user can explore extremely complex problems that cannot be solved in any other manner. This capability is often thought of as the major use of Computer Algebra systems. However, one should not lose sight of the fact that MACSYMA is more often used as an advanced calculator to perform everyday symbolic and numeric problems. It also complements conventional tools such as reference tables or numeric processors.

4. A great deal of knowledge has gone into the MACSYMA knowledge base. Therefore the user has access to mathematical techniques that are not available from any other resources, and the user can solve problems even though he may not know or understand the techniques that the system uses to arrive at an answer.

5. A user can test mathematical conjectures easily and painlessly. One frequently encounters mathematical results in the literature and questions their validity. Often MACSYMA can be used to check these results using algebraic or numeric techniques or a combination of these. Similarly one can use the system to show that some problems do not have a solution.

6. MACSYMA is easy to use. Individuals without prior computing experience can learn to solve fairly difficult problems with MACSYMA in a few hours or less. While MACSYMA is written in a dialect of LISP, the user need never see this base language. MACSYMA itself is a full programming language, almost mathematical in nature, whose syntax resembles versions of ALGOL.

There are two additional reasons for using MACSYMA that are more important than the others.

7. One can concentrate on the intellectual content of a problem leaving computational details to the computer. This often results in accidental discoveries and, owing to power of the program, these occur at a far greater rate than when calculations are done by hand.

8. But the most important reason is that, to quote R. W. Hamming, "The purpose of computing is insight, not numbers." This exemplifies the major benefit of using MACSYMA, and I will demonstrate the validity of this statement by showing not only how one gains insight but also how one uses MACSYMA for theory building. However, a second quotation reputed to be by Hamming is correct as well, namely that "The purpose of computing is not yet in sight."

In the chapter on applications I shall illustrate each of these points with specific examples.

2. Capabilities and Uses of MACSYMA

2.1 Capabilities of MACSYMA

It is not possible to fully indicate the capabilities of MACSYMA in a few lines since the reference manual itself occupies more than 500 pages [MA1]. However, some of the more important capabilities include (in addition to the basic arithmetical operations) facilities to provide analytical tools for

LIMITS	TAYLOR SERIES (SEVERAL VARIABLES)
DERIVATIVES	POISSON SERIES
INDEFINITE INTEGRATION	LAPLACE TRANSFORMATIONS
DEFINITE INTEGRATION	INDEFINITE SUMMATION
ORDINARY DIFFERENTIAL EQUATIONS	MATRIX MANIPULATION
SYSTEMS OF EQUATIONS (NON-LINEAR)	VECTOR MANIPULATION
SIMPLIFICATION	TENSOR MANIPULATION
FACTORIZATION	FORTRAN GENERATION

There are other routines for calculations in number theory, combinatorics, continued fractions, set theory and complex arithmetic. There is also a share library currently containing about 80 subroutines. Some of these perform computations such as asymptotic analysis and optimization while others manipulate many of the higher transcendental functions. In addition one can evaluate expressions numerically, to arbitrary precision, at most stages of a computation. MACSYMA also provides extensive graphic capabilities to the user.

To put the capabilities of MACSYMA in perspective we could say that MACSYMA knows a large percentage of the mathematical techniques used in engineering and the sciences. I do not mean to imply that MACSYMA can do everything. It is easy to come up with examples that MACSYMA cannot handle, and I will present some of these. Perhaps the following quotation will add the necessary balance. It is an exit message from some MIT computers that often flashes on our screens when logging out. It states: "I am a computer. I am dumber than any human and smarter than any administrator." MACSYMA is remarkable in both the questions it can and cannot answer. It will be many years before it evolves into a system that rivals the human in more than a few areas. But until then, it is the most useful tool that any engineer or scientist can have at his disposal.

2.2 Uses of MACSYMA

It is difficult to list the application fields of MACSYMA because users often do not describe the tools that helped them perform their research. However, from Computer Algebra conferences [MUC1,MUC2,MUC3] and publications in the open literature we do know that MACSYMA has been used in the following fields:

ACOUSTICS	FLUID DYNAMICS
ALGEBRAIC GEOMETRY	GENERAL RELATIVITY
ANTENNA THEORY	NUMBER THEORY
CELESTIAL MECHANICS	NUMERICAL ANALYSIS
COMPUTER-AIDED DESIGN	PARTICLE PHYSICS
CONTROL THEORY	PLASMA PHYSICS
DEFORMATION ANALYSIS	SOLID-STATE PHYSICS
ECONOMETRICS	STRUCTURAL MECHANICS
EXPERIMENTAL MATHEMATICS	THERMODYNAMICS

Researchers have reported using MACSYMA to explore problems in:

AIRFOIL DESIGN	MAXIMUM LIKELIHOOD ESTIMATION
ATOMIC SCATTERING CROSS SECTIONS	NUCLEAR MAGNETIC RESONANCE
BALLISTIC MISSILE DEFENSE SYSTEMS	OPTIMAL CONTROL THEORY
DECISION ANALYSIS IN MEDICINE	POLYMER MODELING
ELECTRON MICROSCOPE DESIGN	PROPELLER DESIGN
EMULSION CHEMISTRY	RESOLVING CLOSELY SPACED OPTICAL TARGETS
FINITE ELEMENT ANALYSIS	ROBOTICS
GENETIC STUDIES OF FAMILY RESEMBLANCE	SHIP HULL DESIGN
HELICOPTER BLADE MOTION	SPECTRAL ANALYSIS
LARGE SCALE INTEGRATED CIRCUIT DESIGN	UNDERWATER SHOCK WAVES

3. Examples of MACSYMA

3.1 Polynomial Equations

Here is an elementary example that demonstrates the ability of MACSYMA to solve equations. In MACSYMA, as with most systems, one has user input lines and computer output lines. Below, in the input line (C1), we have written an expression in an ALGOL like syntax, terminated it with a semi-colon, and in (D1) the computer displays the expression in a two dimensional format in a form similar to hand notation. Terminating an input string with $ inhibits the display of the D lines.

```
(C1) X^3+B*X^2+A^2*X^2-9*A*X^2+A^2*B*X-2*A*B*X-

        9*A^3*X+14*A^2*X-2*A^3*B+14*A^4=0;


          3     2    2 2      2     2                    3
(D1) X  + B X  + A  X  - 9 A X  + A  B X - 2 A B X - 9 A  X

                            2       3        4
                      + 14 A  X - 2 A  B + 14 A  = 0
```

In (C2) we now ask MACSYMA to solve the expression (D1) for X and the three roots appear in a list in (D2).

```
(C2) SOLVE(D1,X);

                                 2
(D2)              [X = 7 A - B,   X = - A ,   X = 2 A]
```

Notice that MACSYMA has obtained the roots analytically and that numeric approximations have not been made. This demonstrates a fundamental difference between a Computer Algebra system and an ordinary numeric equation solver, namely the ability to obtain a solution without approximations. I could have given MACSYMA a "numeric" cubic equation in X by specifying numeric values for A and B. MACSYMA then would have solved the equation and given the numeric roots approximately or exactly depending upon the specified command.

MACSYMA can also solve quadratic, cubic and quartic equations as well as some classes of higher degree equations. However, it obviously cannot solve equations analytically in closed form when methods are not known, eg. a general fifth degree equation or one of higher degree. It can however, find approximate real solutions of polynomial equations of any degree to arbitrary precision.

3.2 Differential Calculus

MACSYMA knows about calculus. In (D1) we have an exponentiated function that is often used as an example in a first course in differential calculus.

$$
X^{X^X}
$$

(D1)

We now ask MACSYMA to differentiate (D1) with respect to X, using the DIFF command, to obtain this classic textbook result of differentiation. Notice how fast, 3/100 CPU seconds, MACSYMA computes this derivative.

```
(C2) DIFF(D1,X);
Time= 30 msec.
```

(D2)
$$
X^{X^X} (X^X \ \text{LOG}(X) \ (\text{LOG}(X) + 1) + X^{X-1})
$$

Below is a more complicated function, the error function of the tangent of the arc-cosine of the natural logarithm of X. Notice that MACSYMA does not display the identical input. This is because the input in (C1) passes through MACSYMA's simplifier. MACSYMA recognizes that the tangent of the arc-cosine of a function satisfies a trigonometric identity, namely TAN(ACOS(X)) = SQRT(1-X^2)/X. It takes this into account before displaying (D1).

```
(C1) ERF(TAN(ACOS(LOG(X))));
```

(D1)
$$
\text{ERF}\left(\frac{\text{SQRT}(1 - \text{LOG}^2(X))}{\text{LOG}(X)} \right)
$$

Now when MACSYMA is asked to differentiate (D1) with respect to X, it does so in a straightforward manner and simplifies the result using the canonical rational simplifier RATSIMP. This command puts the expression in a numerator-over-denominator form canceling any common divisors. In (D2) the symbols %E and %PI are MACSYMA's representations for e (the base of the natural logarithms) and π.

```
(C2) DIFF(D1,X),RATSIMP;
Time= 1585 msec.
```

(D2)
$$
- \frac{2 \%E^{1 - \frac{1}{\text{LOG}^2(X)}}}{\text{SQRT}(\%PI) \ X \ \text{LOG}^2(X) \ \text{SQRT}(1 - \text{LOG}^2(X))}
$$

3.3 Trigonometry

MACSYMA knows about elementary mathematics. For example, it can manipulate expressions containing multiple angles of trigonometric functions. In (D1) below is an expression that does not simplify in any obvious way. First we use the TRIGEXPAND command on (D1) which expands out trigonometric (hyperbolic) functions of sums of angles and of multiple angles resulting in (D2). We then EXPAND out (D2) and find that (D2) vanishes identically.

This is an example showing how to use MACSYMA to do "experimental mathematics". I suspected that (D1) vanished but trying to verify this by hand was both time consuming and error prone because of the multitude of signs. This identity does not appear to be known. It would be interesting if someone could find a geometrical interpretation of it.

```
(D1) SIN(Y - X) SIN(Z - X) SIN(Z - Y) + COS(Y - X) COS(Z - X) SIN(Z - Y)

    - COS(Y - X) SIN(Z - X) COS(Z - Y) + SIN(Y - X) COS(Z - X) COS(Z - Y)

(C2) TRIGEXPAND(D1);
Time= 1400.0 msec.

(D2) (COS(X) SIN(Y) - SIN(X) COS(Y)) (SIN(X) SIN(Z) + COS(X) COS(Z))

    (SIN(Y) SIN(Z) + COS(Y) COS(Z)) - (SIN(X) SIN(Y) + COS(X) COS(Y))

    (COS(X) SIN(Z) - SIN(X) COS(Z)) (SIN(Y) SIN(Z) + COS(Y) COS(Z))

    + (SIN(X) SIN(Y) + COS(X) COS(Y)) (SIN(X) SIN(Z) + COS(X) COS(Z))

    (COS(Y) SIN(Z) - SIN(Y) COS(Z)) + (COS(X) SIN(Y) - SIN(X) COS(Y))

    (COS(X) SIN(Z) - SIN(X) COS(Z)) (COS(Y) SIN(Z) - SIN(Y) COS(Z))

(C3) EXPAND(D2);
Time= 1350.0 msec.

(D4)                                    0
```

MACSYMA can also go back the other way, so to speak, beginning with an expression such as (D5) and finding an equivalent form for it in terms of multiple angles using the TRIGREDUCE command.

```
             5        2       3          4
(D5)    2 SIN (X) - 8 COS (X) SIN (X) + 6 COS (X) SIN(X)

(C6) EXPAND(TRIGREDUCE(D5));

(D6)                      SIN(5 X) + SIN(X)
```

3.4 Very Large Problems

A frequent question asks the size of the largest problem that MACSYMA can solve. The size of a problem that can be handled is, of course, to a large extent a function of the underlying computer hardware. I shall go into this question in more depth in Chapter 7. Here is a preliminary answer; here is a very large problem:

$$(D1) \qquad (X - 1)^{100} \quad (2 X - 1)^{100} \quad (3 X - 1)^{700}$$

This is a polynomial in X and if one were to expand it out without collecting terms one would have

```
(C2) NTERMS(D1);
```

```
(D2)                    7150901
```

NTERMS(*expr*) gives the number of terms that *expr* would have if it were fully expanded out and no cancellations or combination of terms occurred. This number, in (D2), is precisely 101^2*701 which one would expect from the binomial expansion formula. We now ask MACSYMA to expand the expression. The command RATEXPAND uses a very efficient algorithm for expanding polynomials. After waiting about 13 CPU minutes MACSYMA returns:

```
(C3) RATEXPAND(D1);
Time = 764150 msec.
```

```
(D3)   122427186804066208416145044327206249704010297
       794616687779023569766751053609544125677306330
       282403068164313282073928984176895647340305797
       701184895851976317431643022541219564575431932
       361445842322265611989432234712383670344350435
       476671159706350681167253578453360950549918549
       395826481540244545237495806672865354745228669         900
       444407778719348014578179086676206584907014351093 76 X     + ...
```

$$... - 2295846600 X^3 + 2876600 X^2 - 2400 X + 1$$

For obvious reasons I have included only the first term and the last 4 terms of (D3). Notice the leading coefficient of X^{900} in (D3). MACSYMA can manipulate integers of arbitrary size, and floating point numbers to arbitrary precision. In fact this very large integer is exactly $2^{100}*3^{700}$ which is also expected from the binomial expansion.

3.5 Factorization

MACSYMA can factor expressions. Below is a multivariate polynomial in four variables.

$$
\begin{aligned}
(D1) \quad & - 36\, W^2 X^7 Y^4 Z^8 + 3\, W^2 X^6 Y^3 Z^8 - 24\, W^3 X^7 Y^4 Z^6 \\
& + 2\, W^3 X^6 Y^3 Z^6 + 96\, W^2 X^8 Y^6 Z^5 - 168\, W^4 X^7 Y^6 Z^5 \\
& + 12\, W^2 X^7 Y^6 Z^5 - 216\, W^2 X^{10} Y^5 Z^5 - 8\, W^2 X^7 Y^5 Z^5 + 9\, X^7 Y^5 Z^5 \\
& + 14\, W^4 X^6 Y^5 Z^5 - W^2 X^6 Y^5 Z^5 + 18\, W^2 X^9 Y^4 Z^5 + 87\, X^7 Y^3 Z^5 \\
& - 3\, W^2 X^6 Y^3 Z^5 + 6\, W X^7 Y^5 Z^3 + 58\, W X^7 Y^3 Z^3 - 2\, W X^3 Y^6 Z^3{}^3 \\
& - 24\, X^8 Y^7 Z^2 + 42\, W^2 X^7 Y^7 Z^2 - 3\, X^7 Y^7 Z^2 + 54\, X^{10} Y^6 Z^2 \\
& - 232\, X^8 Y^5 Z^2 + 414\, W^2 X^7 Y^5 Z^2 - 29\, X^7 Y^5 Z^2 - 14\, W^4 X^6 Y^5 Z^2 \\
& + W^2 X^6 Y^5 Z^2 + 522\, X^{10} Y^4 Z^2 - 18\, W^2 X^9 Y^4 Z^2
\end{aligned}
$$

We now call the function FACTOR on (D1) and

```
(C2) FACTOR(D1);
Time= 111998 msec.
```

$$
\begin{aligned}
(D2) \quad & - X^6 Y^3 Z^2 (3\, Z^3 + 2\, W Z - 8\, X Y^2 + 14\, W^2 Y^2 - Y^2 + 18\, X^3 Y) \\
& (12\, W^2 X Y Z^3 - W^3 Z^{23} - 3\, X Y^2 - 29\, X^2 + W^2)
\end{aligned}
$$

MACSYMA factors this massive expression in about 2 CPU minutes. One can also extend the domain of factorization to the Gaussian integers or other algebraic fields [Wal].

The time required to factor expressions on a particular machine depends strongly upon the number of variables and the degree of the exponents. It is a simple matter to construct examples that look simple yet that take inordinate amounts of time to factor. For example $X^{2026} - X^{1013} + 1$ factors into two parts. One factor is $X^2 - X + 1$ but the second is an irreducible polynomial with 1351 terms. The current factorization algorithms do not return after several CPU hours on this example. It is possible to enhance the factorization program to run faster on this

case, but one could always invent other "simple looking" examples that would push
CPU times to excess.

3.6 Simplification

A very important feature of MACSYMA is its ability to simplify expressions. When I
studied plane-wave metrics for a new gravitation theory [Pa6,Man1], one particular
calculation produced an expression with several hundred thousand terms. From
geometrical arguments I knew the expression must simplify and indeed, using
MACSYMA, the expression collapsed to a small number of pages of output. The
following expression occurred repeatedly in the course of the calculation and caused
the collapse of the larger expression during simplification.

```
              2   2              2   2
      (SQRT(R  + A ) + A) (SQRT(R  + B ) + B)
(D1)  ---------------------------------------
                        2
                        R
                        2   2        2   2
               SQRT(R  + B ) + SQRT(R  + A ) + B + A
      -  -------------------------------------------
                        2   2        2   2
               SQRT(R  + B ) + SQRT(R  + A ) - B - A


(C2) RATSIMP(D1);
Time= 138 msec.
(D2)                           0
```

When the canonical simplifier RATSIMP is called on (D1) above it returns zero. At
first I did not believe that (D1) is zero, and I spent 14 minutes verifying it by hand
(almost exceeding my 15 minute limit). It is not easy to prove. Combining the
expressions over a common denominator results in a numerator that contains 20
terms when fully expanded, and one must be very careful to assure cancellation. Try
it by hand!

3.7 Evaluating Numeric Expressions and Simplification of Radicals

MACSYMA is also useful in proving that numeric expressions vanish. Below is a
nested radical. Evaluating it numerically in (D2) to 8 decimal places, with the
NUMER command, shows only that it is small. It was known that it had to vanish
but the numeric evaluation routines did not prove this. How does one prove it is
zero?

(D1) 173 SQRT(34) SQRT(2 SQRT(34) + 35) +

1394 SQRT(2 SQRT(34) + 35) - 1567 SQRT(34) - 7276

(C2) D1,NUMER;
(D2)   - 0.00048828125

It is difficult to deal with nested expressions by hand or by computer, but
MACSYMA has facilities for dealing with some problems of this kind. We first call
the function SQRTDENEST on (D1), and this simplifies the radicals by denesting
them. Comparing (D1) with (D3), we see that SQRTDENEST tells us that
SQRT(2*SQRT(34)+35) = SQRT(34)+1. This is far from obvious by inspection
although it is easy to verify by hand. Now, when we expand (D3), MACSYMA
returns zero.

(C3) SQRTDENEST(D1);

(D3) 173 SQRT(34) (SQRT(34) + 1) + 1394 (SQRT(34) + 1)

- 1567 SQRT(34) - 7276

(C4) EXPAND(D3);

(D4)   0

3.8 FORTRAN Generation

Here is an example of the FORTRAN capability in MACSYMA.

(C1) FORTRAN(EXPAND((X+3*Y+7*Z)^8));

```
     5764801*Z**8+19765032*Y*Z**7+6588344*X*Z**7+29647548*Y**2*Z**
 1   6+19765032*X*Y*Z**6+3294172*X**2*Z**6+25412184*Y**3*Z**5+2
 2   5412184*X*Y**2*Z**5+8470728*X**2*Y*Z**5+941192*X**3*Z**5+1
 3   3613670*Y**4*Z**4+18151560*X*Y**3*Z**4+9075780*X**2*Y**2*Z
 4   **4+2016840*X**3*Y*Z**4+168070*X**4*Z**4+4667544*Y**5*Z**3
 5   +7779240*X*Y**4*Z**3+5186160*X**2*Y**3*Z**3+1728720*X**3*Y
 6   **2*Z**3+288120*X**4*Y*Z**3+19208*X**5*Z**3+1000188*Y**6*Z
 7   **2+2000376*X*Y**5*Z**2+1666980*X**2*Y**4*Z**2+740880*X**3
 8   *Y**3*Z**2+185220*X**4*Y**2*Z**2+24696*X**5*Y*Z**2+1372*X*
 9   *6*Z**2+122472*Y**7*Z+285768*X*Y**6*Z+285768*X**2*Y**5*Z+1
 :   58760*X**3*Y**4*Z+52920*X**4*Y**3*Z+10584*X**5*Y**2*Z+1176
 ;   *X**6*Y*Z+56*X**7*Z+6561*Y**8+17496*X*Y**7+20412*X**2*Y**6
 <   +13608*X**3*Y**5+5670*X**4*Y**4+1512*X**5*Y**3+252*X**6*Y*
 =   *2+24*X**7*Y+X**8
```

(D1)   DONE

In (C1) above we ask MACSYMA to convert the expanded form of $(X+3*Y+7*Z)\char94 8$ into legal fortran with 6 spaces at the beginning of the line, continuation lines and ** for exponentiation.

Some sites have actually generated thousands of lines of FORTRAN in this way, written the output to a file and taken it over to their large numeric systems for processing. More powerful FORTRAN functions in MACSYMA will create optimized FORTRAN and allow one to generate code specific to particular computer systems.

3.9 Indefinite Integration

One of the most powerful features in MACSYMA is the ability to perform integration. This is useful, of course, in and of itself. But it is also necessary because some of the commonly used integral tables have error rates as high as 25% [Pa5,Ri1,Ri2,Kl1,Ro1]. One must ponder how many disasters have occurred because engineers have used faulty tables or made errors in hand calculations.

Below, for example, is an error that was found in one of the most popular tables [Gr1].

```
 /                                    2  2              2
 [          1                    2 (- 8 C  X  + 4 B C X - B )
 I ---------------- dX  =  ----------------------------
 ]     2        3/2              3  3/2
 / X (C X  + B X)               3 B  X    SQRT(C X + B)
```

The correct integral, found by MACSYMA, is

```
 /                                    2  2              2
 [          1                    2 (8 C  X  + 4 B C X - B )
 I ---------------- dX  =  --------------------------
 ]     2        3/2              3  3/2
 / X (C X  + B X)               3 B  X    SQRT(C X + B)
```

While we cannot know the cause of this error, it is apparent after looking at many errors that a remarkably large percentage of them in the tables are actual blunders rather than misprints.

Some difficulties in the integral tables are far more obscure. For example, the following integral has its form given correctly [Gr2] as

```
                                                         2
                                                         P
                                                        ----
                                                         2
       INF                                    P      4 Q
       /                           (1 - ERF(---)) %E
       [   - P T                          2 Q
       I %E        ERF(Q T) dT = ---------------------
       ]                                   P
       /
       0
```

However, it is missing the following conditions that are not obvious:

$$Re(P) > 0 \text{ and } |arg(Q)| < \pi/4 \ .$$

Above, ERF(T) is the probability integral defined by

```
                              T
                              /        2
                              [    - X
        SQRT(π) ERF(T)  =  2 I  %E       dX
                              ]
                              /
                              0
```

MACSYMA can handle integrals involving rational functions and combinations of rational, algebraic functions, and the elementary transcendental functions. It also has knowledge about error functions and some of the higher transcendental functions.

Below is an integral that is quite difficult to do by hand. It is not found in standard tables in its given form although it may transform to a recognized case. It is especially difficult to do by hand unless one notices a trick that involves performing a partial fraction decomposition of the integrand with respect to LOG(X). However, MACSYMA handles it readily.

```
                        /
                        [   LOG(X) - 1
       (D1)             I   ------------ dX
                        ]     2        2
                        / LOG (X) - X
```

```
(C2) INTEGRATE(D1,X);
Time= 744 msec.
```

```
                 LOG(LOG(X) + X)     LOG(LOG(X) - X)
       (D2)      ---------------  -  ---------------
                        2                   2
```

Below is another integral that is not found in standard tables and that MACSYMA could not handle until recently owing to a bug. I needed the answer and could not wait for the bug to be fixed. I will use this opportunity to illustrate a point. MACSYMA could not integrate (D1) directly and returned the integral in noun form with the integral sign around it in (D1). This normally means that MACSYMA cannot perform the computation. I assumed this was because the system had a difficulty with error functions. Therefore, I got rid of the error function by differentiating (D1) with respect to B resulting in (D2). This is often the same procedure one tries by hand when confronted by difficult integrals.

```
(C1) INTEGRATE(X*EXP(-Q^2*X^2)*ERF(B*X-A),X);

            /        2  2
            [     - Q  X
(D1)        I X %E          ERF(B X - A) dX
            ]
            /


(C2) DIFF(D1,B);

            /                 2    2 2
            [  2  - (B X - A)  - Q  X
          2 I X  %E                      dX
            ]
            /
(D2)        -------------------------------
                        SQRT(%PI)
```

In this form MACSYMA was able to perform the integration with respect to X in (D2). I then asked MACSYMA to integrate that newly found integral with respect to B to obtain the desired result, namely

```
      /       2 2
      [    - Q X
      I X %E       ERF(B X - A) dX =
      ]
      /
              2 2
             A Q
        -  -------
           2    2
          Q  + B         2    2
     B %E         (Q  + B ) X - A B             2 2
                ERF(-----------------)       - Q X
                          2    2              %E       ERF(B X - A)
                    SQRT(Q  + B )
     ------------------------------------  -  ----------------------
              2       2   2                             2
            2 Q  SQRT(Q + B )                         2 Q
```

While MACSYMA can now integrate (D1) directly there is a point to be stressed. MACSYMA sometimes believes it cannot solve a problem, but the user should not always take this computer program (or any other) at its word!

3.10 Definite Integration

Definite integration is far more difficult to code than indefinite integration because the number of known techniques is much larger. One often has the added complication of taking limits at the endpoints of the integral. MACSYMA has impressive capabilities for definite integration. Here is an example of a function whose definite integral does not appear in any of the standard tables:

```
                  INF
                  /              2
                  [     2   - U X
(D1)              I    X  %E          LOG(X) dX
                  ]
                  /
                  0
```

```
(C2) D1,INTEGRATE;
Time= 138442 msec.
```

```
            SQRT(%PI) (LOG(U) + 2 LOG(2) + %GAMMA - 2)
(D2)      - -------------------------------------------
                              3/2
                           8 U
```

In (C2) above we ask **MACSYMA** to perform the integration seen in (D1) with respect to X from 0 to infinity. In the answer, %GAMMA is the **MACSYMA** syntax for the Euler-Mascheroni constant = 0.577215664.. .

We can now generate new integrals by differentiating. From (D1) and (D2) we have,

```
(C3) 'INTEGRATE(D1,X)=D2;
```

```
        INF
        /              2
        [     2   - U X
(D3)    I    X  %E          LOG(X) dX =
        ]
        /
        0
                    SQRT(%PI) (LOG(U) + 2 LOG(2) + %GAMMA - 2)
                  - -------------------------------------------
                                        3/2
                                     8 U
```

and we now differentiate (D3) four times with respect to U and factor to obtain

18

```
(C4) DIFF(D3,U,4),FACTOR;
```

$$(D4) \quad \int_0^{INF} X^{10} \, \%E^{-U X^2} \, LOG(X) \, dX =$$

$$-\frac{3 \, SQRT(\%PI) \, (315 \, LOG(U) + 630 \, LOG(2) + 315 \, \%GAMMA - 1126)}{128 \, U^{11/2}}$$

In addition to definite integration, MACSYMA can perform numeric integration using the Romberg numeric integration procedure. There are a number of other numeric techniques available. And, one has the ability to evaluate expressions numerically to arbitrary precision.

3.11 Laplace Transformations

MACSYMA contains a large number of special purpose routines. For example, there is a package for solving some classes of simultaneous differential equations using Laplace transformations. Below we express our equations in (D1) and (D2) where the subscripts denote differentiation.

```
(C1) EQ1;
```

$$(D1) \qquad F(X)_{X X} - 6 \, G(X)_X - 6 \, SIN(X) = 0$$

```
(C2) EQ2;
```

$$(D2) \qquad 6 \, G(X)_{X X} + A^2 \, F(X)_X - 6 \, COS(X) = 0$$

Next we specify initial values for F(X) and G(X) and their first derivatives, while inhibiting the display of the D-lines by terminating the commands with $.

```
(C3) ATVALUE(F(X),X = 0,0)$

(C4) ATVALUE(G(X),X = 0,1)$

(C5) ATVALUE('DIFF(F(X),X),X = 0,0)$

(C6) ATVALUE('DIFF(G(X),X),X = 0,1)$
```

Finally we use the DESOLVE command which employs Laplace transforms to find the particular integral for the system of equations.

```
(C7) DESOLVE([EQ1,EQ2],[F(X),G(X)]);
```

$$(D7) \quad [F(X) = -\frac{12\ SIN(A\ X)}{A\ (A^2 - 1)} - \frac{6\ COS(A\ X)}{A^2} + \frac{12\ SIN(X)}{A^2 - 1} + \frac{6}{A^2},$$

$$G(X) = \frac{SIN(A\ X)}{A} - \frac{2\ COS(A\ X)}{A^2 - 1} + \frac{(A^2 + 1)\ COS(X)}{A^2 - 1}]$$

3.12 Functions of a Complex Variable

MACSYMA has some capabilities of dealing with functions of a complex variable. For example it is quite good at computing residues:

$$(D1) \qquad \frac{1 - \%E^{2\ Z}}{Z^4\ SIN(Z)}$$

```
(C2) RESIDUE(D1,Z,0);
```

$$(D2) \qquad\qquad\qquad - 1$$

Above it found the residue of (D1) at $Z = 0$. Below MACSYMA finds the real and imaginary parts of (D3) where %I is MACSYMA's syntax for SQRT(-1). The function TRIGSIMP employs the identities $COS(X)^2 + SIN(X)^2 = 1$ and $COSH(X)^2 - SINH(X)^2 = 1$ to simplify, when possible, expressions containing trigonometric or hyperbolic functions.

$$(D3) \qquad \frac{A\ \%E^{\%I\ Z} + B}{C\ \%E^{\%I\ Z} + D}$$

```
(C4) TRIGSIMP(REALPART(D3));
```

$$(D4) \qquad \frac{(A\ D + B\ C)\ COS(Z) + B\ D + A\ C}{2\ C\ D\ COS(Z) + D^2 + C^2}$$

```
(C5) TRIGSIMP(IMAGPART(D3));

                      (A D - B C) SIN(Z)
(D5)                  ----------------------
                                  2     2
                      2 C D COS(Z) + D  + C
```

This example is quite difficult to do by hand.

3.13 Combinatorial Functions

MACSYMA has the capability of manipulating and simplifying combinatorial expressions. For example, in (D1), we have an expression involving binomial coefficients where BINOMIAL(N,M):= N!/(M!*(N-M)!).

```
(C1) BINOMIAL(N,M)*BINOMIAL(N+3,M+2)/BINOMIAL(N+3,M+4);

            BINOMIAL(N, M) BINOMIAL(N + 3, M + 2)
(D1)        ------------------------------------
                   BINOMIAL(N + 3, M + 4)
```

We now call the function **MAKEFACT** on (D1), and this converts binomial coefficients to factorials.

```
(C2) MAKEFACT(D1);

            (M + 4)! N! (N - M - 1)!
(D2)        -------------------------------
            M! (M + 2)! (N - M)! (N - M + 1)!
```

Calling MAKEGAMMA on (D2) converts this expression of factorials to Gamma functions.

```
(C3) MAKEGAMMA(D2);

            GAMMA(M + 5) GAMMA(N + 1) GAMMA(N - M)
(D3) -------------------------------------------------------
     GAMMA(M + 1) GAMMA(M + 3) GAMMA(N - M + 1) GAMMA(N - M + 2)
```

We can also simplify expressions involving factorials. Using the **MINFACTORIAL** function on (D2) results in

```
(C4) MINFACTORIAL(D2);

            (M + 3) (M + 4) N!
(D4)        ------------------------------------
                       2
            M! (N - M)  (N - M + 1) (N - M - 1)!
```

3.14 Taylor/Laurent Series

The Taylor (Laurent) series capability is very impressive. Below we ask for the first 21 terms of the series of (D1) about the point X = 0. Notice that MACSYMA computes this expression in less than 1 CPU second.

$$(D1) \qquad A \, SIN(X^3) + \frac{B \, LOG(X^2 - X + 1)}{X^5}$$

```
(C2) TAYLOR(%,X,0,21);
Time= 900.0 msecs.
```

$$(D2)/T/ \quad -\frac{B}{X^4} + \frac{B}{2X^3} + \frac{2B}{3X^2} + \frac{B}{4X} - \frac{B}{5} - \frac{BX}{3} - \frac{BX^2}{7} + \frac{(B + 8A)X^3}{8}$$

$$+ \frac{(2B)X^4}{9} + \frac{BX^5}{10} - \frac{BX^6}{11} - \frac{BX^7}{6} - \frac{BX^8}{13} + \frac{(3B - 7A)X^9}{42}$$

$$+ \frac{(2B)X^{10}}{15} + \frac{BX^{11}}{16} - \frac{BX^{12}}{17} - \frac{BX^{13}}{9} - \frac{BX^{14}}{19} + \frac{(6B + A)X^{15}}{120}$$

$$+ \frac{(2B)X^{16}}{21} + \frac{BX^{17}}{22} - \frac{BX^{18}}{23} - \frac{BX^{19}}{12} - \frac{BX^{20}}{25} + \frac{(2520B - 13A)X^{21}}{65520}$$

$$+ \ \ldots$$

The program can also compute Taylor (Laurent) series in several variables.

3.15 Simultaneous Linear Equations

MACSYMA solves sets of simultaneous linear equations. Below the expressions M1, M2, ... M5 are assigned to (D1), (D2), ... (D5).

```
(C1) M1;
(D1)            3 A + 5 B + 7 C + 11 D + 13 E = 17 R


(C2) M2;
(D2)            19 A + 23 B + 29 C + 31 D + 37 E = 41 S


(C3) M3;
(D3)            43 A + 47 B + 53 C + 59 D + 61 E = 67 T


(C4) M4;
(D4)            71 A + 73 B + 79 C + 83 D + 89 E = 97 X


(C5) M5;
(D5)        101 A + 103 B + 107 C + 109 D + 113 E = 127 Y
```

To solve the system of equations one calls the SOLVE command in (C6) below. The solution (D6) is given in terms of rational numbers, exactly. Notice the near mathematical syntax in the MACSYMA commands. To use MACSYMA, one communicates in a kind of mathematical language rather than a strict programming language.

```
(C6) SOLVE([M1,M2,M3,M4,M5],[A,B,C,D,E]);
Time= 1355 msec.

                10922 Y - 14259 X + 2706 S + 1207 R
(D6) [   A = - -----------------------------------------
                                 284

            64008 Y - 64505 X - 14271 T + 7667 S + 10863 R
         B = -----------------------------------------------
                                 852

            8128 Y - 4365 X - 4757 T - 1845 S + 3621 R
         C = - ---------------------------------------------
                                 284

            3048 Y - 776 X - 4757 T + 2583 S
         D = - --------------------------------
                                 284

            3810 Y + 5917 X - 14271 T + 2501 S + 3621 R
         E = --------------------------------------------- ]
                                 852
```

MACSYMA can also deal with systems of non-linear algebraic equations. Examples are given in sections 6.1 and 6.2.

3.16 Linear Difference Equations and Fibonacci Numbers

The Fibonacci numbers, 1,1,2,3,5,8,13,21,34,55,89,144,233,... , satisfy the recurrence $F[N] = F[N-1] + F[N-2]$. The Nth term in the sequence is the sum of the two preceding terms. MACSYMA has techniques for solving linear difference equations and systems of simultaneous coupled linear difference equations. Below these routines are used to find an explicit formula for the Nth Fibonacci number.

We first initialize the problem by stating that both the 0th and 1st terms in the sequence are unity.

```
(C1) F[0]:1;
(D1)                              1

(C2) F[1]:1;
(D2)                              1
```

We now define the recurrence relation in (C3),

```
(C3) F[N+2] = F[N+1]+F[N];
(D3)                    F      = F      + F
                         N + 2    N + 1    N
```

and ask MACSYMA to compute the explicit formula for the Nth term.

```
(C4) DIFFERENCE(D3,F[N]);

                         N
            (SQRT(5) + 1)  (SQRT(5) + 5)
(D4) F   = ----------------------------
      N                 N
                      10 2
                                          N
                         (1 - SQRT(5))  (SQRT(5) - 5)
                       - ----------------------------
                                      N
                                    10 2
```

This is the answer we seek and MACSYMA now computes the 6th and 1000th term in the sequence.

```
(C5) D4,N = 6,RATSIMP;

(D5)                            F  = 13   .
                                 6

(C6) D4,N = 1000,RATSIMP;

(D6) F      = 70330367711422815821835254877183549770181269836358
      1000

            73274260490508715453711819693357974224949456261173

            34877504492417659910881863632654502236471060120533

            74121273867339111198139373125598767690091902245245

            323403501
```

3.17 Matrix Analysis

MACSYMA has powerful capabilities for manipulating matrices and performs most of the common operations. An example of matrix inversion is given in section 6.11. Here, the steps leading to the eigenvalues of a matrix are presented. We first enter the matrix in (C1), and in (D1) is MACSYMA's graphical representation of the matrix.

```
(C1) MATRIX([0,6,-10,8],[6,0,8,10],[-10,8,15*A,6],[8,10,6,15*A]);
```

$$
\text{(D1)} \quad
\begin{bmatrix}
0 & 6 & -10 & -8 \\
6 & 0 & 8 & 10 \\
-10 & 8 & 15\,A & 6 \\
-8 & 10 & 6 & 15\,A
\end{bmatrix}
$$

We then ask MACSYMA to compute the characteristic polynomial of (D1) for the parameter "L" and then solve the resulting polynomial for L. The four eigenvalues are shown below.

```
(C2) SOLVE(CHARPOLY(D1,L),L);
```

$$
\left[L = \frac{15\,A - 3\,\mathrm{SQRT}(5)\,\mathrm{SQRT}(5\,A^2 + 8\,A + 32)}{2}, \right.
$$

$$
L = \frac{3\,\mathrm{SQRT}(5)\,\mathrm{SQRT}(5\,A^2 + 8\,A + 32) + 15\,A}{2},
$$

$$
L = \frac{15\,A - \mathrm{SQRT}(5)\,\mathrm{SQRT}(45\,A^2 - 72\,A + 32)}{2},
$$

$$
\left. L = \frac{\mathrm{SQRT}(5)\,\mathrm{SQRT}(45\,A^2 - 72\,A + 32) + 15\,A}{2} \right]
$$

3.18 Ordinary Differential Equations

Another powerful feature is the MACSYMA program ODE. ODE is a collection of algorithms for solving ordinary differential equations. It was built over several years by E. L. Lafferty, J. P. Golden, R. A. Bogen and B. Kuipers, and its capabilities are described in the MACSYMA reference manual [MA1] in V2-4-14.

In (C1), we first declare that Y is a function of X. This assures that the derivative (2nd) of Y with respect to X will not vanish when (C2) is evaluated.

```
(C1) DEPENDS(Y,X)$
```

```
(C2) (1+X^2)*DIFF(Y,X,2)-2*Y=0;
```

```
                        2
(D2)                (X  + 1) Y     - 2 Y = 0
                             X X
```

We now ask MACSYMA to solve (D2) for Y as a function of X using the ODE command. The general solution with the two integration constants, %K1 and %K2 is found in (D3) in about 2 CPU seconds. The program can also find powerseries solutions for some differential equations when it can solve the recurrence relation. It does this in (D4). MACSYMA can be used to check the answer (D3). In (C5) we tell the system to substitute (D3) into (D2), differentiate the result and simplify.

```
(C3) ODE(D2,Y,X);
Time= 2068 msec.
                2      ATAN(X)      X              2
(D3)    Y = %K2 (X  + 1) (------- + --------) + %K1 (X  + 1)
                             2          2
                                     2 X  + 2
```

```
(C4) ODE(D2,Y,X,SERIES);
Time= 8766 msec.
                                    INF
                                    ====         I  2 I
                        2           \       (- 1)  X
(D4)        Y = %K1 (X  + 1) - %K2 X  >    ---------------
                                    /           1        1
                                    ====  (I - -) (I + -)
                                    I = 0       2        2
```

```
(C5) D2,D3,DIFF,RATSIMP;
Time= 2051 msec.
(D5)                        0 = 0
```

The ODE package in MACSYMA incorporates a technique in Computer Algebra that has the computer search for patterns in a given problem. When there is a pattern match the computer generates the answer by means of a procedure (algorithm). Many good mathematicians do, in fact, use this technique. However, the computer is often better due to its thoroughness and the fact that it does not make the normal human errors.

ODE now has capabilities for solving a large class of differential equations analytically. While it is difficult to quantify capability of this kind, it could be stated that it is as capable as a particular set of books, but not as capable as the authors of those books. For example, it has all the elementary methods in the textbook of Rainville [Ra1] and the tables of Kamke [Ka1]. Perhaps a better way to measure its capabilities is to say that virtually all the known methods for the solution of first and second-order ordinary differential equations are embodied in the program.

However, solving differential equations is really a kind of scientific art. Given a differential equation whose form or structure is not immediately recognizable, one looks for transformations that will convert the given problem into one that is already known. ODE has some modest heuristic capability in this area. It is in this direction that we can expect future developments in MACSYMA. Over the next several years, we expect that clever expression recognizers will be added that will simplify and transform problems in very much the same way as do mathematicians who specialize in the field. The rules are fairly well known; the task is to build them within the resources of MACSYMA and provide enough user control to allow reasonably rapid convergence either to a solution or a decision that the problem cannot be solved in terms of known functions.

3.19 Indefinite Summation

Often, algorithms themselves are discovered by combining Computer Algebra systems with the procedures of experimental mathematics. This means that one gains empirical experience with a heuristic method that works often enough so that one gets an idea of what the answers to a particular problem look like. An example of this technique is the algorithm for indefinite (finite) summation found by Gosper with MACSYMA [Go1]. The restrictions for the procedure are that both the summand and the answer must be expressible as products of powers, factorials, binomials, and/or rational functions. Even with these restrictions, it is possible to handle a large class of the indefinite sums one encounters. The history of the more general problem of indefinite summation contains the names of Newton, Euler, Bernoulli and Boole, and despite the many years these distinguished mathematicians devoted to the problem, the results are surprisingly few in number. One reason for believing in the existence of an algorithm is the success achieved in solving the indefinite integral problem and the close parallels between the two conceptually. Gosper's approach involved testing many cases with MACSYMA until a pattern emerged that worked for all cases and led to the discovery of the algorithm. Recently, using abstract algebra techniques similar to those Risch used for integration, Karr has extended Gosper's work [Ka1]. Some parts of Gosper's algorithm have been implemented in MACSYMA. Here is an example of its power:

$$(D1) \qquad \sum_{N=0}^{M} A^{N} N^{4}$$

The expression (D1) is the indefinite sum we wish to evaluate. The command, NUSUM, that performs this computation, is given in (C2).

```
(C2) NUSUM(A^N*N^4,N,0,M);
Time= 24801 msec.
```

$$
\begin{aligned}
(D2)\quad & A^{M+1}(A^4 M^4 - 4 A^3 M^4 + 6 A^2 M^4 - 4 A M^4 + M^4 - 4 A^4 M^3 \\
& + 12 A^3 M^3 - 12 A^2 M^3 + 4 M^3 + 6 A^4 M^2 - 6 A^2 M^2 - 6 A M^2 + 6 M^2 \\
& - 4 A^4 M - 12 A^3 M + 12 A M + 4 M + A^4 \\
& + 11 A^2 + 11 A + 1)/(A - 1)^5 - \frac{A (A + 1)^2 (A^2 + 10 A + 1)}{5}
\end{aligned}
$$

It takes about 25 CPU seconds to find the result, and we can check the answer by taking the first backward difference of (D2) using the UNSUM command. This is analogous to checking an indefinite integral by differentiating it.

```
(C3) UNSUM(D2,M);
Time= 6581 msec.
```

$$
(D3) \qquad\qquad A^M M^4
$$

3.20 Finding Extremums a Function using Lagrange Multipliers

Here, in an elementary interactive session, we combine a number of MACSYMA commands to find the extremum of a function with constraints using the Lagrange multiplier technique.

In (C1) we assign to F the expression shown in (D1). We shall find the extremum of F subject to two constraints with respective assignments K1 and K2 shown in (D2) and (D3). In this example [X,Y,Z,T] are variables and [A,B] are constants.

```
(C1) F:X^2+2*Y^2+Z^2+T^2;
```

$$
(D1) \qquad\qquad Z^2 + 2 Y^2 + X^2 + T^2
$$

```
(C2) K1:A+X+3*Y-Z+T;
```
$$
(D2) \qquad\qquad - Z + 3 Y + X + T + A
$$

```
(C3) K2:B+2*X-Y+Z+2*T;
```
$$
(D3) \qquad\qquad Z - Y + 2 X + 2 T + B
$$

The Lagrange multiplier technique tells us to form a new function G defined by adding to F the product of the Lagrange multipliers, L1 and L2, into the constraints. This is shown in (C4).

```
(C4) G:F+L1*K1+L2*K2;
```

$$(D4) \quad G = Z^2 + 2 Y^2 + X^2 + T^2$$

$$+ L2 (Z - Y + 2 X + 2 T + B) + L1 (- Z + 3 Y + X + T + A)$$

The next step is to form all partial derivatives of G as shown in (C5).

```
(C5) [DIFF(G,X),DIFF(G,Y),DIFF(G,Z),DIFF(G,T),DIFF(G,L1),DIFF(G,L2)];
```

```
(D5) [2 X + 2 L2 + L1, 4 Y - L2 + 3 L1, 2 Z + L2 - L1,

     2 T + 2 L2 + L1, - Z + 3 Y + X + T + A, Z - Y + 2 X + 2 T + B]
```

We then solve the system (D5) for each of the six parameters.

```
(C6) SOLVE(D5,[X,Y,Z,T,L1,L2]);
```

$$(D6) \quad \left[\left[X = -\frac{27 B + 13 A}{138}, \; Y = -\frac{5 A - 2 B}{23}, \; Z = \frac{11 A - 9 B}{69}, \right.\right.$$

$$\left.\left.T = -\frac{27 B + 13 A}{138}, \; L1 = \frac{19 A - 3 B}{69}, \; L2 = -\frac{A - 5 B}{23}\right]\right]$$

For the final step we substitute (D6) into F as defined in (D1). The expression (D7) is the extremum of (D1) subject to the constraints.

```
(C7) D1,D6,RATSIMP;
```

$$(D7) \quad \frac{15 B^2 - 6 A B + 19 A^2}{138}$$

4. Extending the Capabilities of MACSYMA

4.1 Block Programs and Hessians

One can easily extend the capabilities of MACSYMA and add to the knowledge base by defining new functions. For example, MACSYMA does not know about Hessians although the example below actually appears in the MACSYMA reference manual. Hessians are determinants of symmetric matrices whose elements are the second partial derivatives of a function [Gou1].

To illustrate some of the features of MACSYMA programs, we define the Hessian using a program block. A block in MACSYMA is similar to a subroutine in FORTRAN, or a procedure in ALGOL or PL/1. A block contains local variables that will not conflict with variables having the same names outside or global to the block. For example, in (C1) the atomic variables DFXX ... DFZZ are locally assigned the values of the various mixed second partial derivatives. We then compute the determinant and simplify it through the RATSIMP and FACTOR commands within the block statement. In fact, any MACSYMA command can be included within a block.

```
(C1) HESSIAN(F,X,Y,Z):=BLOCK([DFXX,DFXY,DFXZ,DFYY,DFYZ,DFZZ],
          DFXX:DIFF(F,X,2),
            DFXY:DIFF(F,X,1,Y,1),
              DFXZ:DIFF(F,X,1,Z,1),
                DFYY:DIFF(F,Y,2),
                  DFYZ:DIFF(F,Y,1,Z,1),
                    DFZZ:DIFF(F,Z,2),
                      FACTOR(RATSIMP(DETERMINANT(MATRIX
    ([DFXX,DFXY,DFXZ],[DFXY,DFYY,DFYZ],[DFXZ,DFYZ,DFZZ])))))$
```

This functional definition is only appropriate for the three variable case, but it can easily be extended for any number of dimensions. Below, in (D2), we choose a function of three variables, and the routine computes the Hessian in (D3).

```
(C2) COS(X/(Y+Z))*EXP(X-2*Z);

                        X - 2 Z        X
(D2)                  %E        COS(-----)
                                     Z + Y
```

```
(C3) HESSIAN(D2,X,Y,Z);
Time= 19127 msec.

                3 X - 6 Z        X          2    X
            4 %E          COS(-----) SIN (-----)
                               Z + Y          Z + Y
(D3)        - ------------------------------------
                                  4
                              (Z + Y)
```

4.2 Recursive Definition for Chebyshev Polynomials

While adding features to the knowledge base is fairly easy, careful programming is essential. Suppose MACSYMA did not know about Chebyshev polynomials. A recursive definition of these is given in (C1) and (C2).

```
(C1) T[N]:=2*X*T[N-1]-T[N-2];

(D1)          T  := 2 X T      - T
               N         N - 1    N - 2
```

```
(C2) (T[0]:1, T[1]:X)$
```

To compute the 20th polynomial, one simply calls T[20] as below. Of course it must be simplified, and we expand it with the command (see example 3.4) in (C4).

```
(C3) T[20]$
Time= 2880.0 msecs.

(C4) RATEXPAND(D3);
Time= 27083 msec.

            20              18              16              14
(D4) 524288 X    - 2621440 X   + 5570560 X    - 6553600 X

            12              10             8            6           4
   + 4659200 X   - 2050048 X   + 549120 X  - 84480 X  + 6600 X

          2
   - 200 X  + 1
```

This method is slow because we were not clever in our approach. We generated a huge expression in (D3) (10,946 terms, see example 3.4), as seen in (D5), that we later expanded to simplify it.

```
(C5) NTERMS(D3);

(D5)                        10946
```

We now repeat the same problem, adding a RATEXPAND command to the recursive definition. We change the name of the array to S because we have already used the array name T. If we had redefined T, but had not removed array elements already computed, then the old values would have been used. Alternately we could remove the definition of T from memory.

```
(C6) S[N]:=RATEXPAND(2*X*S[N-1])-S[N-2];

(D6)          S   := 2 X S      - S
               N         N - 1    N - 2

(C7) (S[0]:1, S[1]:X)$

(C8) S[20];
Time= 1700.0 msecs.

            20              18              16              14
(D8) 524288 X    - 2621440 X   + 5570560 X    - 6553600 X

            12              10             8            6           4
   + 4659200 X   - 2050048 X   + 549120 X  - 84480 X  + 6600 X

          2
   - 200 X  + 1
```

We now find the same (correct) expression for the 20th polynomial in 1/22 the CPU time. One can generate many examples where careful programming can save orders of magnitude of processing time. Other examples of saving processing time in real applications will be shown later.

5. Special Purpose Packages

There are special purpose packages in MACSYMA for use in General Relativity, high energy physics and several other fields of study. These auxiliary packages often provide features that are useful to non-specialists.

5.1 Component Tensor Manipulation

The Component Tensor Manipulation Program, CTENSR, provides the user with tools for computing many of the geometrical objects encountered in General Relativity and Differential Geometry. For a given metric tensor, CTENSR can compute the Christoffel symbols, the curvature tensor, the Weyl tensor, the Einstein and Ricci tensors, the geodesic equations of motion, and other useful geometrical objects.

In the example below, CTENSR is used to reproduce Dingle's formulas [Di1]. For thirty years these were the cornerstone of hand computations in relativity. It required a team effort of several months in 1933 to compute these formulas by hand. Below, MACSYMA does the identical calculation in less than one CPU minute.

CTENSR is an out of core file that is loaded with the command (C1). In addition, some flags must be preset to control the simplification processes, but these are not shown here. In (C2) we specify the dimension of the manifold to be 4 and then assign to OMEGA the list of coordinate labels.

```
(C1) LOAD("CTENSR")$
```

```
(C2) DIM:4$
```

```
(C3) OMEGA:[X,Y,Z,T]$
```

The metric tensor LG is now specified as a matrix, and the functional dependence of the elements of the metric is declared in (C4).

```
(C3) LG:MATRIX([A,0,0,0],[0,B,0,0],[0,0,C,0],[0,0,0,-D]);
```

$$
(D3) \quad
\begin{bmatrix}
A & 0 & 0 & 0 \\
0 & B & 0 & 0 \\
0 & 0 & C & 0 \\
0 & 0 & 0 & -D
\end{bmatrix}
$$

```
(C4) DEPENDS([A,B,C,D],[X,Y,Z,T])$
```

The program now runs itself. The command (C5) tells the program to compute and simplify the 10 unique components of the mixed Einstein tensor, and it takes about

40 CPU seconds to accomplish this. The argument "FALSE" inhibits the display of the components.

```
(C5) EINSTEIN(FALSE)$
Time= 40200.0 msecs.
```

We now ask it to print G[1,4], which is the factored XT component of the mixed Einstein tensor. The subscripts denotes partial differentiation.

```
(C6) G[1,4];

             2                   2         2
(D6)  - (A B  C C  D  + A B B  C  D  + A B  C  C  D
            T X          T    X         T    X

       2              2                      2
   + A  B   C C  D - 2 A B  C C   D + A B B  B  C  D
       T     X            T X        T    X

       2              2               2 2 2
   + A  B B  C  D - 2 A B B   C  D)/(4 A B  C  D )
       T    X            T X
```

5.2 Coordinate Transformations

The CTENSR package offers the user the ability to transform objects between coordinate systems. The interaction is shown below:

```
(C1) OMEGA:[X,Y,Z]$

(C2) IDENT(3);
                              [ 1  0  0 ]
                              [         ]
(D2)                          [ 0  1  0 ]
                              [         ]
                              [ 0  0  1 ]
```

In (C1) we assign to OMEGA the coordinate labels as a list. We want the metric tensor to correspond to a rectangular Cartesian 3-space so we define it to be the identity matrix. The command TTRANSFORM begins a routine that prompts the user to enter the three transformations shown below. The algebra is completed by MACSYMA and results in the new line element in spherical coordinates (D3).

```
(C3) TTRANSFORM(D2);
```

Transform # 1
x*sin(y)*sin(z);

Transform # 2
x*sin(y)*cos(z);

Transform # 3
x*cos(y);

$$
(D3) \quad
\begin{bmatrix}
1 & 0 & 0 \\
0 & X^2 & 0 \\
0 & 0 & X^2 \, SIN^2 (Y)
\end{bmatrix}
$$

5.3 Indicial Tensor Manipulation

The tensorial capabilities in **MACSYMA** include a program called ITENSR [Pa7,Pa8] that allows the user to manipulate indexed objects. One of the fundamental identities in General Relativity, and differential geometry is Bianchi's identity. It involves the covariant derivative of the Riemann tensor cycled with two of the non-skew-symmetric tensor indices. One normally proves this identity by choosing a coordinate system in which the connection coefficients vanish. There is nothing wrong with that method, but to illustrate the power of the program we shall carrying out the computation without choosing geodesic coordinates.

$$
C^{U}_{RST;W} + C^{U}_{RTW;S} + C^{U}_{RWS;T}
$$

One normally writes the Bianchi identity as above where the 4-indexed object C is the curvature tensor, and the semi-colon is covariant differentiation.

In ITENSR we express this combination of covariant derivatives with the command (C1) where COVDIFF is a function which performs covariant differentiation. In (C2) we expand (C1) and then call a function RENAME that relabels dummy indices in a consistent way so that each term in the sum has at most one pair of dummy indices with a particular label.

```
(C1)  COVDIFF(CURVATURE([R,S,T],[U]),W)+
      COVDIFF(CURVATURE([R,T,W],[U]),S)+
      COVDIFF(CURVATURE([R,W,S],[U]),T)$
```

```
(C2) RENAME(EXPAND(D1))$
```

The SHOW command in (C3) displays (D2) with covariant and contravariant indices, where CHR2 is the Christoffel symbol of the second kind and %1, %2, ... are dummy indices.

```
(C3) SHOW(D2);

(D3)
          U        %1       %2           U        %1       %2
   2 CHR2     CHR2     CHR2     - 2 CHR2     CHR2     CHR2
          %1 S      %2 T     R W          %1 T     %2 S     R W


          %2       U        %1           %2       U        %1
 + 2 CHR2     CHR2     CHR2     - 2 CHR2     CHR2     CHR2
          %1 S      %2 T     R W          %1 T     %2 S     R W


          U        %1       %2           U        %1       %2
 - 2 CHR2     CHR2     CHR2     + 2 CHR2     CHR2     CHR2
          %1 S      %2 W     R T          %1 W     %2 S     R T


          %2       U        %1           %2       U        %1
 - 2 CHR2     CHR2     CHR2     + 2 CHR2     CHR2     CHR2
          %1 S      %2 W     R T          %1 W     %2 S     R T


          U        %1       %2           U        %1       %2
 + 2 CHR2     CHR2     CHR2     - 2 CHR2     CHR2     CHR2
          %1 T      %2 W     R S          %1 W     %2 T     R S


          %2       U        %1           %2       U        %1
 + 2 CHR2     CHR2     CHR2     - 2 CHR2     CHR2     CHR2
          %1 T      %2 W     R S          %1 W     %2 T     R S
```

In (C4) the CANFORM command simplifies (D3). It takes into account the symmetry of the CHR2s in the covariant indices and also changes the labels of dummy indices. It finds that the expression vanishes, as we had anticipated.

```
(C4) CANFORM(D3);

(D4)                              0
```

MACSYMA provides the user with an atmosphere for testing conjectures. This enabled me to use ITENSR to test and thereby discover a new algorithm for performing multiple covariant differentiation that has been implemented in MACSYMA [Pa9]. The standard method tells one to use the methods of covariant differentiation on tensorial objects only when performing multiple covariant differentiation. It turns out that one can take multiple covariant derivatives on the basis of covariant and contravariant indices, by treating all objects as tensorial, and arrive at the correct answer in a far more direct and simple way.

One can become dizzy trying to simplify objects such as (D3) by hand. While this MACSYMA example is impressive, it is the case that trying to simplify tensorial objects, based upon their symmetry properties, by hand or by computer is very difficult. There is no known algorithm for performing this simplification. Another package, STENSR, written by Lars Hornfeldt at the Institute of Theoretical Physics in Stockholm is a very advanced program and can manipulate tensors and as well as spinors. It was implemented in MACSYMA on the MIT-MC machine a few years ago. While ITENSR and STENSR are better than humans in simplifying expressions containing symmetric objects, I do not believe that any computer program can yet outperform people when the symmetry properties become sufficiently complicated.

6. Applications of MACSYMA

Over the last 10 years I have used MACSYMA on a large number of problems submitted by researchers who have not had access to Computer Algebra systems. I have chosen a few of these problems to illustrate the power and utility of MACSYMA. These examples should give the reader some ideas for adapting the system to meet their own needs. Unfortunately, many of these problems were sent without any indication to the user's motivation for studying them. Users often tend to shy away from describing the significance of their problems, especially when the problems have commercial or military implications. When the problems have come from commercial sites, I have left out references. Readers interested in detail may contact with me privately.

6.1 Emulsion Chemistry/Try This by Hand!

Two years ago I gave a MACSYMA talk to a group from a large manufacturer of photographic equipment. As is usual after the talk I gave a demonstration of MACSYMA and took questions from the attendees. One of the scientists produced the following problem reported to be in the field of emulsion chemistry. It is a system of nonlinear algebraic equations given by

```
(C1)  [R1*(R2+R3) = 2*A, R2*(R1+R3) = 2*B, R3*(R1+R2) = 2*C]$

(C2) SOLVE(D1,[R1,R2,R3]);
Time= 93830 msec.

                    (- C + B + A) (C - B + A)
(D2) [ R1 = SQRT(-------------------------)
                          C + B - A

                 (- C + B + A) (C + B - A)
       R2 = SQRT(-------------------------)
                          C - B + A

                  (C - B + A) (C + B - A)
       R3 = SQRT(-------------------------) ]
                        - C + B + A
```

MACSYMA solved it readily while we waited (there is a second solution that is the negative of (D2)). The person then showed me his other problem that is a generalization of (D1) in four variables.

$$R1*(R2+R3+R4) = A, \quad R2*(R1+R3+R4) = B,$$
$$R3*(R1+R2+R4) = C, \quad R4*(R1+R2+R3) = D$$

MACSYMA ran for several minutes without returning an answer so I took it back to MIT and let it run. It turns out that there are 16 solutions involving the roots of quartic equations. The expressions are far too massive to display here.

6.2 Commodities Market/Cannot be Solved by Hand

The following system of nonlinear equations occurred in a study of the commodities market. The person tried to solve the system by hand over a period of several months, and gave up. He tried numeric analysis and came to the incorrect conclusion that the system was singular. After suspending work on the equations for some time he contacted me to determine whether MACSYMA could help.

The system of equations is shown in (D1). The variables are [S,T,U,V,W,X,Y,Z].

```
(D1) [  U + S - 2 A = 0,

        V + T - 2 B = 0,

        V - U - T + S = 0,

        V Z - T Y - U X + S W = 0,

        V Z + T Y + U X + S W - 2 C E = 0,

        - V Z - T Y + U X + S W - 2 C D = 0,

        - S U (Z + Y) + S T (Z + X) + U V (Y + W) - T V (X + W)

                    - 2 PO (U - T) (V - S) = 0,

        U (Z - W) + T (W - Z) + S (Y - X) + V (X - Y)

                    - RO (U - T) (V - S) = 0 ]
```

An analysis of about one hour with MACSYMA resulted in some useful results for the proposer. We discovered that a general solution of this system cannot be found unless certain restrictions are placed upon the parameters. In particular, unless R0 is zero, closed form solutions cannot be found. With R0 = 0 and some other restrictions, there are two solutions, one of which is:

```
(C2) SOLVE(D1,[S,T,U,V,W,X,Y,Z]);

            B + A      3 B - A     3 A - B     B + A        C E
(D2) [S = -----, T = -------, U = -------, V = -----, W = -----,
            2            2            2          2        B + A

              C (E + 2 D)     C (E - 2 D)     C E
        X = - -----------, Y = -----------, Z = -----]
              B - 3 A          3 B - A         B + A
```

In fact this solution was of sufficient interest that the proposer was able to continue with his research. A check of the solution is given in (D3) by substituting (D2) into (D1) and calling the canonical simplifier RATSIMP.

```
(C3) D1,D2,RATSIMP;
```

```
(D3)    [0 = 0, 0 = 0, 0 = 0, 0 = 0, 0 = 0, 0 = 0, 0 = 0, 0 = 0]
```

6.3 Quantum Optics/Impossible Without MACSYMA

Here is a problem that dramatically demonstrates how essential Computer Algebra systems are for certain kinds of problems. It is unlikely that the associated research program could be approached in any other manner. The basic problem involves optical transitions between two orbitals on different atoms. For a particular geometry, this involves three-dimensional integration of the product of Gaussians as shown in (D1)

$$(D1) \quad (2X^2 - P^2) \, \%E^{-\frac{(Z-C)^2 + (Y-B)^2 + (X-A)^2}{S^2} - \frac{Z^2 + Y^2 + X^2}{P^2}}$$

The problem is now broken into three steps, first integrating (D1) over Z from MINF (minus infinity) to INF (infinity), and similarly for Y and X.

```
(C2) INTEGRATE(D1,Z,MINF,INF)$
Time= 61300.0 msec.
```

```
(C3) INTEGRATE(D2,Y,MINF,INF)$
Time= 383983.34 msec.
```

```
(C4) INTEGRATE(D3,X,MINF,INF)$
Time= 253766.66 msec.
```

To obtain the final form of the answer we factor (D4).

```
(C5) FACTOR(D4);
Time= 2400.0 msec.
```

$$(D5) \quad -\frac{P^{5/2} S^{3/2} (S^2 + P^2 - 2A^2) \, \%E^{-\frac{C^2 + B^2 + A^2}{S^2 + P^2}}}{(S^2 + P^2)^{7/2}}$$

The research program involves performing hundreds of integrations of this kind for different geometries. The integrands are similar to (D1) and the computations equally complex. Although the answers look fairly simple when factored, the integration times are quite large.

6.4 Quantum Theory/Gaining Insight

Here is a problem in quantum theory that was proposed several years ago. The analysis with MACSYMA shows how the user can gain insight by using the system. The problem involves a matrix of the kind shown in (D3) with terms adjacent to and along the main diagonal. The problem is to find the determinant for the NxN generalization of (D3). I do not know whether this problem has actually been solved, but setting it up in MACSYMA is interesting and demonstrates one of my introductory points about gaining insight. A DELTA function is defined in (C1) and we use this in (C2) to define the array F[I,J]. Then we use the GENMATRIX command to generate the matrix from the array by specifying the number of rows and columns. For the 5x5 case we obtain (D3).

```
(C1) DELTA(I,J):= IF I = J THEN 1 ELSE 0$

(C2) F[I,J]:= L*Q[I]*DELTA(I,J) + P[I]*DELTA(I,J+2) + R[I]*DELTA(I,J-2)$

(C3) GENMATRIX(F,5,5),DELTA;
```

$$
(D3) \quad
\begin{bmatrix}
Q_1 L & 0 & R_1 & 0 & 0 \\
0 & Q_2 L & 0 & R_2 & 0 \\
P_3 & 0 & Q_3 L & 0 & R_3 \\
0 & P_4 & 0 & Q_4 L & 0 \\
0 & 0 & P_5 & 0 & Q_5 L
\end{bmatrix}
$$

Next, we define a new function MAKEDET(N) that, for a particular value of N, takes the array F and generates the corresponding NxN matrix, computes the determinant of the matrix and factors it. In (C5) MACSYMA is asked to display the corresponding factored determinant for N = 3, 4,.. 7.

```
(C4) MAKEDET(N):=FACTOR(DETERMINANT(EV(GENMATRIX(F,N,N),DELTA)))$

(C5) FOR I:3 THRU 7 DO PRINT(I = MAKEDET(I))$
```

$$3 = Q_2 L (Q_1 Q_3 L^2 - R_1 P_3)$$

$$4 = (Q_1 Q_3 L^2 - R_1 P_3)(Q_2 Q_4 L^2 - R_2 P_4)$$

$$5 = L(Q_2\,Q_4\,L - R_2\,P_4)^2\,(Q_1\,Q_3\,Q_5\,L - R_1\,P_3\,Q_5 - Q_1\,R_3\,P_5)^2$$

$$6 = L^2\,(Q_1\,Q_3\,Q_5\,L - R_1\,P_3\,Q_5 - Q_1\,R_3\,P_5)^2$$

$$(Q_2\,Q_4\,Q_6\,L - R_2\,P_4\,Q_6 - Q_2\,R_4\,P_6)^2$$

$$7 = L(Q_2\,Q_4\,Q_6\,L - R_2\,P_4\,Q_6 - Q_2\,R_4\,P_6)^2$$

$$(Q_1\,Q_3\,Q_5\,Q_7\,L^4 - R_1\,P_3\,Q_5\,Q_7\,L^2 - Q_1\,R_3\,P_5\,Q_7\,L^2 - Q_1\,Q_3\,R_5\,P_7\,L^2$$

$$+ R_1\,P_3\,R_5\,P_7)$$

Notice that there are patterns appearing in the expressions above. For each index N, the last term on the right also appears in the succeeding expression. There are both parity relations in these expressions in addition to relations involving permutations of the matrix elements. Other relations may be found here too, and insight into the nature of this problem is established by looking at several cases in this way. This insight can be traced to MACSYMA's ability in factoring these expressions. This is something that might not happen if one were to attempt the problem by hand.

6.5 Quantum Gravity/New Frontiers

Another problem that illustrates the power and interactive nature of MACSYMA occurred a couple of years ago when a colleague working in quantum gravity at Los Alamos phoned me at home one evening. He was excited believing he had found a new solution to Einstein's equations. This would indeed be exciting, and while he was on one telephone, I used my second phone to dial up MACSYMA at the MIT-MC machine. As we spoke, I gave MACSYMA (see CTENSR in 5.1) his proposed solution. For the diagonal metric shown below

```
  2  1  2       2        2        2
dS   =  - dx + A*C  dy + B*C  dz  - A*C  dt
  C
```

his solution took the form in (D1) below where the components of the metric are the following complicated expressions in y and t:

$$\text{(D1)} \quad \left[A = \frac{(\text{COSH}^2(t) - \text{SIN}^2(y))^2}{\text{COSH}^2(t)}, \ B = \text{COSH}^2(t) \ \text{SIN}^2(y), \right.$$

$$\left. C = \text{\%E}^{-2 \ \text{ATAN}(\text{SINH}(t))} \right]$$

MACSYMA found the field equations, and the three independent components of the Ricci tensor are shown below in (D2). The subscripts are partial derivatives.

$$\text{(D2)} \quad [2 \ B \ C \ C_{t \ t} \ - 2 \ B \ (C_t)^2 + B \ C \ C_t,$$

$$2 \ A^2 \ B \ C^2 \ C_{t \ t} \ - 2 \ A^2 \ B \ (C_t)^2 + A^2 \ B \ B_t \ C \ C_t + 2 \ A^2 \ B \ B_{y \ y} \ C^2$$

$$- A \ (B_y)^2 \ C - A \ A_y \ B \ B_y \ C + A \ A_t \ B \ B_t \ C + 2 \ A \ A_{y \ y} \ B^2 \ C$$

$$- 2 \ (A_y)^2 \ B^2 \ C + 2 \ A \ A_{t \ t} \ B^2 \ C - 2 \ (A_t)^2 \ B^2 \ C,$$

$$2 \ A \ B^2 \ C_{t \ t} + A \ B \ B_t \ C_t + A \ A_y \ B \ B_y \ C$$

$$+ 2 \ A \ B \ B_{t \ t} \ C - A \ (B_t)^2 \ C - A \ A_t \ B \ B_t \ C + 2 \ A \ A_{y \ y} \ B^2 \ C$$

$$- 2 \ (A_y)^2 \ B^2 \ C + 2 \ A \ A_{t \ t} \ B^2 \ C - 2 \ (A_t)^2 \ B^2 \ C]$$

If the solution is correct, substituting (D1) into (D2), differentiating and simplifying should result in zero. Indeed this is the result shown in (D3).

```
(C3) D2,D1,DIFF,RATSIMP;
Time= 58892 msec.
```

$$\text{(D3)} \qquad \qquad [0, \ 0, \ 0]$$

The entire interaction on the telephones took about 20 minutes and the proposer was happy to learn that his solution was correct. However, his reaction was subdued when I pointed out that the solution he found was one of Harrison's metrics (found by hand) and known since 1959 [Ha1].

6.6 Boundary Layers/FORTRAN Generation

Having MACSYMA is like having the world's most powerful sliderule at your side. Three years ago we received a call from a researcher at an aircraft company who had heard about MACSYMA. He wished to evaluate an integral arising from a turbulence and boundary layer problem. We were told that an engineer had been assigned to the task, and had spent three weeks on the problem, always obtaining a different answer. Could MACSYMA help? The deceptively simple looking integral is given by (D1). It is complicated by the exponent 4 in the integrand. Numeric methods would not suffice in this case because of the two parameters in the integrand. We gave him an account to use MACSYMA on the MIT-MC machine, he typed in the expression, and 10 seconds later had the answer. If you count the terms, after expanding out (D2), you will find 70. This explains why it is so tricky to correctly evaluate the integral by hand.

```
                     /
                     [                3      2      4
        (D1)         I  (K LOG(N) - 2 N  + 3 N  + B) dN
                     ]
                     /

(C2) D2,INTEGRATE;
Time= 10488 msec.
                        3      2      4
(D2) N (K LOG(N) - 2 N  + 3 N  + B)

        3      2                          3      2      3
 + 2 N (3 N  - 4 N  - 2 K) (K LOG(N) - 2 N  + 3 N  + B)

        6        5         4           3         2         2
 + N (1080 N  - 2940 N  + 2016 N  - 1575 K N  + 2240 K N  + 840 K )

         3      2      2
(K LOG(N) - 2 N  + 3 N  + B) /70
            9            8            7            6             6
 + N (272160 N  - 1125600 N  + 1561140 N  - 631800 K N  - 725760 N

            5            4          2 3            2 2
 + 1808100 K N  - 1298304 K N  + 694575 K  N  - 1019200 K  N

          3              3      2
 - 352800 K ) (K LOG(N) - 2 N  + 3 N  + B)/14700
```

```
     9        4          3            2
   N  (42768 N  - 237952 N  + 498966 N  - 467181 N + 164736)
 + ---------------------------------------------------------
                            5005


     7        3          2
   K N  (2437776 N  - 10510000 N  + 15149673 N - 7299072)
 - -------------------------------------------------------
                          88200


   2 5           2
   K N  (23996250 N  - 70529375 N + 51894528)
 + -------------------------------------------
                        514500
     3 3
   5 K  N  (1377 N - 2048)           4
 - ----------------------- + 24 K  N
            144
```

Actually, the researcher needed the FORTRAN expression for the integral in (D2). MACSYMA has the facility for converting MACSYMA expressions in FORTRAN as shown below.

```
(C3) FORTRAN(D2);
     N*(K*LOG(N)-2*N**3+3*N**2+B)**4+2*N*(3*N**3-4*N**2-2*K)*(K*LOG(N)-
   1  2*N**3+3*N**2+B)**3+N*(1080*N**6-2940*N**5+2016*N**4-1575*K*N**
   2  3+2240*K*N**2+840*K**2)*(K*LOG(N)-2*N**3+3*N**2+B)**2/70.0+N*(2
   3  72160*N**9-1125600*N**8+1561140*N**7-631800*K*N**6-725760*N**6+
   4  1808100*K*N**5-1298304*K*N**4+694575*K**2*N**3-1019200*K**2*N**
   5  2-352800*K**3)*(K*LOG(N)-2*N**3+3*N**2+B)/14700.0+N**9*(42768*N
   6  **4-237952*N**3+498966*N**2-467181*N+164736)/5005.0-K*N**7*(243
   7  7776*N**3-10510000*N**2+15149673*N-7299072)/88200.0+K**2*N**5*(
   8  23996250*N**2-70529375*N+51894528)/514500.0+(-5.0)*K**3*N**3*(1
   9  377*N-2048)/144.0+24*K**4*N
Time= 2250.0 msecs.
```

Some users have actually generated thousands of lines of FORTRAN and transferred the code by networks to their large numeric processors for number crunching.

6.7 Nuclear Chemistry/Exact Answers

The following integral appears in a study of the work diagram for Plutonium. In his original paper, the proposer [Si1] made numeric approximations not certain whether (D1) could be integrated in closed form. Having learned about MACSYMA, the proposer wanted to know whether his numeric approximations were accurate. It turns out that MACSYMA integrates (D1) readily. With this result in hand, the person was able to verify the accuracy of his earlier results.

```
        /  4       3      2     2
        [ (M  + 4 C M  + 9 D M  + C M  + 4 D M + C D) LOG(M)
(D1)    I -------------------------------------------------- dM
        ]                     3    2          2
        /                   (M  + M  + C M + D)
```

```
(C2) D1,INTEGRATE;
Time = 6880 msec.
```

```
              2
         M (3 M  + 2 M + C) LOG(M)           3    2
(D2)     -------------------------- - LOG(M  + M  + C M + D)
               3    2
             M  + M  + C M + D
```

```
(C3) DIFF(D2,M),FACTOR;
```

```
          4       3      2     2
        (M  + 4 C M  + 9 D M  + C M  + 4 D M + C D) LOG(M)
(D3)    -------------------------------------------------
                          3    2          2
                        (M  + M  + C M + D)
```

There was the added benefit of being able to check the answer in (C3) by differentiating (D2) to retrieve to the original integrand in (D1).

6.8 Fluid Flow/Theory Building

MACSYMA's capabilities in rational arithmetic can help in theory building. In the following problem the authors were studying steady, inviscid, incompressible, two-dimensional flow with a free surface, that is, water waves [Hu1]. Using MACSYMA, they were able to derive sequences of large rational numbers, an example of which is shown in (D1).   However, the largest number in the sequence exhausted the time/memory of their computer, so they could not continue.

These rational numbers led to an approximate result that was orders of magnitude better than anyone had ever produced before, but the authors were still limited in the analysis. Sometimes numeric results can be extended using MACSYMA, in this case to guess the Nth term in the sequence. To do this we first factor (D1), and this results in (D2). The game now involves examining these seven fractions and trying to construct a function that generates them.

```
           9   625  117649  4782969   25937424601   23298085122481
(D1)  [1, -,  ---, ------, -------, -----------, --------------]
           8   192   9216    81920     88473600      14863564800
```

```
(C2) FACTOR(D1);
```

```
           2    4      6      14        10            12
           3    5      7       3        11            13
(D2)  [1, --, ----, ------, -----, ---------, -----------]
           3    6    10 2    14     17 3  2    20  4  2
           2    2 3   2  3    2  5   2   3  5    2  3  5  7
```

From the form of the denominators in (D2) one gets the impression that factorials are present, and it is not difficult to guess at a possible form. Such a function is given below in (D3).

```
(C3) F(L):=(L-1/2)^(2*L-2)*L/L!^2;
```

$$(D3) \qquad F(L) := \frac{\left(L - \dfrac{1}{2}\right)^{2L-2} L}{L!^{2}}$$

For this function the first nine values of the argument give:

```
(C4) [F(1),F(2),F(3),F(4),F(5),F(6),F(7),F(8),F(9)];
```

```
           9   625  117649  4782969   25937424601   23298085122481
(D4)  [1, -,  ---, ------, -------, -----------, -------------- ,
           8   192   9216    81920     88473600      14863564800


      14416259765625   48661191875666868481
      --------------, --------------------]
         1644167168      958878292377600
```

In (D4) we find all the terms in (D1), along with two additional terms. One does not immediately know whether the generating function F(L) is either unique or correct, and some additional analysis, perhaps induction, will have to be done to verify this. As an alternative, one might go back to the original analysis of the problem, try to force an additional couple of terms into the sequence, and see if they fit the formula. But in this case, and many others, I have found that this rather simplistic approach can lead to significant results.

6.9 Solitons/"It can be shown that"

Everyone who reads technical articles in the sciences finds expressions supported by statements such as "it is evident that" or "it can be shown that". Before embarking on research involving such results, it is essential to know whether a relation is

45

correct or not. Often MACSYMA allows one to painlessly verify such statements.
(D1) below gives a shock wave solution of Burger's equation involving stability of
Korteweg-de Vries soliton [Je1]. We are told, without proof, that the solution (D2)
satisfies the differential equation (D1) subject to the constraint (D3).

$$
\text{(D1)} \quad -F\,(24\ \mathrm{SECH}^2(Y)\ \mathrm{TANH}(Y) + A) - 4\,\frac{dF}{dY}\,(1 - 3\ \mathrm{SECH}^2(Y)) + \frac{d^3F}{dY^3} = 0
$$

(C2) SOLUTION;

$$
\text{(D2)} \quad F = 4\,\left(\frac{d^2}{dY^2}(\%E^{(G-1)Y}\ \mathrm{SECH}(Y))\right) + (G-2)^2\ G\ \%E^{G\,Y}
$$

(C3) CONSTRAINT;

$$
\text{(D3)} \qquad A = G^3 - 4\,G
$$

To see if this is correct, we first set the switch **EXPONENTIALIZE** to **TRUE** so
that all trigonometric and hyperbolic expressions are expressed in exponential form.
Now we need not not be concerned about making trigonometric simplifications. In
(C5) we take (D2), differentiate it, and substitute this into (D1). We take the
resulting expression and simplify and factor it to obtain (D6). For this to vanish, one
of the factors must vanish, and the first factor is precisely the constraint (D3). The
other factors cannot vanish for arbitrary values of Y. Thus, the result given in the
paper is correct.

(C4) EXPONENTIALIZE:TRUE$

(C5) D1,EV(D2,DIFF),DIFF$
Time= 6139 msec.

(C6) FACTOR(RATSIMP(D5));
Time= 16336 msec.

$$
\text{(D6)} \quad (G^3 - 4\,G - A)\,(G\,\%E^{2Y} - 2\,\%E^{2Y} + G + 2)
$$

$$
(G^2\,\%E^{4Y} - 2\,G\,\%E^{4Y} + 2\,G^2\,\%E^{2Y} - 16\,\%E^{2Y} + G^2 + 2\,G)\,\%E^{G\,Y}
$$

$$
/(\%E^{2Y} + 1)^3 = 0
$$

6.10 Antenna Theory/Impossible Problems

MACSYMA cannot do the impossible, yet such requests are often made. The
following is an indefinite integral as it was shown to me. It arises from an antenna
problem studied by the Air Force. The solution of the problem involves a combination
of numeric and algebraic capabilities to prove a null result.

Below, (D1) is an awful looking expression involving %I = SQRT(-1), several
parameters [K,F,P,X0,Y0,Z0], and variables X and Y and U(X,Y) given by (D2). The
exercise is to find the indefinite integral of (D1) over X and Y. The researchers spent
many CPU hours trying to coerce MACSYMA to return an answer, but MACSYMA
would only return the noun form. That is the integrand is returned with an integral
sign, and it normally implies that MACSYMA cannot perform the integration.

```
                                                  2    2
            2   2                      %I K (F COS(P) (Y  + X ) + SIN(P) X - U)
(D1) U (K  U  - 3 %I K U - 3) %E


                  2    2
  (COS(P) (2 F Y Y0 + F (Y  + X ) - Z0) + 2 F SIN(P) X) (Z0 - F (Y  + X ))


       2    2
 /SQRT(Y  + X  + 1)



                2    2
                Y  + X      2           2           2
    (D2)     U = (------- - Z0)  + (Y - Y0)  + (X - X0)
                   4 F
```

How can one tell if this expression can be integrated in closed form in terms of a
finite series of elementary functions (integrated in finite terms)? A more reasonable
question might be whether it is even worth looking for the closed form. For this
example we can make a good deal of progress. In (C3) we substitute (D2) into (D1),
and in (C4) we take this expression and evaluate it for some particular values of the
constants (and Y). The purpose is to simplify the double integral as much as possible
to a particular end that will be seen. We then factor the expression in (D4) resulting
in (D5).

```
(C3) D1,D2$

(C4) D3,F=1,Y=0,X0=0,Y0=0,Z0=0,K=16*%I,PSI=%PI/2$

(C5) FACTOR(D4)$
                                                       4       2
     5    2        8      6        4       2          X  + 16 X  - 16 X
    X  (X  + 16) (X  + 32 X  + 253 X  - 48 X  + 3) %E
(D5) -----------------------------------------------------------------
                                    2
                            8 SQRT(X  + 1)
```

It can be shown that (D5) cannot be integrated in finite terms. However, even if
one did not know this, searching for the integral of (D5) in tables or using

MACSYMA would lead one to suspect that it is very difficult to integrate. The horror of the original integral was obscured by all of the parameters present, but by removing them its complexity becomes more transparent. The main argument we can now use is that since we cannot integrate in finite terms over X for a particular value of the parameters, it means that we cannot integrate in finite terms over X in general. Hence, the original expression cannot be integrated in finite terms, and one should look to other methods to obtain the desired result.

6.11 Resolution of Closely Spaced Targets/Numeric Systems Cannot Cope

To assess the parameter estimation performance associated with closely spaced object problems, a theoretical lower bound (Cramer-Rao) is often used. In computing this bound, one must invert the Fisher Information matrix (shown below). In the case of interest the elements of (D1) were numbers. The matrix was almost singular and the numeric inversion routine, to double precision, failed to return an answer. The way we approached the problem was to compute the analytic form for the inverse and then insert numerical values. The actual calculation is shown below:

```
(C1) MATRIX([1,A,0,B,0,C],[A,1,-B,0,-C,0],[0,-B,-R,-D,-E,-F],
     [B,0,-D,-R,-F,-E],[0,-C,-E,-F,-G,-H],[C,0,-F,-E,-H,-G]);
```

```
          [ 1    A    0    B    0    C ]
          [                            ]
          [ A    1  - B    0  - C    0 ]
          [                            ]
          [ 0  - B  - R  - D  - E  - F ]
(D1)      [                            ]
          [ B    0  - D  - R  - F  - E ]
          [                            ]
          [ 0  - C  - E  - F  - G  - H ]
          [                            ]
          [ C    0  - F  - E  - H  - G ]
```

```
(C2) D1^^(-1)$
Time= 44927 msec.
```

The inversion of (D1) with the command (C2) takes less than 1 CPU minute, and below is the factored form of the 1-1 element of the inverse matrix. The problem was basically solved and numeric values could now be inserted, but notice that we received more than we asked for. The denominator of (D3), the determinant of the matrix, factors into two parts. The near singularity of the matrix can be attributed to the value of the determinant being near zero. The factorization of the determinant into two parts allowed us to look at these parts separately and gain physical insight into the singularity of the matrix.

(C3) FACTOR(D2[1,1]);

$$(D3)\ (H^2R^2 - G^2R^2 - C^2GR + BH^2R - 4EFHR - 2BCFHR$$

$$-\, B^2G^2R + 2F^2GR + 2E^2GR + 2BCEGR + C^2F^2R + C^2E^2R$$

$$-\, D^2H^2 + 2DF^2H - 2B^2EFH + 2DE^2H + 2BCDEH + D^2G^2$$

$$+\, B^2F^2G - 4DEFG - 2BCDFG + B^2E^2G + C^2D^2G - F^4$$

$$+\, 2E^2F^2 + 2BCE^2F - 2C^2DEF - E^4 - 2BCE^3)/$$

$$((AHR - HR + AGR - GR - C^2R + ADH - DH - B^2H + ADG$$

$$-\, DG - B^2G - AF^2 + F^2 - 2AEF + 2EF + 2BCF - AE^2 + E^2$$

$$+\, 2BCE - C^2D)(AHR + HR - AGR - GR - C^2R - ADH - DH$$

$$+\, B^2H + ADG + DG - B^2G + AF^2 + F^2 - 2AEF - 2EF$$

$$-\, 2BCF + AE^2 + E^2 + 2BCE + C^2D))$$

6.12 Optical Discrimination/Saving Time and Money

To discriminate an object by viewing its radiated brightness at various angles, one can use the energy flux on the object above a diffusely emitting earth. For a spherical object, we encountered the following integral that does not appear to be tabulated:

$$(D1)\qquad \int \cos(P)\,\mathrm{ATAN}\!\left(\frac{A}{\cos(P)}\right)\,dP$$

This integral was embedded within a four-dimensional integral that was being evaluated numerically. We ran the problem on a large CYBER system, and after 3 CPU hours obtained only 15 points. At several hundred dollars/CPU minute, this is an expensive way to obtain a result (or not obtain a result). We traced the slow convergence of the four-dimensional integral to the slow convergence of (D1) over the range of interest. We then evaluated (D1) with MACSYMA, although it takes a bit of work to get the integral into the form shown in (D2). Once found we replaced (D1) with (D2) in the FORTRAN program, and the numeric program then returned several hundred points after only 2 CPU minutes. Thus, a problem that initially was intractable owing to time and budgetary constraints was easily solved by preprocessing with MACSYMA.

```
(C2) D1,INTEGRATE;

                2             A TAN(P)                    SIN(P)
(D2)  - SQRT(A  + 1) ATAN(------------) + 2 A ATAN(----------)
                              2                     COS(P) + 1
                         SQRT(A  + 1)


                                                        A
                                      + SIN(P) ATAN(------)
                                                     COS(P)
```

6.13 Resolving Closely Spaced Optical Targets/Combining MACSYMA with Reference Tables

The CSO resolution problem is usually studied in the spatial domain, but can be looked at in the frequency domain as well. To evaluate the estimation performance, in a particular case, the following integral occurred:

```
        INF
        /
        [          SIN(%PI (X - M)) SIN(%PI (X - N))
(D1)    I        ---------------------------------------------- dX
        ]                            2                2
        /         (X - M) (X - N) (1 - (X - M) ) (1 - (X - N) )
        MINF
```

This integral is not tabulated. However, we can perform a partial fraction decomposition of the terms containing M in the integrand to find

```
                    1                    1            1
(D2)    ---------------------- = (- ------------- + -----
                        2            2 (X - M + 1)   X - M
        (X - M) (1 - (X - M) )


                                        1
                                - -------------)
                                   2 (X - M - 1)
```

with a similar expression for N. Combining these under the integral sign and expanding allows one to write (D1) as the sum of nine integrals, each of which is tabulated.

We have used MACSYMA in this example as an advanced calculator to do a partial-fraction decomposition. Left to a hand calculation, one might not even try this computation owing to the messy algebra. But, with MACSYMA no more than a few seconds of the user's time is needed to try it.

6.14 Approximation to the Planck Integral/New Results in Old Fields

One would think that almost everything of interest or utility is known about the Planck integral. We shall see that this is not the case. It is known that the Planck integral cannot be evaluated in closed form in terms of a finite series of elementary functions [Pa10]. Therefore, one must choose between a numeric integration (slow) or numeric approximations to the integral (fast). In the infrared range, approximations to the Planck integral converge very slowly. Recently, MACSYMA was used to find a new approximation to the Planck integral that converges many times faster in this range. Hundreds of CPU hours were saved in a particular research project.

Below are equivalent expressions (setting constants to unity) for the Planck integral in the wavelength and frequency domains.

```
     /                          /    3
     [        1                 [    V
     I -------------- dL  =  I ------- dV
     ]  5    1/L                ]    V
     / L  (XE    - 1)           / XE  - 1
```

We next rewrite the integrand as a powerseries.

```
                                       INF              INF
       3         3   - V                ====             ====
       V         V  XE         3   - V \       - N V   3 \       - N V
     ------- = ---------  =  V  XE      >    XE       = V  >   XE
       V           - V                 /                   /
     XE   - 1   1 - XE                 ====             ====
                                       N = 0            N = 1
```

We now integrate the last term on the right above, with respect to V, to find the known powerseries expansion for the Planck integral.

```
                          INF
     /    3               ====        3     2
     [    V               \          V    3 V    6 V   6     - N V
     I ------- dV  =  -     >      (-- + ---- + --- + --) XE
     ]    V               /          N     2     3    4
     / XE  - 1            ====             N     N    N
                          N = 1
```

The first sum on the right hand side of the expression above

```
      INF
      ====   - N V
      \     %E                    - V
       >    ------- = - LOG( 1 - %E   )
      /       N
      ====
      N = 1
```

can be evaluated in closed form as shown. It turns out that this is the term (of the four) that converges most slowly in the infrared. We now have a new approximation to the Planck integral:

```
                 INF
   /    3        ====      2
   [    V        \      3 V    6 V    6      - N V
   I ------- dV = - >   (---- + --- + --) %E
   ]    V        /      2     3     4
   / %E  - 1     ====    N    N     N
                 N = 1
                                      3           - V
                                  + V  LOG( 1 - %E   )
```

Using this relation, rather than the relation with four sums, is clearly preferable for all ranges of the frequency but it is particularly efficient in the infrared. This relation for the Planck integral does not appear in the standard literature although the closed form for the sum is known. Apparently nobody used it in this context before or if they did it was not publicized very well.

After finding this result with MACSYMA in 1979, I performed a literature search to try to find the approximation and was unsuccessful. It would be surprising if the result is really new. However, it makes more sense to derive a relation in a few minutes with MACSYMA than to spend several hours searching for the relation (that one believes should exist) in the literature.

Finally, the form above also leads to a new representation of the Planck integral in terms of Polylogarithms:

```
    3       - V     2       - V          - V           - V
  - V  LI (%E   ) - 3 V  LI (%E   ) - 6 V LI (%E   ) - 6 LI (%E   )
       1                 2                 3               4
```

6.15 Group Theory/Testing Conjectures

A problem in the parametrization of the special orthogonal group led to a conjecture that circulated among the MACSYMA hackers at MIT: If M and N are antisymmetric matrices, then the determinant of (ID - M.N) is a perfect square where ID is the identity matrix. Several of us became interested in this problem and immediately asked MACSYMA whether it is true in certain cases. It is easy to verify the conjecture by hand for the 2x2 and 3x3 case. The 4x4 can almost be done by hand but for the 5x5 and 6x6 cases MACSYMA is needed. Below is the 5x5 verification:

```
(C1) M5;
                    [    0      X12      X13      X14      X15  ]
                    [                                           ]
                    [ - X12     0        X23      X24      X25  ]
                    [                                           ]
(D1)                [ - X13   - X23      0        X34      X35  ]
                    [                                           ]
                    [ - X14   - X24    - X34      0        X45  ]
                    [                                           ]
                    [ - X15   - X25    - X35    - X45      0    ]

(C2) N5;
                    [    0      Y12      Y13      Y14      Y15  ]
                    [                                           ]
                    [ - Y12     0        Y23      Y24      Y25  ]
                    [                                           ]
(D2)                [ - Y13   - Y23      0        Y34      Y35  ]
                    [                                           ]
                    [ - Y14   - Y24    - Y34      0        Y45  ]
                    [                                           ]
                    [ - Y15   - Y25    - Y35    - Y45      0    ]
```

We compute the determinant below in (D3), where M5.N5 is the MACSYMA syntax for matrix multiplication and IDENT(5) is the 5x5 identity matrix. The determinant is very large as we see from (D4) (the NTERMS command is described in example 3.4). Even after simplification in (C5) it is large, but at least it is manageable.

```
(C3) DETERMINANT(IDENT(5)-M5.N5)$

(C4) NTERMS(D3);

(D4)                            56792

(C5) RATSIMP(D3)$

(C6) NTERMS(D5);

(D6)                            1506
```

To solve the problem, we take the square root of the square free part (SQRF) of (D5). SQRF is an algorithm similar to FACTOR except for the condition that the polynomial factors are "square free". The result is shown below:

(C7) SQRT(SQRF(D5));

(D7) X23 X45 Y23 Y45 - X24 X35 Y23 Y45 + X25 X34 Y23 Y45

+ X13 X45 Y13 Y45 - X14 X35 Y13 Y45 + X15 X34 Y13 Y45 + X45 Y45

+ X12 X45 Y12 Y45 - X14 X25 Y12 Y45 + X15 X24 Y12 Y45 + X35 Y35

- X23 X45 Y24 Y35 + X24 X35 Y24 Y35 - X25 X34 Y24 Y35 + X34 Y34

- X13 X45 Y14 Y35 + X14 X35 Y14 Y35 - X15 X34 Y14 Y35 + X25 Y25

+ X12 X35 Y12 Y35 - X13 X25 Y12 Y35 + X15 X23 Y12 Y35 + X24 Y24

+ X23 X45 Y25 Y34 - X24 X35 Y25 Y34 + X25 X34 Y25 Y34 + X23 Y23

+ X13 X45 Y15 Y34 - X14 X35 Y15 Y34 + X15 X34 Y15 Y34 + X15 Y15

+ X12 X34 Y12 Y34 - X13 X24 Y12 Y34 + X14 X23 Y12 Y34 + X14 Y14

- X12 X45 Y14 Y25 + X14 X25 Y14 Y25 - X15 X24 Y14 Y25 + X13 Y13

- X12 X35 Y13 Y25 + X13 X25 Y13 Y25 - X15 X23 Y13 Y25 + X12 Y12

+ X12 X45 Y15 Y24 - X14 X25 Y15 Y24 + X15 X24 Y15 Y24

- X12 X34 Y13 Y24 + X13 X24 Y13 Y24 - X14 X23 Y13 Y24

+ X12 X35 Y15 Y23 - X13 X25 Y15 Y23 + X15 X23 Y15 Y23

+ X12 X34 Y14 Y23 - X13 X24 Y14 Y23 + X14 X23 Y14 Y23 + 1

6.16 Estimation Theory/A Research Problem

Until now I have shown applications that are rather easily stated and solved. None of them required more than a few hours to analyze with MACSYMA. The example below is both difficult and time consuming. The problem was proposed by a mathematician at MIT's Lincoln Laboratory. He had a research program that required solving a problem in estimation theory. He worked on it by hand for a long time because it could not be solved by numeric systems. Eventually he put aside the problem because the analysis seemed hopeless. When he showed it to me it was not clear that MACSYMA could help. It required several days before I came up with an idea for using MACSYMA that worked.

The recursive differential relation below arises in a study of maximum likelihood estimation applied to multiple-pulse monopulse problems. The mathematical problem involves finding the limit of the array element V[N,Y,L] as L approaches 1 where V[N,Y,L] satisfies the differential recursion relation

$$V_{N\ Y\ L} = V_{N-1\ Y\ L} + \frac{1}{2*Y} \frac{d\ V_{N-1\ Y\ L}}{d\ L}$$

The first element in the array is known and is given as

$$V[1,Y,L] = 2\ \%E^{-L\ Y}\ (SQRT(L^2 - 1)\ Y\ SINH(SQRT(L^2 - 1)\ Y)$$

$$+ (L^2\ Y - 1)\ COSH(SQRT(L^2 - 1)\ Y))/Y + \frac{\%E^{-2\ L\ Y} + 1}{Y^2} - \frac{2\ L}{Y} + 1$$

It is not efficient, in the current version of **MACSYMA**, to try compute the array elements and use the LIMIT package directly. Rather, it is better to compute the limit as the zeroth-order term of the Taylor series. This procedure can easily be substantiated.

Let us examine the computational difficulties by writing T(L) for V[1,Y,L]. In (C1) we enter the recursion relation and in (C2) assign T(L) to V[1,Y,L].

    (C1) V[N,Y,L]:=V[N-1,Y,L]+DIFF(V[N-1,Y,L],L)/(2*Y);

$$(D1) \qquad V_{N,\ Y,\ L} := V_{N-1,\ Y,\ L} + \frac{DIFF(V_{N-1,\ Y,\ L}, L)}{2\ Y}$$

    (C2) V[1,Y,L]:T(L)$

Next we evaluate V[2,Y,L] and substitute for T(L) in (C4).

    (C3) EV(V[2,Y,L],RATSIMP,EXPAND);

$$(D3) \qquad \frac{T(L)_L}{2\ Y} + T(L)$$

    (C4) D3,T(L)=2*%E^(-L*Y)*(SQRT(L^2-1)*Y*SINH(SQRT(L^2-1)*Y)+
         (L*Y-1)*COSH(SQRT(L^2-1)*Y))/Y^2+(EXP(-2*L*Y)+1)/Y^2 -2*L/Y+1$

We now compute the limit as the zeroth-order term in the Taylor series about the expansion point L = 1. This looks too simple.

    (C5) V[2,Y,1]= TAYLOR(D4,L,1,0);

$$V_{2,\ Y,\ 1} = \frac{(Y + 2)\ \%E^{-Y}}{Y} - \frac{2}{Y} + 1$$

To truly appreciate the computational horrors were one to do this by hand, consider the ninth member of the array V[9,Y,L]:

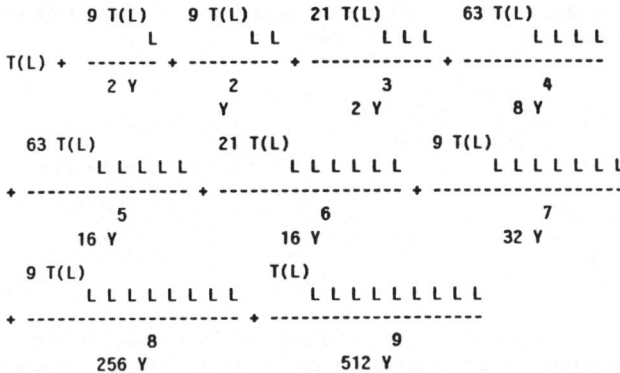

Into this one must substitute the value of T(L) as given above, differentiate as indicated, and then compute the limit as L approaches 1. I used MACSYMA to compute the limit of the array elements for the first 11 cases, and some of these are shown below.

$$V_{3,\,Y,\,1} = \frac{(Y^3 + 9Y^2 + 18Y + 6)\,XE^{-Y}}{6Y^2} - \frac{2}{Y} - \frac{1}{Y^2} + 1$$

$$V_{4,\,Y,\,1} = \frac{(Y^4 + 20Y^3 + 120Y^2 + 240Y + 120)\,XE^{-Y}}{60Y^2} - \frac{2}{Y} - \frac{2}{Y^2} + 1$$

. . . . .
. . . . .

$$V_{11,\,Y,\,1} = (Y^{11} + 209Y^{10} + 18810Y^9 + 959310Y^8 + 30697920Y^7$$
$$+ 644656320Y^6 + 9025188480Y^5 + 83805321600Y^4 + 502831929600Y^3$$
$$+ 1843717075200Y^2 + 3687434150400Y + 3016991577600)\,XE^{-Y}$$
$$/(335221286400\,Y^2) + \frac{Y^2 - 2Y - 9}{Y^2}$$

The game now begins. One tries to relate the large integers in the array elements to the array index. This can be very time consuming and for this case it took me more than two days. I eventually deduced an expression

```
                                 N
                                ====              J
                          - Y \              Y
        (N - 2)! N! XE      >    -------------------------
                          /      J! (N - J)! (N + J - 3)!
                                ====
    2   N - 2                   J = 0
1 - - - ----- + ------------------------------------------------
    Y     2                              2
        Y                              Y
```

that reproduces the elements of the arrays for each of the 11 cases. I returned this expression to the mathematician who posed the problem. He believed I had done the analysis by hand, and he spent the next six weeks proving that it was correct. I can report that his research is now moving along well, and that he is also writing several papers on the subject.

7. Limits of Computational Power

Many people are interested in the size of problems that can be solved now and in speculations about the size of solvable problems in the future. Here are a few examples that will give some insight into this.

7.1 Geometric Series

The first example is based on the elementary geometric series defined in (D1) for an arbitrary integer M. This expression vanishes identically. To generate a more complicated problem we substitute 1 + 1/X for X in (D1) resulting in (D2)

```
               M
              ====              M
              \       N      X (X  - 1)
      (D1)     >     X   -   ----------  =  0
              /                X - 1
              ====
              N = 1
```

(C2) D1, X = 1+1/X;

$$
(D2) \quad \sum_{N=1}^{M} \left(\frac{1}{X}+1\right)^{N} - \left(\left(\frac{1}{X}+1\right)^{M}-1\right)\left(\frac{1}{X}+1\right) X = 0
$$

This expression (D2) must also vanish for all integer M.

In (C3) MACSYMA is asked to sum (D2) from 1 to 1000. The resulting truncated expression (D3) has a large number of terms. The MACSYMA function NTERMS gives one a handle on how large an expression would be if it were fully expanded out and no simplification of like terms occurred. MACSYMA finds that there are 503504 terms before simplification. From the binomial theorem it is not difficult to show that this number is precisely $(M^2 + 7{*}M + 8)/2$ for M = 1000. MACSYMA simplifies (D3) after almost 2 CPU hours, and the expected answer is returned.

(C3) D2,M = 1000,SUM;

$$
(D3) \quad \left(\frac{1}{X}+1\right)^{1000} + \left(\frac{1}{X}+1\right)^{999} + \ldots + \left(\frac{1}{X}+1\right)^{2} + \left(\frac{1}{X}+1\right)^{1} + 1
$$

$$
- \left(\left(\frac{1}{X}+1\right)^{1000} - 1\right)\left(\frac{1}{X}+1\right) X = 0
$$

(C4) NTERMS(LHS(D3));
(D4)   503504

(C5) RATSIMP(D3);
Time= 7226000 msec. = (2 hrs)

(D5)   0 = 0

7.2 Taylor Series

In this example (see example 3.14) in which MACSYMA computes the first 500 terms of the Taylor series of (D1) about X = 0.

$$
(D1) \quad A \, SIN(X^3) + \frac{B \, LOG(X^2 - X + 1)}{X^5}
$$

(C2) TAYLOR(D1,X,0,500)$
Time= 590000.0 msec.

This expression is huge, completely useless, and it is not displayed. However, MACSYMA does find it in about 10 CPU minutes, and the example does say something about symbolic computing power.

7.3 Einstein's Field Equations in Four Dimensions

Here is a problem, with more physical motivation, that I proposed four years ago. It better illustrates the question of limits on computational power today. For two, three and four dimensions we have computed and completely simplified the components of the Ricci tensor (the object that is the source of the vacuum field equations of Einstein's General Theory of Relativity) using the most general metric tensor. In four dimensions the components of the metric tensor can be expressed by the matrix

```
[ A(X,Y,Z,T)   B(X,Y,Z,T)   C(X,Y,Z,T)   D(X,Y,Z,T) ]
[                                                    ]
[ B(X,Y,Z,T)   E(X,Y,Z,T)   F(X,Y,Z,T)   G(X,Y,Z,T) ]
[                                                    ]
[ C(X,Y,Z,T)   F(X,Y,Z,T)   H(X,Y,Z,T)   I(X,Y,Z,T) ]
[                                                    ]
[ D(X,Y,Z,T)   G(X,Y,Z,T)   I(X,Y,Z,T)   J(X,Y,Z,T) ]
```

For this case the covariant Ricci tensor consists of 10 second-order coupled partial differential equations in the four space-time variables. The diagonal components have 9990 terms each and the off-diagonal components have 13280 terms each. Some researchers have speculated that these equations will eventually be solved in this most general case. Pondering the complexity of the expressions, one wonders if this could really come to pass. Each component, when printed, is the size of a 200 page book.

The object of the exercise is to compute and simplify the components of the covariant Ricci tensor. Here is a table giving the result of the computation:

DIMENSION	DIAGONAL TERMS	OFF DIAGONAL TERMS	CPU SECONDS MACSYMA	CPU SECONDS CAMAL
2	17	17	5	1
3	416	519	30	3
4	9990	13280	350	45

Several Computer Algebra systems attempted to generate the table above. Not all were successful. The CPU times to complete the three and four dimensional calculations (on the systems that succeeded) ranged from an average of 30 and 350 CPU seconds/component respectively by L. S. Hornfeldt and me using MACSYMA, to an average of 3 and 45 seconds/component respectively by R. G. McLenaghan and G. J. Fee using CAMAL [Fil]. MACSYMA was run on the MIT Multics operating system on a Honeywell 68/DPS and CAMAL on an IBM 4341 at the University of Waterloo. The time difference is mainly due to the machine speeds. Special algorithms were constructed that reduced the computation times.

I would challenge any system to compute the Ricci tensor in 5 dimensions and completely simplify the components. I doubt that any computer in the 1980s will have that capacity. With this challenge offered I will likely be proven wrong.

7.4 Computational Power in 2084?

It seems appropriate to conclude this paper by speculating about the power of Computer Algebra systems 100 years from now. Below is a 14x14 matrix:

$$
\begin{vmatrix}
M_{aa} & M_{ab} & M_{ac} & M_{ad} & M_{ae} & M_{af} & M_{ag} & M_{ah} & M_{ai} & M_{aj} & M_{ak} & M_{al} & M_{am} & M_{an} \\
M_{ba} & M_{bb} & M_{bc} & M_{bd} & M_{be} & M_{bf} & M_{bg} & M_{bh} & M_{bi} & M_{bj} & M_{bk} & M_{bl} & M_{bm} & M_{bn} \\
M_{ca} & M_{cb} & M_{cc} & M_{cd} & M_{ce} & M_{cf} & M_{cg} & M_{ch} & M_{ci} & M_{cj} & M_{ck} & M_{cl} & M_{cm} & M_{cn} \\
M_{da} & M_{db} & M_{dc} & M_{dd} & M_{de} & M_{df} & M_{dg} & M_{dh} & M_{di} & M_{dj} & M_{dk} & M_{dl} & M_{dm} & M_{dn} \\
M_{ea} & M_{eb} & M_{ec} & M_{ed} & M_{ee} & M_{ef} & M_{eg} & M_{eh} & M_{ei} & M_{ej} & M_{ek} & M_{el} & M_{em} & M_{en} \\
M_{fa} & M_{fb} & M_{fc} & M_{fd} & M_{fe} & M_{ff} & M_{fg} & M_{fh} & M_{fi} & M_{fj} & M_{fk} & M_{fl} & M_{fm} & M_{fn} \\
M_{ga} & M_{gb} & M_{gc} & M_{gd} & M_{ge} & M_{gf} & M_{gg} & M_{gh} & M_{gi} & M_{gj} & M_{gk} & M_{gl} & M_{gm} & M_{gn} \\
M_{ha} & M_{hb} & M_{hc} & M_{hd} & M_{he} & M_{hf} & M_{hg} & M_{hh} & M_{hi} & M_{hj} & M_{hk} & M_{hl} & M_{hm} & M_{hn} \\
M_{ia} & M_{ib} & M_{ic} & M_{id} & M_{ie} & M_{if} & M_{ig} & M_{ih} & M_{ii} & M_{ij} & M_{ik} & M_{il} & M_{im} & M_{in} \\
M_{ja} & M_{jb} & M_{jc} & M_{jd} & M_{je} & M_{jf} & M_{jg} & M_{jh} & M_{ji} & M_{jj} & M_{jk} & M_{jl} & M_{jm} & M_{jn} \\
M_{ka} & M_{kb} & M_{kc} & M_{kd} & M_{ke} & M_{kf} & M_{kg} & M_{kh} & M_{ki} & M_{kj} & M_{kk} & M_{kl} & M_{km} & M_{kn} \\
M_{la} & M_{lb} & M_{lc} & M_{ld} & M_{le} & M_{lf} & M_{lg} & M_{lh} & M_{li} & M_{lj} & M_{lk} & M_{ll} & M_{lm} & M_{ln} \\
M_{ma} & M_{mb} & M_{mc} & M_{md} & M_{me} & M_{mf} & M_{mg} & M_{mh} & M_{mi} & M_{mj} & M_{mk} & M_{ml} & M_{mm} & M_{mn} \\
M_{na} & M_{nb} & M_{nc} & M_{nd} & M_{ne} & M_{nf} & M_{ng} & M_{nh} & M_{ni} & M_{nj} & M_{nk} & M_{nl} & M_{nm} & M_{nn}
\end{vmatrix}
$$

Let us call this matrix MATR. MACSYMA computes the determinant in about 4 CPU hours. The number of terms in the determinant of a general NxN matrix is N! and no cancellations are possible. Indeed, simply attempting to print one single term exhausts the memory of the machine. A reason for this is that the algorithm MACSYMA uses for computing determinants involves an expansion by co-factors so that the determinant is a deeply nested product. Of course, as we also see, 14! is a very large number. Perhaps 100 years from now (or perhaps in less than 10 years), one will be able to compute and manipulate such expressions on a hand-held device. Of course it had better have a large display!

```
(C1) DETERMINANT(MATR)$
Time= 13672000 msec.

(C2) 14!;

(D2)                  87178291200
```

60

REFERENCES

[Di1] Dingle, H. *Proc. Nat. Acad. Sci.* **19** (1933) 559.

[Fi1] Fitch, J. P. *CAMAL User's Manual* Univ. Cambridge Computation Lab. (1975).

[Go1] Gosper, R. W. *Proc. Nat. Acad. Sci.* **75** (1978) 40 (see also Reference MA1, 1-6-18).

[Gou1] Goursat, E. *A Course in Mathematical Analysis* Vol.1 Dover, New York, (1904) Sec. 30.

[Gr1] Gradshteyn, I. S. and Ryzhik, I. M. *Table of Integrals, Series and Products* 4th edition, Academic Press, New York (1965) Pg. 85, Sec. 2.269.4. The error is corrected in the 4th [sic] edition (1980).

[Gr2] Gradshteyn, I. S. and Ryzhik, I. M. *Table of Integrals, Series and Products* 4th edition, Academic Press, New York (1980) Pg. 648, Sec. 6.282.1.

[Ha1] Harrison, B. K. *Phys. Rev.* **116** (1959) 1285.

[Hu1] Hui, W. H. and Tenti, G. *J. Applied Math. and Physics* **33** (1982) 569-589.

[Je1] Jeffery, A. and Kakutani, T. *Ind. U. Math. J.* **20** (1970) 463.

[Ka1] Karr, M. *J. Assoc. Comp. Mach.* **28** (1981) 305.

[Kam1] Kamke, E. *Differentialgleichungen: Losungsmethoden und Losungen* Akademische Verlagsgellschaft, Leipzig (1943).

[Ki1] Kilmister, C. W., and Newman, D. *J. Proc. Camb. Phil. Soc.* **57** (1961) 851.

[Kl1] Klerer, M. and Grossman, F. *A New Table of Indefinite Integrals (Computer Processed)* Dover, New York (1971). The authors actually used a numerical computation system to check errors in the indefinite integral tables.

[MA1] *The MACSYMA Reference Manual, (Version 10)*, Massachusetts Institute of Technology and Symbolics, December (1984).

[Man1] Mansouri, F. and Chang, L. N. *Phys. Rev. D* **13** (1976) 3192.

[Mo1] Moses, J. "MACSYMA - the fifth year". In *Proceedings Eurosam 74 Conference*, Stockholm, August (1974).

[MUC1] *Proceedings of the 1977 MACSYMA Users' Conference*, (R. J. Fateman, Ed.), University of California, Berkeley, CA, July 27-29, NASA CP-2012, Washington, D.C. (1977).

[MUC2] *Proceedings of the 1979 MACSYMA Users' Conference*, (V. E. Lewis, Ed.), Washington, D.C. June 20-22, (1979).

[MUC3] *Proceedings of the 1984 MACSYMA Users' Conference*, (V. E. Golden, Ed.), General Electric Corporate Research and Development, Schenectady, NY, July 23-25, (1984).

[Pa1] Pavelle, R. *J. Math. Phys.* **16** (1975) 1199.

[Pa2] Pavelle, R. *Phys. Rev. Lett.* **33** (1974) 1461.

[Pa3] Pavelle, R. *Phys. Rev. Lett.* **34** (1975) 1114.

[Pa4] Pavelle, R. *Phys. Rev. Lett.* **37** (1976) 961.

[Pa5] Pavelle, R., Rothstein, M. and Fitch, J. P. "Computer Algebra," *Scientific American* **245** (December 1981).

[Pa6] Pavelle, R. *Phys. Rev. Lett.* **40** (1978) 267.

[Pa7] Pavelle, R. and Bogen, R. A. *Lett. Math. Physics* **2** (1977) 55 (erratum **2** (1978) 255).

[Pa8] See MA1, Vol. 2, Sec. 3.

[Pa9] Pavelle, R. *Gen. Rel. Grav.* **7** (1976) 383.

[Pa10] Pavelle, R. *J. Math. Phys.* **21** (1980) 14.

[Ra1] Rainville, E. D. *Elementary Differential Equations* Macmillan, New York, N.Y. (1964).

[Ri1] Risch, R. *Trans. Amer. Math. Soc.* **139** (1969) 167.

[Ri2] Risch, R. *Bull. Amer. Math. Soc.* **76** (1970) 605.

[Ro1] Rosenlicht, M. *Amer. Math. Monthly* November (1972) 963. The author provides a readable discussion of the Risch algorithm.

[Si1] Silver, G. L. *J. Inorg. Nuc. Chem.* **43** No. 11 (1981) 2997.

[va1] van Hulzen, J. A. and Calmet, J. "Computer Algebra Systems" in Buchberger, B. Collins, G. E. and Loos, R. *Computer Algebra, Symbolic and Algebraic Manipulation* Springer-Verlag, Wien, New York (1983) 220.

[Wa1] Wang, P. S. *Math. Comp.* **32** (1978) 1215.

[Ya1] Yang, C. N. *Phys. Rev. Lett.* **33** (1974) 445.

[Yu1] Yun, D. Y. Y. and Stoutemyer, D. "Symbolic Mathematical Computation" in Belzer, J., Holzman, A. G. (Eds.). *Encyclopedia of Computer Science and Technology* **15** Marcel Dekker, New York-Basel (1980) 235.

2

MODERN SYMBOLIC MATHEMATICAL COMPUTATION SYSTEMS

PAUL S. WANG*
Department of Mathematical Sciences, Kent State University,
Kent, Ohio, 44242

ABSTRACT

An elementary introduction to computer-based symbolic mathematical computation is presented. Features of symbolic computation as contrasted by familiar numerical computation are described. Some insight into how these symbolic systems work, the internal organization, and data structures are given. Two examples of basic algorithms implemented in symbolic computational systems are provided.

INTRODUCTION

Symbolic Mathematical systems compute with numbers, symbols, and formulas. They are characterized by exact computation rather than numerical approximation. These systems computerize the kind of computations students of mathematics are trained to do using pencil and paper. These systems perform mathematical operations with such speed and accuracy that calculations which took decades to do by hand now take mere hours. These computer systems can manipulate complicated formulas and return answers in terms of symbols and formulas, not just numbers. With this revolutionary tool, computing will never be the same again.

To give the novice or new user an introduction, our approach here is to first provide a general discussion about computer-based symbolic mathematical computation, contrasting it with familiar numerical computation. Then we shall present a few glimpses into how these symbolic computation systems work internally. Two examples of basic algorithms implemented in symbolic computation systems will then be described.

---

* Work reported herein has been supported in part by the National Science Foundation under Grant MCS 82-01239, and in part by the Department of Energy under Grant DE-AC02-ER7602075-A010.

SYMBOLIC AND NUMERICAL COMPUTATIONS

There are generally two broad kinds of mathematical computations that we think of when we mention scientific computation on a computer. These are numerical computation and symbolic computation. By numerical computation we mean the computer is used as a number cruncher. Many scientific problems can be cast in such a way that its solution calls for repeated computations with numbers. Therefore, in numerical computation, the computer system is used to carry out calculations with numbers and numbers only. On the other hand, there is symbolic computation. The purpose of a computer based symbolic computational system is to compute with symbols, equations, polynomials and other formulas. Modern symbolic systems can performs many highly sophisticated computations that will amaze a mathematics expert. Symbolic computation systems emphasize exactness in the calculations while numerical systems operate on numbers that are as nearly correct as the precision of the computer system allows. The two approaches are both important for the mathematical solution of scientific problems.

In the next section we will discuss some drawbacks in numerical computations. This will provide additional motivation for understanding symbolic computational systems. In later sections we shall present the features of symbolic systems and how they work.

ACCURACY OF NUMERICAL COMPUTATIONS

A typical example of numerical computation is the use of Newton's method to find roots of an equation. In such computations on a computer one computes with floating-point numbers. For example the fraction 1/3 would be 0.333333 in floating-point representation. The number of places retained after the decimal point is dependent on the computer system used. To maintain speed of operation with these floating-point numbers, the precision, or the number of digits kept after the decimal point, is usually fixed. In examples, we shall use six digits. For many applications, these numerical calculations are simple, fast and accurate enough. However, because of the limited precision used, numbers such as 1/3 are represented with a built-in error. Consider

$$0.333333 + 0.333333 + 0.333333$$

This will yield 0.999999 instead of 1.0. But we would like to get 1 if we add three 1/3 together. If we look at 2/3, it is represented as 0.666667. Now what is 0.666667 - 0.333333 ? If it is 0.333334 then we suddenly realize that 1/3 is 0.333334 as well as 0.333333. Thus, when large amount of computation is done with floating-point numbers, one must be very careful.

PROBLEMS OF ASSOCIATIVITY WITH FLOATING-POINT NUMBERS

The familiar mathematical operations plus, "+", and times, "*" satisfy a property called "associativity". Simply this means that if we want to compute A + B + C we can do either (A + B) + C or A + (B + C) and the result obtained is always the same. Similarly for the times operator "*". This is true for most numbers A, B and C. However, for floating-point numbers, this all-important property is no longer there for either "+" or "*".

For example,

$$(111113. - 111111.) + 5.51111 = 2.00000 + 5.51111$$
$$= 7.51111$$

$$111113. + (- 111111. + 5.51111) = 111113. - 111105.$$
$$= 8.00000$$

FEATURES OF SYMBOLIC COMPUTATION SYSTEMS

One of the features of modern symbolic computation systems is exact computation with integers, rational numbers and symbolic formulas involving one or more unknow symbols. Thus, in a symbolic computation system, one can deal with integers of arbitrary size. The integer 200! (two hundred factorial) can be handled just as easily as 20. Fractions as 1/3 and irrational numbers such as SQRT(2) are all kept exact rather than turned into floating-point numbers.

Expressions can be composed from operators, numbers and unknown symbols. Built-in procedures are provided for manipulation of polynomials, quotient of polynomials (rational functions), and other expressions that may involve SIN, COS etc. Many high-level operations are provided as a simple command. For example differentiation, factoring, partial fraction expansion, solution of a system of linear equations, integration, series expansion of functions matrix operations, etc.

Another aspect of computer-based symbolic computation systems is that they usually also provide floating-point operations and the added ability to carry out floating-point operations under a user defined precision. This means that a user first decide how many digits should be kept after the decimal point, then the system can carry out all floating-point calculations in accordance with the user defined precision. Many people feel this is much more convenient than the fixed precisions provided by most numerical systems.

In the next sections, we shall look into some of the basic operations of symbolic systems and other important features of such systems.

EXACT COMPUTATION WITH NUMBERS

In symbolic systems arithmetic with numbers are kept exact. This is done by using software controlled structures to represent the numbers that we are computing with rather than turning them into floating-point numbers. Let us consider arithmetic with fractions. The fraction 1/3 can be represented by a list

(fraction 1 3)

The list has three elements. The first element is the name fraction which identifies what the list is representing. The second and third element is the numerator and denominator of the fraction being represented, respectively. Here are a few more examples,

$$1/4 \quad \text{(fraction 1 4)}$$
$$5/2 \quad \text{(fraction 5 2)}$$
$$-7/11 \quad \text{(fraction -7 11)}$$

Note, the last fraction could be (fraction 7 -11) but we disallow that. The reason is that it is best to represent quantities that are equal with an identical representation. This allows for faster comparisons and cancellations when carrying out computations. Thus we restrict the denominator in a fraction representation to positive integers. We may also want to represent both 2/4 and 1/2 as (fraction 1 2). So we say the fraction representation is always reduced to lowest terms. But the greatest common divisor between the numerator and the denominator is not always obvious. Consider 34/255 or 24140/40902. To reduce these fractions to lowest terms a symbolic system needs to have fast build-in programs for the computation of the greatest common divisor between any pairs of integers. We will return to this point later.

Once the fractions are represented in such a fashion then simple programs can be included in a symbolic system to add, subtract, multiply and divide fractions by manipulating the parts in the fractional representation rather similar to what one would do by hand. In software terminology, such structures for representing quantities or data are referred to as data structures.

Another aspect of exact computation with numbers is handling large integers. Instead of having a limit as to how large an integer can become, modern symbolic computation system allow arbitrary large integers. Thus, it is possible to compute with

200! = 78865786736479050355236321393218506229513597768717326329474253324435944996340334292030428401198462390417721213891963883025764279024263710506192662495282993111346285727076331723739698894392244562145166424025403329186413122742829485327752424240757390324032125740557956866022260319041703240623517008585879617892222278962370389737472000000000000000000000000000000000000000000000000000000

DATA STRUCTURES FOR MATHEMATICAL EXPRESSIONS

The simplest kind of mathematical expressions that involve unknowns are polynomial. Here are several polynomials involving one or more unknown symbols.

$$x^4 + 1 \tag{1}$$

$$b^2 - 4\,a\,c \tag{2}$$

$$(x + 1)^3 \tag{3}$$

How do we represent such symbolic expressions inside a computer? Well, we can use the idea for representing fractions. That is, we will use a data structure. For example, (1) may be represented as

(plus (expt x 4) 1)

Similarly, (2) and (3) can be represented as

(plus (expt b 2) (times -4 a c))

(expt (plus x 1) 3)

We see that inside a computer we can use a list to represent a symbolic expression. The first element on the list is always an "operator" such as "plus" or "times" or "expt". The rest of the list gives the terms under this operator. The representation of each term is again a list with the same structure. One can see that this method, known as the prefix notation, is quite general and it can be used to represent all mathematical expressions, not just polynomials. For example the expression

$$log\,(x\,) + \,cos\,(y^{2}) \qquad (4)$$

is represented as

(plus (log x) (cos (expt y 2)))

Mathematical quantities can thus be represented and programs can be written to manipulate these "lists" to perform well-defined operations such as addition, subtraction, multiplication, division, division with remainder, and so on.

SPECIAL DATA STRUCTURES FOR POLYNOMIALS

While the prefix notation is quite general and useful, it may not be the best choice for representing certain narrow classes of expressions that are much more structured and restricted. For a polynomial, we can consider alternative representations that leads to more efficient programs for many operations involving polynomials. For instance, the list

(x 4 8 2 9 1 10 0 11)

can be used to represent the polynomial

$$8\,x^{4} + 9\,x^{2} + 10\,x\, + 11.$$

We see that the list starts with the variable x followed by the leading exponent then by the leading coefficient, etc. This representation is much more concise than the general prefix notation we just discussed. For programs dealing exclusively with polynomials, this representation is a good choice for many applications.

The above discussed representation is especially effective for polynomials with many terms missing, e.g. $x^{20} + 1$. This type of polynomials are said to be "sparse". Another representation is better suited for dense polynomials. A polynomial is dense if it has almost all terms. In other words, a polynomial with few missing terms is dense. For such a polynomial, we can use a list of just the coefficients, leaving out the exponents. For example,

$$(x\ 1\ \text{-3}\ 8\ 9\ 20)$$

represents

$$x^4 - 3\,x^3 + 8\,x^2 + 9\,x + 20$$

For polynomial with more than one variable, many possibilities exit for their representation. A natural extension of the sparse representation is to write a multivariate polynomial as a polynomial in one variable with coefficients being polynomials in the remaining variables. Thus,

$$(x\ 2\ 3\ 1\ (y\ 1\ 1\ 0\ 1)\ 0\ (y\ 1\ 1\ 0\ \text{-1}))$$

represents

$$3\,x^2 + (y + 1)\,x + y - 1$$

COMPUTING WITH RATIONAL FUNCTIONS

A symbolic computation system deals also with rational functions. A rational function is a quotient where the numerator and the denominator are both polynomials. In computations involving rational functions, a frequent operation is canceling the greatest common divisor (gcd) between the numerator and denominator. A simple example is,

$$\frac{1}{(x - 1)^2} + \frac{1}{x^2 - 1} = \frac{2\,x}{x^3 - x^2 - x + 1}$$

It turns out to be an interesting research subject to design and implement fast programs for polynomial gcd. Modern symbolic mathematical systems all contain very fast routines for polynomial gcd computations.

Another operation often performed with rational functions is partial fraction expansion. This involves the factorization of the denominator and then a sequence of division with remainder operations. The factorization of a polynomial is not an easy task. It is almost impossible to do by hand for a slightly complicated polynomial. Research in the past decade has lead to amazingly fast algorithms for the factorization of polynomials.

INTERACTIVE PROBLEM SOLVING

Most modern symbolic mathematical systems provide a user environment in which the user solves a problem through interaction with the computer system. The user would issue a command to perform a specific operation. The systems carries out the operation and displays the result of the operation on the user's terminal. After seeing the display, the user then decides what command to issue next. This cycle is usually referred to as the command interpretation cycle. It depicted graphically here.

Command execution cycle

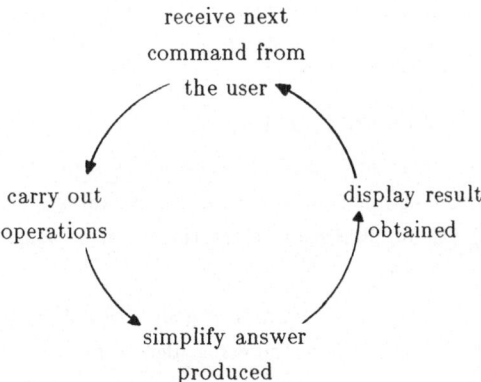

receive next
command from
the user

carry out
operations

display result
obtained

simplify answer
produced

In this manner, the computer is used rather like an assistant to perform tedious and error prone calculations allowing the user to think and to concentrates on the solution strategy. The speed with which operations can be carried out also means that one can try many alternate ways of attacking a given problem without wasting too much time. In many situations, this speed up will translate into improvements in the quality of the work at hand.

PRINTING MATHEMATICAL FORMULAS

Most current systems display mathematical expressions in two dimensional form rather like formulas in a text book. Numerical systems do not have this problem because they deal exclusively with numbers.

To display any mathematical expression in a text book fashion requires ingenious programs for positioning superscripts, subscripts, division bars; for aligning rows and columns of a matrix; and for breaking up a large expression into several sections for display on a terminal which has a limited width, etc. There are many examples of such display later in this article. As high-resolution bit-mapped display terminals become widely used, these display programs are certain to become even more sophisticated.

BUILT-IN ALGORITHMS

A symbolic computation system contains many built-in algorithms for performing mathematical operations. One of the most active research areas in support of building modern symbolic mathematical systems is the development of new or improved mathematical algorithms which can solve practical problems with reasonable speed. Let us examine a couple of basic algorithms.

THE EUCLIDEAN ALGORITHM

As mentioned before, there are many places within a symbolic mathematical system where the greatest common divisor (gcd) of two integers is needed, the most obvious place being the simplification of fractions. Let us consider computing the gcd of two integers.

If m and n are integers, not both zero, then gcd(m,n) is the greatest integer that evenly divides both m and n. The following algorithm, known as the Euclidean Algorithm, can be used to quickly arrive at gcd(m,n) given m and n.

Euclidean Algorithm:

Given: integers m and n, not both zero

Result: gcd(m,n)

STEP E0:   set m = abs(m), n = abs(n)

STEP E1:   If (n=0) then the algorithm terminates
              with m as the result computed.

STEP E2:   Compute r = n mod m

STEP E3:   Set n = m, m = r, and go back to STEP E1.

In the above algorithm, abs(m) means the absolute value of m and n mod m means the remainder of n divided by m. To see how the Euclidean Algorithms works, the best way is perhaps to look at an example. The gcd(40902,24140) computation is shown in figure 1.

Example for Euclidean gcd Algorithm

m	n	m mod n
90902	24140	16762
24140	16762	7378
16762	7378	2006
7378	2006	1360
2006	1360	646
1360	646	68
646	68	34
68	34	0

Figure 1   gcd(40902,24140) = 34

SYMBOLIC DIFFERENTIATION

One of the most fundamental operations in mathematical computation is differentiation. Modern symbolic mathematical system can perform differentiation very well indeed. This seemingly difficult task is actually quite simple once we know the algorithm used.

Consider differenting an arbitrary elementary function g(x) with respect to x.

$$\text{diff}(g(x),x)$$

The algorithm shown in figure 2 is the heart of a symbolic differentiation algorithm.

Differentiation Algorithm diff(g,x):
given: any elementary function g(x) and the variable x
result: the derivative of g(x) with respect to x

STEP D1: if g = x then 1
STEP D2: if g is a constant then 0
STEP D3: if g = a + b then diff(a,x) + diff(b,x)
STEP D4: if g = a * b then b*diff(a,x) + a*diff(b,x)
STEP D5: if g = a ** b then (a**b * log(a))diff(b,x) + b * diff(a,x)/a
STEP D6: if g = f(x) where f is sin, cos, tan, log, exp, and so on then
          the result is retrieved from a table where the
          derivatives of these functions are stored.
STEP D7: if g = a(b(x)) then apply the chain rule.

Figure 2   Heart of differentiation algorithm

This differentiation algorithm recursively applies itself to sub-expressions in the given function g(x). A table stores the derivatives of the basic functions that are handled by the algorithm. If we implement this algorithm all by itself it will not work very well. It will compute the correct answer, however the answer is not simplified. This means zero terms are not dropped in a sum, like terms are not combined, etc. However, in a symbolic mathematical system, a simplifier is always there to make intermediate results and the final answer simplified.

It is possible to extend the table containing basic derivative information and allow the algorithm to handle a larger class of functions. Some symbolic mathematical system allows the user to provide extensions to this table.

CONCLUSIONS

As computer hardware become more powerful and inexpensive, symbolic mathematical computation systems will become more affordable and widely used. Their application potential is great. One day people will take access to a "symbolic calculator" for granted. It is hoped that this elementary introduction will aid the introduction of these systems to people who can benefit from their use.

REFERENCES

van Hulzen, J. A. and Calmet, J.: "Computer Algebra Systems" in Buchberger, B., Collins, G. E., and Loos, R., ed. *Computer Algebra, Symbolic and Algebraic Manipulation* Springer-Verlag, Wien, New York (1983).

Knuth, D. E.: *Seminumerical Algorithms, The Art of Computer Programming Vol. 2*, 2nd ed., Addison-Wesley, Reading, Mass. (1980).

Pavelle, R., Rothstein, M. and Fitch J. P.: "Computer Algebra," *Scientific American* 245 (December 1981).

3

USING VAXIMA TO WRITE FORTRAN CODE†

STANLY STEINBERG

Department of Mathematics and Statistics, University of New Mexico, Albuquerque, NM 87131

PATRICK ROACHE

Ecodynamics Research Associates, Inc., P.O. Box 8172, Albuquerque, NM 87198

ABSTRACT

This paper describes the symbol manipulation aspects of a project that produced a large FORTRAN program that is now used to model lasers and other physical devices. VAXIMA (MACSYMA) was used to write subroutines that were combined with standard software to produce the full program.

INTRODUCTION AND PROJECT OVERVIEW

The purpose of this paper is to describe the symbol manipulation aspects of a project that used VAXIMA (MACSYMA) [8] to write FORTRAN subroutines that are part of a finite difference code that is now used to model lasers and other physical devices. Some of the results of this project were reported in [10, 11, 12, 13, 15, 16] and some substantial improvements will be reported in [18]. Without the help of a symbol manipulator, portions of the project would have been impossible. With the help of VAXIMA, useful code was produced in a few weeks. Thus, as users of a symbol manipulator, we had a reduction in code development time that was infinite, a fact of importance for anyone considering using a symbol manipulator in a code development project. Unfortunately, not all of our report is accolades.

The physical devices that are being modeled are assumed to be in a steady state and consequently it is assumed that the physics of interest can be modeled by a partial differential equation (called the hosted equation) that involves the Laplace operator or a generalization of this operator. Mathematically, the equations must be elliptic. Such differential equations have been well studied both analytically and numerically. The difficulties come not from the differential equations, but from the fact that the physical devices have an irregular shape. This means that the full model

---

† This work was partially supported by the U.S. Army Research Office, by the U.S. Air Force Office of Scientific Research, by the National Science Foundation Grant #MCS-8102683, and by System Development Foundation. This paper was also presented at the Third MACSYMA User's Conference, July 23, 1984 in Schenectady, New York.

will involve a boundary value problem in an irregular three dimensional region, or if the device has sufficient symmetry, then a boundary value problem in an irregular two dimensional region.

There are several different finite difference or finite element methods available for handling such problems. We are interested in a technique called *Boundary-Fitted Coordinates* that involves finite difference techniques. This subject has become a field of study in its own right as evidenced by the proceedings [2, 6, 19]. The basic idea is to find a transformation (or change of coordinates) that maps the given region (called *physical* space) into a rectangular region (called *logical* space), see Figure 1. In the rectangular region it is easy to produce finite difference schemes. Although the idea is simple, there are several complications and it is these complications that made using a symbol manipulator so helpful.

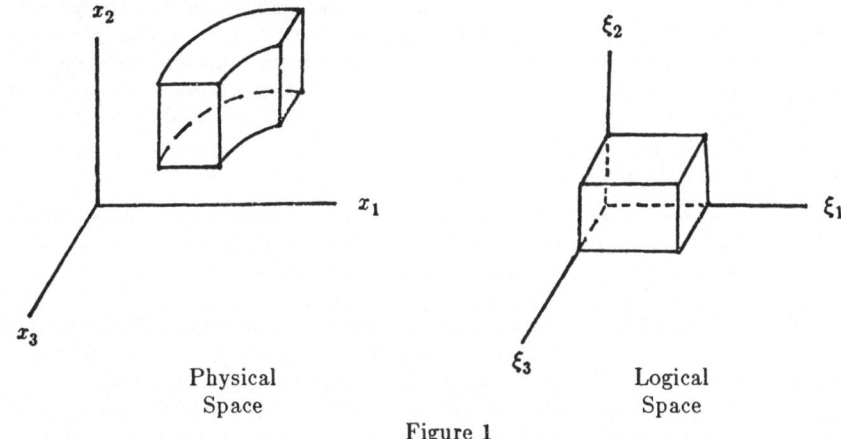

Physical
Space

Logical
Space

Figure 1

One complication is that the given regions are usually so irregular that it is impossible to find analytic transformations; the transformations must be determined numerically. The geometric idea that underlies the numerical methods is that the transformation should be smooth. Historically this is translated into requiring the transformation be harmonic: that each component of the transformation satisfy Laplace's equation. If the transformations are required to be harmonic as mapping from logical space to physical space, then converting the differential equations to finite difference schemes is easy. However, it was discovered that such a formulation leads to poorly behaved numerical methods while if the transformations are required to be harmonic as mappings from physical space to logical space, then the numerical methods are better behaved. In this formulation, the Laplace equations are called the smoothness equations. The fact that the smoothness equations must now be transformed to equations on logical space is an important complication in this approach. This is done using the yet to be determined transformation from logical to physical space which results in a coupled system of quasi-linear elliptic differential equations (also called the smoothness equations) for determining the the transformation from logical to physical space. The prescription of the transformation on the boundary of the region provides the boundary conditions for the smoothness equations. We used VAXIMA to do the algebra and calculus in this step. The condition that the transformation be harmonic has been generalized, and we included these generalizations in our work. More recently, the harmonic condition has been replaced by a variational principle which we will describe in Section 5.

Another complication is that the hosted equations (the equations describing the physical process) must be transformed to logical space. Because the transformation is not known analytically, a general transformation must be used. To give the reader some idea of the size of these problems, we note that the Laplacian in three dimensions in general coordinates and in fully expanded form contains 1611 terms. Again, we used VAXIMA to do this algebra and calculus although we did not always use the fully expanded form. Once the smoothness and the hosted equations are known in logical space, they must be converted to finite difference form. This was also done with VAXIMA. Finally VAXIMA was used to write all of the formulas into a file in a form appropriate for the FORTRAN compiler. In fact, we had VAXIMA write two complete subroutines, one for the hosted equation and one for the smoothness equations. A few minutes work with the editor and the subroutines were ready to compile.

After the subroutines were written, they were combined with other standard numerical software to produce a program that could model physical devices. Although the level of confidence in the resulting code was high there was no guarantee that it was correct. Consequently, the program was checked using convergence rate testing on a set of examples that would exercise all parts of the program. The program has now been distributed to several universities and laboratories for production modeling. At the end of this paper we have included three coordinate systems generated by our programs. Each figure represents a coordinate system in the interior of a laser cavity. Figures 6 and 7 represent regions that have one axis of symmetry while Figure 8 represents the surface of a three dimensional region. We did not label the axes in these figures because such labeling is arbitrary and should be chosen for the convenience of the user. The details of the devices being modeled are described in [11, 13].

We now describe some parts of this project in more detail. In Section 2 we will describe the mathematical formulation of coordinate changes, in Section 3 we will describe how to introduce the finite difference schemes, and in Section 4 we will describe how to write the FORTRAN code. Section 5 is devoted to describing a variational formulation of the grid problem while Section 6 is devoted to a summary of what was accomplished. This project provided us with experiences we believe are relevant to general applied mathematics and symbol manipulations problems so some of our opinions are presented in Section 7. Finally, some of the basic ideas used in this project we used in a project to develop a program that performs analytic changes of coordinate for partial differential equation, so a brief introduction to this material is given is Section 8.

MATHEMATICAL FORMULATION

Here we will give a brief introduction to the mathematics involved in our problem. The mathematical formulation is done in n-dimensions, not because we need the formulation for dimensions other than 2 and 3, but because this allows us to write VAXIMA code that works for all dimensions including 2 and 3. Thus a point in space will be denoted

$$\vec{x} = (x_i) = (x_1, x_2, \ldots, x_n) \tag{1}$$

where n is a positive integer parameter and we think of $\vec{x}$ as a column vector. The simplest hosted equation is the Laplacian,

$$\Delta f = \sum_i \frac{\partial^2 f}{\partial x_i^2} \; , \tag{2}$$

where $f = f(\vec{x})$ and we assume that all sums run from 1 to n. In general, the hosted equation will have the form

$$Lf = \sum_{ij} a_{ij} \frac{\partial^2}{\partial x_i \partial x_j} f + \sum_i b_i \frac{\partial}{\partial x_i} f + cf + d \tag{3}$$

where a_{ij} , b_i , c , d and f depend on $\vec{x}$. The smoothness equations have a similar form where the coefficients depend on the unknown transformation, that is, the equations are quasi-linear rather than linear.

In our applications we will be doing numerical calculations in logical space so we will want to write all of our formulas in terms of the $\vec{\xi}$ variables. Thus we write

$$\vec{x} = \vec{x}(\vec{\xi}), \tag{4}$$

and choose the Jacobian matrix J to be

$$J = (J_{ij}) = (\frac{\partial x_j}{\partial \xi_i}) = \begin{Bmatrix} \dfrac{\partial x_1}{\partial \xi_1} & \dfrac{\partial x_2}{\partial \xi_1} & \cdots \\ \dfrac{\partial x_1}{\partial \xi_2} & \cdots & \\ \cdots & & \cdots \end{Bmatrix} \tag{5}$$

and K to be the cofactor matrix of J. Thus if $\Delta = determinant(J)$, then

$$J^{-1} = \frac{K}{\Delta} \; . \tag{6}$$

The chain rule gives

$$\frac{\partial}{\partial \xi_i} = \sum_j J_{ij} \frac{\partial}{\partial x_j} \tag{7}$$

and multiplying by K/Δ gives

$$\frac{\partial}{\partial x_i} = \frac{1}{\Delta} \sum_j K_{ij} \frac{\partial}{\partial \xi_j} . \tag{8}$$

The formula for $\partial/\partial x_i$ can be used to compute the second derivatives,

$$\frac{\partial^2}{\partial x_i \partial x_j} = \sum_{rs} \frac{1}{\Delta} K_{ir} \frac{\partial}{\partial \xi_r} (\frac{1}{\Delta} K_{js} \frac{\partial}{\partial \xi_s})$$

$$= \sum_{rs} \frac{1}{\Delta^2} K_{ir} K_{js} \frac{\partial^2}{\partial \xi_r \partial \xi_s} + \sum_{rs} \frac{1}{\Delta} K_{ir} \frac{\partial}{\partial \xi_r} (\frac{1}{\Delta} K_{js}) \frac{\partial}{\partial \xi_s} . \tag{9}$$

The next step is to use the above formulas to transform the hosted and smoothness equations to logical space. In our first approach, this is exactly what we did. We believe that this is a very natural approach to the problem and the fact that this leads one to a trap points out that symbol manipulation is not as easy as it may seem. Using this approach, VAXIMA required about 60 cpu hours to write the subroutine for the hosted equation in three dimensions and produced about 1800 lines of very dense FORTRAN code. This work was done on a VAX 11/780 computer with 4 megabytes of RAM memory. After some minor adjustments to the f77 compiler, approximately one cpu hour was required to compile the subroutine. As it turned out, the subroutine was correct the first time it was written, which is not the same as saying that no errors were made in the symbol code.

If a less obvious approach is taken, then VAXIMA can write an equivalent subroutine in about 8 cpu minutes (60 cpu hours over 8 cpu minutes equals 450). This subroutine contains only 180 lines of code (1800 lines over 180 lines equals 10). As we proceed with our discussion we will point out the differences between the first and the second approaches that make such a great difference in the VAXIMA run time and the size of the subroutine.

Before we proceed we need to point out that we would have liked to derive the above formulas using VAXIMA. This is not practical because VAXIMA does not know about vectors of length n, where n is an integer parameter and does not know about functions of n variables where n is an integer parameter. We believe that none of the existing symbol manipulators can do this type of computation.

A major improvement in our programs was achieved by noting that there is a classical formula for differentiating the inverse of a matrix A whose entries are functions of the ξ variables,

$$\frac{\partial}{\partial \xi_i} \frac{1}{A} = -\frac{1}{A} \frac{\partial A}{\partial \xi_i} \frac{1}{A} \, . \tag{10}$$

Combining this with a previous formula gives

$$\frac{\partial}{\partial \xi_r}(\frac{1}{\Delta} K_{j\bullet}) = -\frac{1}{\Delta^2} \sum_{uv} K_{ju} \frac{\partial}{\partial \xi_r} J_{uv} K_{v\bullet} \, . \tag{11}$$

and thus

$$\frac{\partial^2}{\partial x_i \partial x_j} = \frac{1}{\Delta^2} \sum_{r\bullet} K_{ir} K_{j\bullet} \frac{\partial^2}{\partial \xi_r \partial \xi_\bullet} - \frac{1}{\Delta^3} \sum_{r\bullet uv} K_{ir} K_{ju} K_{v\bullet} (\frac{\partial}{\partial \xi_r} J_{uv}) \frac{\partial}{\partial \xi_\bullet} \, . \tag{12}$$

It is this last formula that was used in the second approach. Moreover, in the first approach we plugged everything into the differential equation that was to be transformed and then expanded out the formula. This produced a rather large expression which, in turn, accounts for part of the excessive time used in the first approach. In the second approach, we introduced some intermediate quantities that are equal to various coefficients in the previous expression. This has the effect of reducing the size of the expressions to be manipulated but makes it very difficult to understand what is the simplest form for expressions defined in terms of the intermediate expressions. In either case, this produces the formulas needed to generate the finite difference equations. For more details see [16].

FINITE DIFFERENCES

The next step is to replace all of the ξ derivatives in the previous formulas by standard centered finite differences (which we do not write here). The portions of the VAXIMA code that deal with finite differences were programmed separately for two and three dimensions because we found it impossible to do this with the dimension n as a formal parameter, see also [22]. The fact that we could not manage a general formulation was disappointing, but VAXIMA was still an indispensable tool for completing this part of our work.

In our first approach we simply substituted these formulas into the expressions for the differential equations. Now our formulas were getting really large but were, in reality, not very complicated. In three dimensions, the resulting difference equations have the form

$$\sum_{|i|+|j|+|k|\leq 3} c_{i,j,k}(\xi_1,\xi_2,\xi_3) \, g(\xi_1+i\Delta\xi_1,\xi_2+j\Delta\xi_2,\xi_3+k\Delta\xi_3) = R(\xi_1,\xi_2,\xi_3) \, . \tag{13}$$

Here i , j , k each run over the set of values $\{-1,0,1\}$. The coefficients $c_{i,j,k}$ are called a stencil. For convenience, we view the elements of the stencil as points on a cube in three space and then label the elements as in Figure 2. Note that not all of the stencil elements have been labeled so as to avoid cluttering the figure. In addition, the ξ_1 axis points to the right, the ξ_2 axis points up, and the ξ_3 axis points forward.

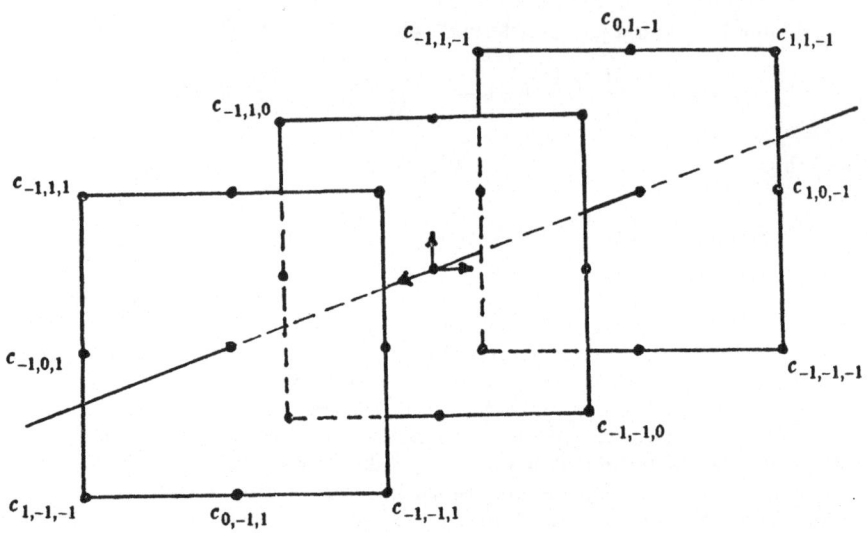

The Stencil
Figure 2

It is easy to check that, because we used centered differences, there are many relationships among various stencil elements, for example,

$$c_{1,1,1} = 0 \ , \ c_{-1,0,1} = c_{1,0,1} \ .$$

These relationships reduce the 27 stencil elements to 10 independent stencil elements.

Because of these relationships, the formulas generated by the first method contain substantial redundant information. In the second method the difference formulas were never substituted into the differential equations. Instead, various coefficients of the differential equation were collected and combined to give the formulas for the stencil elements. This can be thought of as reversing the order of the substitution and the collecting of the coefficients. We did not find this convenient in VAXIMA. It is our impression that decisions concerning the order in which the steps of a computation are to be done are very important. Shouldn't symbol manipulators provide facilities to help the user with such decisions?

At this point the stencil elements are defined in terms of intermediate expressions which are, in turn, defined in terms of the parameters and derivatives of parameters appearing in the original differential equations. These expressions also contain derivatives of the coordinate change. The situation for the smoothness equations is a bit different from that for the hosted equation but the computations are similar. For more details, see [16]. In either case, these parameter functions and derivatives must be replaced by a notation that the FORTRAN compiler understands. In the three dimensional case, we created 31 atomic variables (variable names) and substituted these into the formulas needed to write the FORTRAN subroutine.

In the first formulation this was a disaster. At this point we were dealing with a very large formula and every time a substitution was made the simplifier rearranged the terms in the formula in an attempt to find simplifications where we knew that none existed. We estimate that roughly 50 of the 60 cpu hours used in the first method went into this step, that is, by stopping this irrelevant work we reduced the computation to about 10 cpu hours. The remaining savings required some mathematics to be done.

At this point our VAXIMA program had produced all of the formulas needed to write the FORTRAN subroutines. These formulas were recorded in several lists in the form of VAXIMA equations that corresponded to FORTRAN assignment statements. We had also collected lists of all of the variables used in the formulas that were to be used in the subroutines. So let us write the subroutines.

WRITING FORTRAN

At this point we had all of the formulas needed to write the FORTRAN code. We had also saved some lists that contained all of the variable used in these formulas. The next thing that needed to be done was to create the subroutine header, some comments, some variable declarations, and some loops over the grid in logical space. We could have written these things into a file and then used the VAXIMA *fortran* function to convert the formulas to FORTRAN syntax and then written these expressions into a file. A few minutes work with an editor would have produced the subroutine. Instead, we decided to write a VAXIMA function to produce a complete subroutine. This is not a particularly efficient way to do things given the state of the facilities in VAXIMA or, for than matter, any other symbol manipulator. However, we now have some opinions on what is needed to make writing floating point programs more efficient and less irritating.

When we are writing FORTRAN code we would like our symbol manipulator to be a good programmer's assistant. We are not interested in converting symbol manipulation programs or functions to floating point programs (although some may be interested in this). One step in writing a program is creating the assignment formulas which is the forte of symbol manipulators. These formulas need to be in or converted to FORTRAN syntax. The arithmetic operations seem to cause no difficulty whereas the logical operators do cause problems. First note that the VAXIMA *fortran* function converts the VAXIMA assignment operator ":" to the FORTRAN assignment operator "=" while converting the VAXIMA logical "=" to the FORTRAN ".eq.". Here is a short sample of a VAXIMA run that illustrates the point.

```
(c_1)  x:a^2+b^2;
```

$$
(d\_1) \qquad\qquad a^2 + b^2
$$

```
(c_2)  y:x+c*d;
```

$$
(d\_2) \qquad\qquad a^2 + b^2 + c\,d
$$

```
(c_3)  fortran(y);

(d_3)        a**2+b**2+c*d

(c_4)  fortran(y:x+c*d);

(d_4)        y=a**2+b**2+c*d

(c_5)  fortran(z=a^2+b^2);

(d_5)        z.eq.a**2+b**2
```

VAXIMA Output
Figure 3

Now imagine that the the values being assigned are very complicated and we don't want VAXIMA to do the substitutions, we want FORTRAN to do floating point evaluations and substitutions. Now things become a bit convoluted. We opted to use the VAXIMA "=" in our assignment formulas and then use the editor to change the ".eq." to "=". We believe a good resolution to this problem is to have a data type that is a FORTRAN assignment statement so that the meaning of "=" is clear.

A good assistant could help with many other chores. It would be nice to have the manipulator check that we had declared all of the variables used in the subroutine, that the formulas were in a consistent order, that all of the values used in the assignment formulas were computed locally or passed into the subroutine through the calling sequence or a common block, and that all of the values passed out of the subroutine were computed; it also would have been nice to have the manipulator read into the program a file that contained some comments. No doubt, there are many other programming tasks that a good assistant could do for the user. Again, we believe that it would be helpful if the symbol manipulator had a data type called FORTRAN program and could manipulate such an object. One thing that we could do with VAXIMA was very useful: we wrote all of the assignment formulas used in the subroutine into a file in the VAXIMA two dimensional format. This made reading the FORTRAN code much easier. In fact, much of our FORTRAN code is not very readable by humans. However, we hope the the VAXIMA code and the two dimensional format formulas are readable.

Of course, we have saved the real bad news to the last. All of the subroutines that we have written using VAXIMA contain an outrageous amount of redundant arithmetic. This can be corrected with a large amount of *at the terminal* work with either VAXIMA or a text editor, but there are many chances for mistakes. Because the subroutines that we wrote account for a small percentage of the total run time of our programs, we have not yet optimized the formulas. Anyone planning on using a symbol manipulator to write code should be aware of this problem. There is a function in VAXIMA called "optimize" that will correct this problem for small formulas, although "optimize" does not change any of our formulas and was not designed to work on a list of formulas where the formulas contain common expressions. It does not seem that it will be difficult to improve this situation. However, it is not clear to us how to optimize both the arithmetic count and the stability of formulas. We are currently working on these problems. To give the reader some idea of what we are talking about we have included, see Figure 4, one assignment statement from the three dimensional code.

```
u1 = s33*vk31*z33+2*s23*vk31*z23+s22*vk31*z22+2*s13*vk31*z13+2*s1
1    2*vk31*z12+s11*vk31*z11+s33*vk21*y33+2*s23*vk21*y23+s22*vk21*y2
2    2+2*s13*vk21*y13+2*s12*vk21*y12+s11*vk21*y11+s33*vk11*x33+2*s23
3    *vk11*x23+s22*vk11*x22+2*s13*vk11*x13+2*s12*vk11*x12+s11*vk11*x
4    11
```

Sample of VAXIMA Written FORTRAN Code
Figure 4

Note that almost any rewriting of this expression in a form analogous to that given by Horner's rule will improve the operation count and presumably the stability of evaluating the expression. The problem is that there are many ways of rewriting the formula and so it is not obvious how to automate such a procedure.

THE VARIATIONAL FORMULATION

Recently, there has been an interest in formulating variational problems for determining coordinate systems in physical space [3, 14, 18]. A problem with the previous methods is that they are ad hoc. There is some geometric intuition but the parameters in the method have no direct geometric interpretation. With the variational methods the parameters do have a geometric interpretation and consequently this geometric intuition can be used to help determine the parameters. Previously, numerical experimentation and experience were the best guides to choosing the parameters. In the simplest cases the variational methods yield the previous methods. However, the variational method provides direct control over many aspects of the grid including the smoothness, the angles between the grid lines, and the area or volume of the grid cells. For more details see [3, 14, 18].

In this section we will describe how to convert a variational problem into a FORTRAN subroutine. We will see that this type of problem is very appropriate for a symbol manipulator. On the other hand, the derivation of a variational problem from geometric intuition seems the proper domain for human thinking so we leave describing how this is done to another paper [18]. The simplest variational problem is the one that is related to smoothness, so we now describe that problem and its conversion to a FORTRAN subroutine for generating a grid. The more general variational problems involve the same mathematics; they are just more complicated and complicated enough to warrant using a symbol manipulator.

We will use the notation of the Section 2 and formulate an n-dimensional version of the variational problem. The variational problem is to find a transformation $\vec{x} = \vec{x}(\vec{\xi})$ mapping a rectangle S in logical space to the given region in physical space which minimizes the integral

$$I(\vec{x}) = \int_S \sum_{i,j=1}^{n} \left(\frac{\partial x_i}{\partial \xi_j} \right)^2 d\xi_1 \cdots d\xi_n . \tag{14}$$

The transformation $\vec{x}(\vec{\xi})$ is to be specified on the boundary of S. It is known that if $\vec{x}(\vec{\xi})$ minimizes $I(\vec{x})$ then the components $x_i(\vec{\xi})$ must satisfy the Euler equations which are a system of partial differential equations, that are well known. Instead of writing the Euler equations we will derive them. This is because the method of deriving these equations is far more interesting than the equations themselves.

Let $\vec{c}(\vec{\xi})$ be defined on the region S and be zero on the boundary of S. Then for every ϵ, the transformation $\vec{x} = \vec{x}(\vec{\xi}) + \epsilon\, \vec{c}(\vec{\xi})$ maps the region S in logical space to the given region in physical space. If $\vec{x} = \vec{x}(\vec{\xi})$ minimizes $I(\vec{x})$ then $\epsilon = 0$ must be a minimum of

$$F(\epsilon) = I(\vec{x} + \epsilon\vec{c}) , \tag{15}$$

that is, it must be the case that

$$\frac{dF}{d\epsilon}(0) = 0 . \tag{16}$$

A common way of thinking of the above is to consider $I(\vec{x})$ to be a functional (function) on an infinite dimensional space of smooth functions and then the previous derivative is thought of as being a directional derivative of the functional $I(\vec{x})$ in the direction $\vec{c}$. This type of derivative is a direct generalization of the notion of directional derivative in finite dimensional spaces and is frequently called a Fréchet or

Gâteaux derivative. This derivative has many diverse applications in applied mathematics.

A bit of calculus gives

$$\frac{dF}{d\epsilon}(0) = \int_S \sum_{i,j=1}^{n} \frac{\partial x_i}{\partial \xi_j} \frac{\partial c_i}{\partial \xi_j} d\xi_1 \cdots d\xi_n = 0 .$$ (17)

An integration by parts gives

$$\int_S \sum_{i,j=1}^{n} \frac{\partial^2 x_i}{\partial \xi_j^2} c_i \, d\xi_1 \cdots d\xi_n = 0 .$$ (18)

The integration by parts is easy to do symbolically because what is required is to remove all of the derivatives from the c_i.

The previous integral must be zero for all choices of $\vec{c}$ which are zero on the boundary of S, which implies that the coefficient of each c_i must be zero. Thus

$$\sum_{j=1}^{n} \frac{\partial^2 x_i}{\partial \xi_j^2} = 0 , \; 1 \leq i \leq n ,$$ (19)

which are the usual smoothness equations!

Now we are in a position to apply all that we have learned previously. We would like to note that we are very fond of this approach to grid generation problems: it has geometric insight, straightforward computations, and considerable versatility. In addition, VAXIMA handles the computation easily if we don't try to completely implement the n-dimensional formulation.

SUMMARY

Let us look at what was accomplished. Using our first mathematical formulation it is certainly possible to generate the two dimensional FORTRAN subroutine by hand but probably impossible to generate the three dimensional subroutine by hand. The new mathematical formulation of the problem has probably brought the writing by hand of the three dimensional subroutine within the realm of possibility. For two dimensional codes there probably is no time saved in producing the subroutines using VAXIMA. Even though no time is saved in the production, it is still advantageous to use VAXIMA. The reason is that there is a very small probability of *typo* type errors in the VAXIMA written code. In fact, we did not have to spend any time debugging the FORTRAN subroutines although there was a small problem in combining the subroutines with the elliptic equation solver. This comment is a bit unfair because a reasonable amount of time was spent debugging the symbol codes. However, because the symbol codes are written at a higher mathematical level than the subroutines, they are usually correct or produce garbage, and consequently are considerably easier to debug than FORTRAN code. Thus the VAXIMA project had the advantage of requiring less debugging time and producing a product that we were confident was correct.

As stated before, the mathematical formulation used to write the VAXIMA code was derived for the general n-dimensional case. About half of this was programmed in VAXIMA using n as a formal parameter. The parts of the formulation that could not be programmed for the general case were programmed for the two dimensional case and then the two dimensional subroutines were written and tested. A modest amount of programming produced three dimensional versions of those parts that were not general and then it was possible to write the three dimensional subroutines. The fact that much of the VAXIMA code was used in the two dimensional case or was a direct analog of the code used in the three dimensional case gave us a very high level of confidence in the three dimensional subroutines, and now we had realized a tremendous saving of time!

Certainly VAXIMA is a useful tool. However, this project and other projects have shown that there are problems. Here we do not mean *bug* type problems; in fact this type of problem is rather rare; we mean problems in the fundamental design of VAXIMA. Clearly part of the problem could be that VAXIMA evolved over a number of years and involved a large number of programmers. However, we have looked at several of the new general purpose manipulators, which were certainly designed, and find that the problems are still there. Thus we are led to believe that there is some disparity between what the symbol manipulation community is designing and the needs of the applied mathematicians who use symbol manipulators.

COMMENTS

In this section we will make some general comments about the use of symbol manipulators in applied mathematics.

One of the most important phenomenon in symbol manipulation is that of *intermediate expression swell*. Certainly, our 60 cpu hour run times were a result of this phenomena. It seems reasonably clear that faster hardware is not going to be all that helpful in tackling many problems that have large intermediate expressions. Clearly some problems will have a *best* formulation that is very large and for such problems fast machines are crucial. We believe that improved design of symbol manipulators along with the user community developing more skill in using these programs will have more impact.

The problem of large intermediate expressions is not unique to computer symbol manipulation; the same problem occurs in hand computation. When students do computations in elementary courses we often refer to their approach as *plug and chug* and are clearly aware that more experience may improve their computational abilities: they will start to have an overview of their computations and will start to choose among several computational strategy. We believe [7] that more experienced users of mathematics use abstraction to overcome the intermediate expression swell problem. They tend to introduce symbols to represent large expressions. Such symbols need to be well chosen; they must have nice manipulation properties and represent important parts of the underlying problem. The more abstract symbols may be manipulated in an attempt to find an approach to a problem that has tractable intermediate expressions.

The fact that we could discover a reformulation of our problem that reduced our run time by nearly three orders of magnitude can be interpreted in many ways; perhaps we should have thought more before programming. From an applied mathematics point of view, our original programs were very natural and this is what allowed our rapid progress. We believe that the use of the identity that reduced the run time could have been found by a symbol manipulator. The identity is well known. The use of the identity is indicated because it allows some of the calculations to be done before the messy details are put into the formulas; clearly this a good thing to try.

Another problem is the notion of functional dependencies that is used by the VAXIMA differentiation routines. This notion is not adequate for our need and is probably not adequate to carry out many applied mathematics projects that involve multivariate calculus. This problem caused us to carry out hand derivations of the coordinate transformation formulas rather than doing this work in VAXIMA. As far as we can tell all of the manipulators that are commonly available have problems in this area. This problem is so important that M. Wester and one of the authors have published a paper [20] on this subject and are presenting a separate paper [21] on this subject at this conference.

As we noted above, some of the VAXIMA code could be written using the dimension, n, of physical space as a parameter. The fact that about half of the VAXIMA code could not be written this way was more than a nuisance. One of the problems here is that it is not possible to define vectors of length n where n is an integer parameter and then have VAXIMA know how to manipulate such objects. Stated more mathematically, it is not feasible to teach VAXIMA about abstract vector spaces. We believe it is impossible to over-estimate the importance of abstract vector spaces in applied mathematics. A simpler version of this problem can be found in the fact that it is not easy to define lists of length n where n is a formal integer parameter and then have VAXIMA (or Lisp) know how to manipulate these objects. As far as we know, no other manipulator has this type of facility.

ANALYTIC CHANGES OF COORDINATES

Some of the previous ideas we used in a project [17] that developed a program to perform analytic changes of coordinates for partial differential equations. There are some facilities available [8] for changing the independent variables in partial differential equations: what we wanted was a program that would change both the dependent and independent variables. The program developed will transform up to second order partial differential equations in any number of dependent and independent variables. As the algebra here is quite complicated, let us briefly describe the case of one dependent and one independent variable. In fact, what is needed is a program that will change any partial derivative in one coordinate frame to partial derivatives in a second coordinate frame. These formulas are then substituted into the partial differential equation.

Let x be the independent variable, while y is the dependent variable in the given coordinate frame, and let u be the independent variable and v be the dependent variable in the new coordinate frame, as is shown in Figure 5.

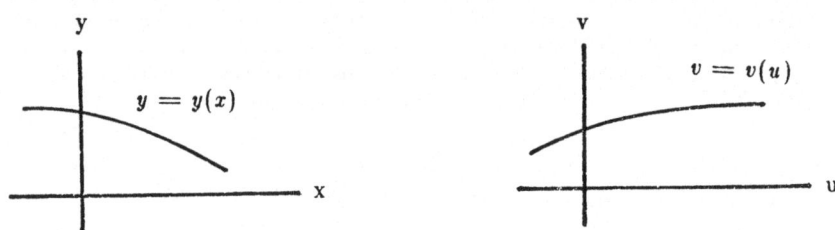

Change of Coordinates
Figure 5

The curves in Figure 5 are given by $y = y(x)$ and $v = v(u)$. We are interested in transforming the derivative $y' = dy/dx$ into an expression involving the derivative $v' = dv/du$.

We assume that the transformations are given implicitly,

$$F(x,y,u,v) = 0 \ , \ G(x,y,u,v) = 0 \ , \tag{20}$$

because this was the case that occurred in our applications. Here we assume that it is possible to solve these equations numerically for x and y in terms of u and v, that is, a certain Jacobian described below is not zero and that the Jacobian of the resulting transformation is nonzero. This will imply that the inverse transformation exists, that is, the equations can be numerically solved for u and v in terms of x and y. We are not assuming that the equations can be solve algebraicly, although if this can be done then the results we obtain can be improved.

Because this case is so simple it can be done in many ways. We found that the ideas from elementary calculus were not powerful enough to allow us to do the more general problem so we opted to use differential forms to solve the problem. It should be noted that differential forms are nice because they convert analytic problems to linear algebraic problems as we will see below. First, calculate the differentials of the transformation:

$$F_x \ dx + F_y \ dy + F_u \ du + F_v \ dv = 0 \ , \tag{21}$$

$$G_x \ dx + G_y \ dy + G_u \ du + G_v \ dv = 0$$

where dx, dy, du, and dv are the differential of the dependent and independent variables and $F_x = \partial F / \partial x$ and so forth. Introduce the matrices

$$M_1 = \begin{pmatrix} F_y & F_x \\ G_y & G_x \end{pmatrix} \ , \ M_2 = \begin{pmatrix} F_v & F_u \\ G_v & G_u \end{pmatrix} \ . \tag{22}$$

Our assumptions on the Jacobians means that the determinants of the matrices M_1 and M_2 must not be zero. Now the system of equations (8.2) can be solved for dy and dx yielding

$$dy = A \ dv + B \ du \ , \ dx = C \ dv + D \ du \tag{23}$$

where the matrix

$$M = \begin{pmatrix} A & B \\ C & D \end{pmatrix} \tag{24}$$

is given by

$$M = M_1^{-1} \ M_2 \ . \tag{25}$$

Note that M depends on x, y, u, and v.

To transform the first derivatives note that

$$\frac{dy}{dx} = \frac{A \ dv + B \ du}{C \ dv + D \ du} = \frac{A \dfrac{dv}{du} + B}{C \dfrac{dv}{du} + D} \tag{26}$$

which implies that

$$y' = \frac{A + B \ v'}{C + D \ v'} \tag{27}$$

As noted above A, B, C, and D depend on x, y, u, and v. If the equations (8.1) can be solved algebraically for x and y in terms of u and v, then this information can be used in the previous equation. This should not be done before computing the second derivative.

The computation of the transformation of higher derivatives is simple: for the second derivative compute, as we did for the first derivative, divide dy' by dx and

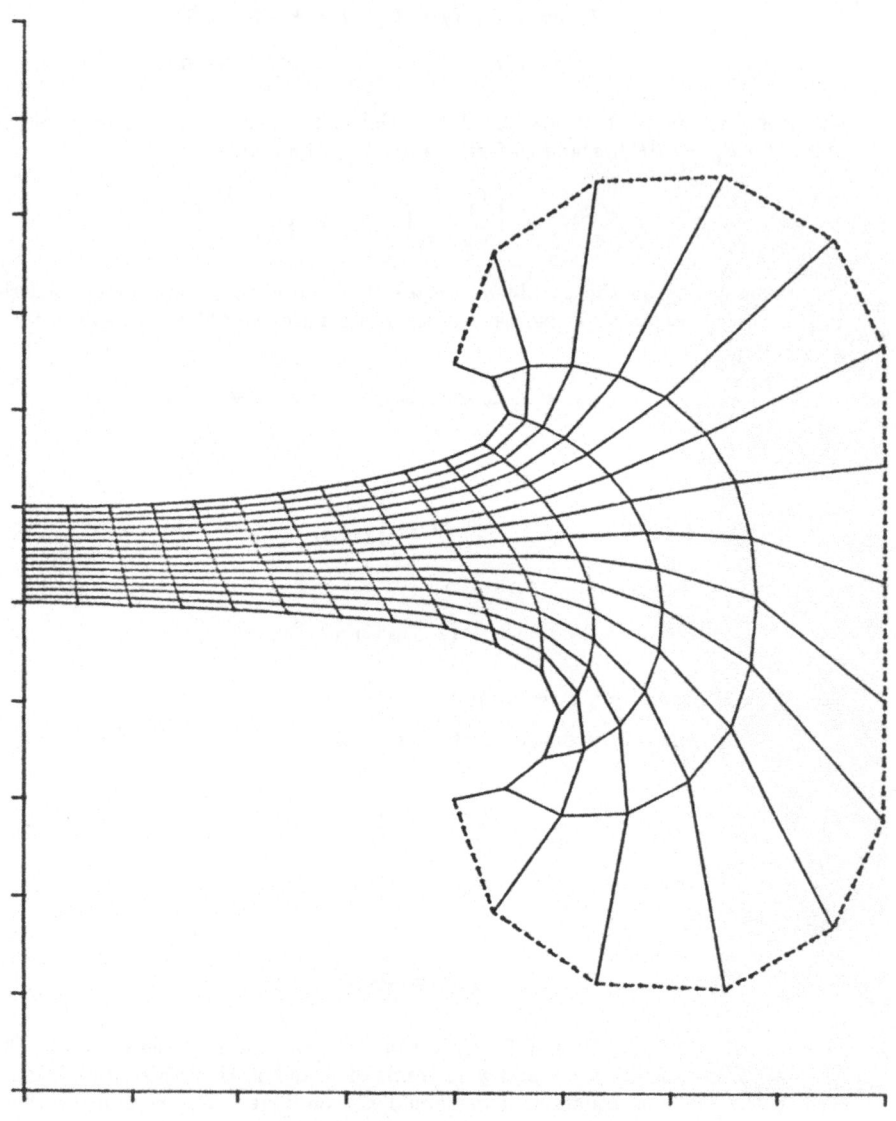

Grid for a 2D Laser Cavity
Figure 6

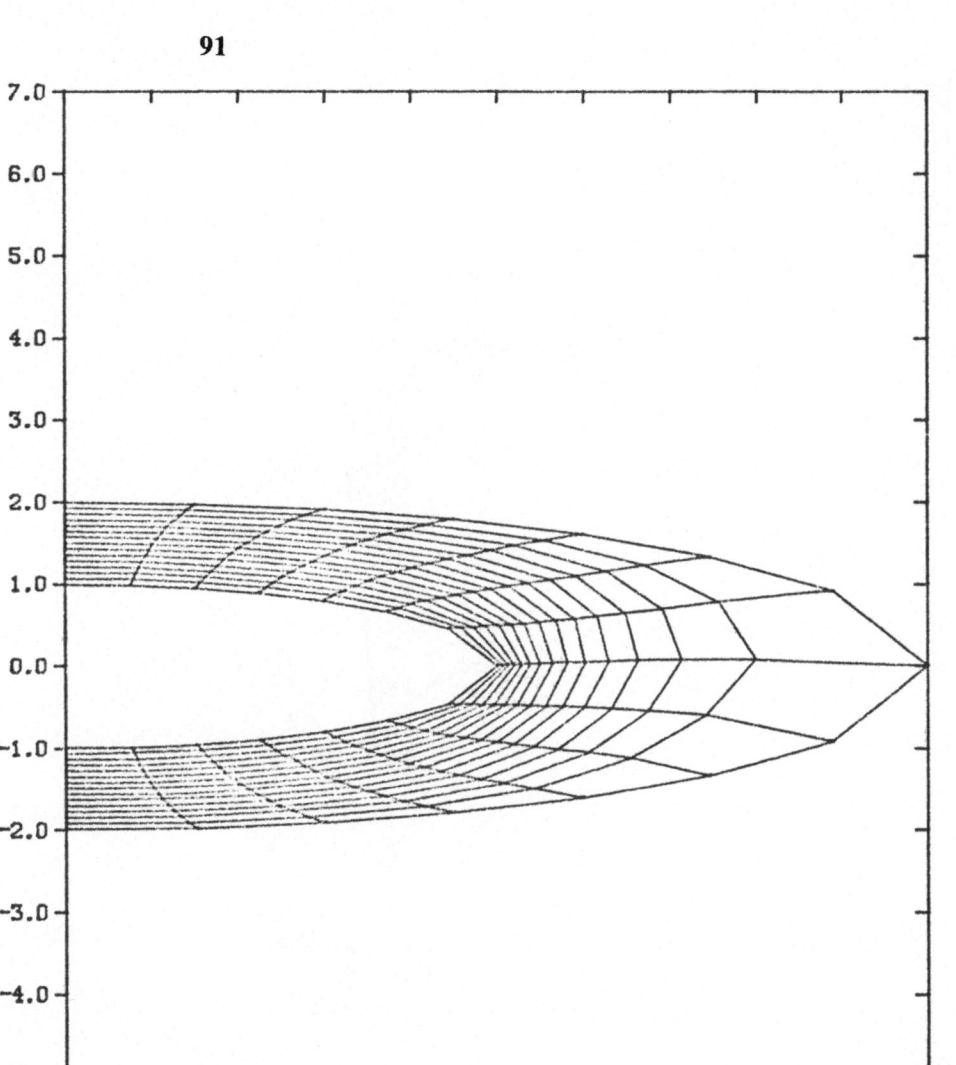

Grid for a 2D Laser Cavity
Figure 7

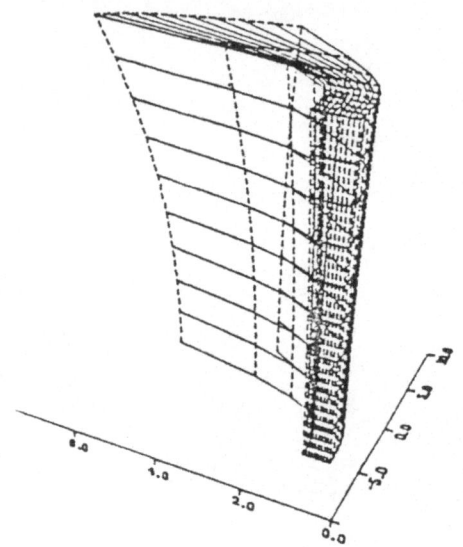

Grid for a 3D Laser Cavity
Figure 8

93

then use the formulas for dy and dx to eliminate these terms from the resulting expression. This computation can be done using the matrix form of the equations, in which case it is important to use the identity that was discussed in Section 2 for differentiating the inverse of a matrix. The computations in the multivariate case are similar but considerably more complicated than what we just did. Since we carried out the derivation of these formulas by hand in [17], we do not believe the computation to be practical in VAXIMA. As can be seen from this note and the paper [17], we believe that it is important for symbol manipulators to know about differential forms.

References

1. Anderson, J.D., Lau, E.L., and Hellings, R.L. *In*: 1979 MACSYMA Users' Conference, (V.E. Lewis, Ed.), (Washington, D.C., 1979), pp. 583-595.

2. Babuska I., Chandra, J., and Flaherty J.E., Eds. Adaptive Computational Methods for Partial Differential Equations, SIAM, (Philadelphia, 1983).

3. Brackbill, J.U. and Saltzman, J.S. J. Computational Physics, *46* (1982), pp. 342-368.

4. Engquist B. and Smedsaas T. SIAM J. Sci. Stat. Comput., *1*, (1980), pp. 249-259.

5. Fateman R. *In*: 1979 MACSYMA Users' Conference, (V.E. Lewis, Ed.), (Washington, D.C., 1979), pp. 563-582.

6. Ghia K.N. and Ghia U., Eds., Advances in Grid Generation, ASME FED, *5*, (1983).

7. Hermann, R. This point was discussed while R. Hermann was visiting the Department of Mathematics and Statistics at the University of New Mexico in April of 1984.

8. MACSYMA Reference Manual, The MATHLAB Group, Laboratory for Computer Science, MIT, Cambridge, MA, 1977.

9. Ng E. and Char B. *In*: 1979 MACSYMA Users' Conference, (V.E. Lewis, Ed.), (Washington, D.C., 1979), pp. 604-621.

10. Roache, P. and Steinberg, S. Talk presented at the Army Research Office - General Electric Corporation Workshop on Symbolic Computations, (Dec. 1982, Schenectady, NY).

11. Roache P.J., Steinberg S., Happ H.J., and Moeny W.M. *In*: Proceedings of the 4th IEEE Pulsed Power Conference, (Albuquerque, NM, June 1983).

12. Roache P.J. and Steinberg S. *In*: Proceedings of the AIAA 6th Computational Fluid Dynamics Conference, (Danvers, Mass., July 1983), pp. 443-462.

13. Roache R.J., Moeny W.M., and Steinberg S. *In*: AIAA 17th Fluid Dynamics, Plasma Dynamics and Lasers Conference, (June 1984, Snowmass, Colo.), pp. 25-27.

14. Saltzman, J.S. and Brackbill, J.U. Numerical Grid Generation, Proceedings of the Symposium on the Numerical Generation of Curvilinear Coordinate Systems and use in the Numerical Solution of Partial Differential Equations, (Thompson J.F., Ed.), (April 1982, Nashville, Tennessee), North-Holland, New York, 1982.

15. Steinberg, S. and Roache, P. Talk presented at the Army Research Office - General Electric Corporation Workshop on Symbolic Computations, (Dec. 1982, Schenectady, NY).

16. Steinberg S. and Roache, P.J. To appear in the Journal of Computational Physics, (1984).

17. Steinberg S. Change of Variables in partial differential equations. Technical report, Dept. of Math. and Stat., Univ. of New Mexico, Albuquerque, March 1983.

18. Steinberg S. and Roache P.J. Variational formulation for numerical coordinate changes, In preparation.

19. Thompson J.F., Ed. Numerical Grid Generation. Proceedings of the Symposium on the Numerical Generation of Curvilinear Coordinate Systems and use in the Numerical Solution of Partial Differential Equations, (April 1982, Nashville, Tennessee), North-Holland, New York, 1982.

20. Wester M. and Steinberg S. ACM SIGSAM Bulletin *17*, (1983), pp. 25-30.

21. Wester M. and Steinberg S. *In*: Third MACSYMA Users' Conference, (V.E. Golden, Ed.), (July 1984, Schenectady, N.Y.).

22. Wirth M.C. *In*: Proc. 1981 ACM Symposium on Symbolic and Algebraic Computation, (Aug., 1981, Snowbird, Utah), pp. 73-78.

4

APPLICATIONS OF SYMBOLIC MATHEMATICS TO MATHEMATICS

A. M. ODLYZKO

AT&T Bell Laboratories, Murray Hill, New Jersey 07974

ABSTRACT

Some of the most interesting applications of symbolic mathematics are in mathematics itself. Areas of both pure and applied mathematics, including coding theory, cryptography, probability theory, analysis, combinatorics, and number theory, have all gained from the availability of the new symbolic manipulation tools. These tools have been used to prove a number of results directly. Their main application, however, has been to obtain insight into behavior of various mathematical objects, which then led to conventional proofs being constructed.

1. INTRODUCTION

This paper is an account of some applications of symbolic algebra systems to mathematics. It is not a general survey of this very wide subject, but rather a selection from my own work, a selection made to illustrate the main point of the paper, which is that the most important applications of symbolic algebra systems are not to compute formulas as such, but to help in formulating and testing hypothesis.

My experience with symbolic mathematics goes back over ten years. Most of this work was with MACSYMA [10], although at various points I or my collaborators used other systems such as ALTRAN [2], MAPLE [5], and SMP [4]. These systems were used in many fields of mathematics, as the examples presented below demonstrate. Still,

95

these extremely varied examples do not cover the full range of applications that have been made, and are a reflection of my research interests. For example, one of the most impressive achievements in the field of computer algebra is the development of algorithms for indefinite integration. However, I have not yet made any serious use of them. The reason is that all of the many integrals I have worked with were either very easy, so that I was able to do them either directly or with the help of a table of integrals, or else they were so hard that even the algorithms in MACSYMA failed on them. On the other hand, I have made use of most of the other capabilities of computer algebra systems at one time or another. Since my problems are usually quite mathematical to start with, I had a strong advantage over people in other fields in not having to devote too much effort into translating these problems into a form suitable for symbolic manipulation programs.

The theme of this paper is reflected in the famous aphorism of R. W. Hamming [8]:

"The purpose of computing is insight, not numbers."

In the context of computer algebra, this maxim could be modified to say that the purpose of symbolic manipulation is insight, not formulas. To make it clear what is meant by this remark, consider some of the most successful applications of symbolic algebra, namely those to physics and celestial mechanics, many of which are listed and summarized briefly in [3]. In those applications, one is usually faced with a tremendously complicated mathematical description of the system under consideration, and the role of the symbolic algebra package is to go through a very long series of complicated transformations which result in a relatively simple formula which can then be used for numerical computation of the desired quantities, such as planetary orbits, for example. The validity of the results depends on the correctness of the program that was used, and little or no insight is provided by those results into why they came out the way they did. Important applications of this type do occur in mathematics, and an example will be presented in Section 2. It is my feeling, though, that such applications are not the most important ones. I see the main role of symbolic algebra systems as that of helping to formulate hypotheses, search for examples and counterexamples, and in general explore ramifications of mathematical models. In other words, the main role of these systems is to obtain mathematical insight. Once that insight is obtained, one can then go on and construct canonical mathematical proofs, in which there might not even be any traces of the use of computer algebra. Several examples of such applications are presented in sections 3-5.

2. CLASSIFICATION OF FINITE SIMPLE GROUPS

This section presents an application of computer algebra to one small chapter in the subject of classification of finite simple groups, namely Bombieri's proof [1] of the uniqueness of simple groups of Ree type. This application was chosen as an example where only raw computing power was needed, and no special insight was gained from the computation.

Thompson had earlier shown that the uniqueness of simple groups of Ree type would follow if one could show that given any automorphism σ of $GF(3^{2n+1})$, $n \geq 1$, $\sigma \neq 3^{n+1}$, and integers a,s such that a is even, s is odd, and $x^{(\sigma+1)} = x^2$ and $x^{(\sigma+2)s} = x$ for all $x \in GF(3^{2n+1})$, there is an element $x \in GF(3^{2n+1})\backslash GF(3)$ such that if

$$z = (x + 1)^s, \quad y = (x - 1)^s, \quad u = (x + 1)^{1-s} - (x - 1)^{1-s},$$

then $u \notin GF(3)$ and

$$z(u-1)^a - (z-y+1)u^a - y(u+1)^2 + (u^2-1)^a \neq 0 . \qquad (2.1)$$

Bombieri succeeded in showing, by a sophisticated argument, that a suitable x for which (2.1) holds can be found except possibly in a finite number of cases. Thus the problem was reduced to a finite computation, namely that of finding, for each case in Bombieri's list, of x that satisfies (2.1).

The computations that completed Bombieri's proof were carried out by D. Hunt and myself (independently) [1]. Given any $x \in GF(3^{2n+1}) \backslash GF(3)$, the chances of (2.1) being false seemed very low, so unless there were some unexpected phenomena (such as a counterexample to the conjecture), it seemed likely that choosing a random x and computing the quantity on the left side of (2.1) would suffice. That, in fact, turned out to be the case.

The computation of the quantity in (2.1) is easily carried out using computer algebra. It is first necessary to generate the field $GF(3^{2n+1})$, which is equivalent to finding an irreducible polynomial of degree $2n+1$ over $GF(3)$, which can be done using known factorization algorithms. Then the computations are carried out in the field $GF(3^{2n+1})$, which is represented as the ring of polynomials over $GF(3)$ modulo the irreducible polynomial. These computations are, in principle, quite straightforward. Their validity (and that of the main result) depend on the correctness of the programs and machines. The fact that (2.1) did hold for the choices of x that were made only confirmed the guess that (2.1) would hold for most x's, and did not provide any insight that would enable one to prove the result without the computations.

As an aside, the computations required to complete Bombieri's proof were in principle perfectly suited to the capabilities of the MACSYMA system. I wrote the necessary program in MACSYMA in a fraction of an hour and ran some small cases successfully through it (and in the process discovered a previously unknown bug in MACSYMA). However, a quick extrapolation of the times needed for those examples showed that the entire computation would have taken on the order of several days of MACSYMA's time. Since I was then a non-paying guest at the MIT MACSYMA machine, I did not feel like straining my hosts' hospitality to such an extent. Therefore I spent a day or two writing a FORTRAN program especially for this job, which then carried out the entire computation in a few hours. Some of the output of that program was checked using MACSYMA, though. Today, with several computer algebra systems available on local computers, I would be much more inclined to use them for the entire computation to save my own time. This is certainly in tune with the modern trend towards the use of higher-level computer languages, which are often less efficient than lower-level ones, but which save programmers' time and yield more portable and more readable code.

3. DISCRIMINANTS OF NUMBER FIELDS

In the previous example, the symbolic algebra application stood out on its own, since the final result depended on the correctness of the computation, and no special insight was provided into the reasons the result turned out the way it did. We next consider another application, where the role of computer algebra was quite different. This example is drawn from my dissertation, which appeared in [11]. The subject of it was improved lower bounds for discriminants of number fields. The methods used were from analytic number theory. Some of the crucial technical lemmas required showing certain rational functions were nonnegative over a wide range. For example, one of these problems was reduced to showing that

$$P(v) = 73v^8 - 20,214v^7 + 2,094,217v^6 - 92,766,496v^5$$
$$+ 1,419,515,855\ v^4 + 3,533,810,602\ v^3 \tag{3.1}$$
$$-2,837,192,781\ v^2 - 29,267,901,572\ v + 237,342,960,316$$

satisfies $P(v) \geqslant 0$ for all $v \geqslant -10$ [11, p. 288]. This is a problem that can be solved easily by some computer algebra systems, since there is an algorithm due to Sturm for determining the exact number of zeros of a polynomial in a given interval. That algorithm is available in MACSYMA. However, the proof presented in [11] does not rely on the use of that algorithm.

It is easy to see, for example, that for $v \geqslant 0$, $P(10v) \geqslant 73 \cdot 10^7 P_2(v)$, where

$$P_2(v) = 10v^8 - 277v^7 + 2,868v^6 - 12,708v^5 + 19,445v^4$$
$$+ 4,840v^3 - 389v^2 - 401v + 325\ .$$

Now for $0 \leqslant v \leqslant 1/2$, say,

$$389v^2 + 401v \leqslant 325, \quad 12,708v^5 \leqslant 19,445v^4, \quad 277v^7 \leqslant 2,868v^6,$$

and so $P_2(v) \geqslant 0$ in that range. Other ranges are treated similarly in [11] to provide the desired proof.

During the preparation of [11], the MACSYMA algorithm for finding the number of zeros of a polynomial in an interval was used (and some bugs in it were found). In fact, I used MACSYMA in the initial analysis that led to the reduction of the problem to the nonnegativity of polynomials like $P(v)$ of (3.1), and I used MACSYMA again in constructing the case-by-case proofs of nonnegativity of the kind outlined above. As a result the proof can be verified even by people without access to computer algebra systems. The role of symbolic manipulation in this case was to speed up my own work by showing which polynomials could be used and in helping to construct paper-and-pencil type proofs.

The application of computer algebra discussed here is roughly intermediate between the case discussed in Section 2, where the output of the computation was the result, and the cases to be discussed next, where computer algebra served only to provide insight. In the case of bounds for discriminants of number fields, the proof is much easier for the reader to check with the use of a computer algebra program than by following the proof given in the paper step-by-step. The point of going to the extra effort was to make the proof verifiable even to those without access to such programs. In the results to be discussed next, even access to a symbolic manipulation system is of no help to the reader in verifying the proofs, although such access was very helpful in figuring out what the results ought be.

4. ENUMERATION OF 2,3-TREES

A 2,3-tree is a rooted, oriented tree each of whose nonleaf nodes has either two or three successors, and all of whose root-to-leaf paths have the same length. 2,3-trees and their generalizations, the B-trees, are used as data structures in situations where it is desirable to be able to insert and delete records in time that is logarithmic in the total number of records present. Let a_n denote the number of 2,3-trees of size (i.e., number of leaves) equal to n. In the paper [12], I showed that

$$a_n \sim n^{-1}\phi^n \, u \, (\log n) \quad \text{as } n \to \infty,$$ (4.1)

where $\phi = (1 + 5^{1/2})/2$ is the "golden ratio," and $u(x)$ is a positive nonconstant continuous function which satisfies $u(x) = u(x + \log(4-\phi))$ for all real x. The proof contains no trace of the use of a symbolic manipulation program. However, such a program (MACSYMA in this case) was very helpful in obtaining this result. The proof uses complex analysis techniques to study the generating function

$$f(z) = \sum_{n=1}^{\infty} a_n z^n .$$ (4.2)

It is a general principle that the behavior of the coefficients a_n of a generating function of the form (4.2) is determined by the analytic behavior of the function $f(z)$ near its singularities. It is easy to show that the generating function $f(z)$ for 2,3-trees is analytic inside the circle $|z| = \phi^{-1}$, and also on that circle with the exception of the point $z = \phi^{-1}$, where it has a singularity. Since it was already known that

$$c_1 n^{-1}\phi^n < a_n < c_2 n^{-1}\phi^n$$ (4.3)

for some constants c_1 and c_2, I started out trying to prove the natural conjecture, namely that

$$a_n \sim cn^{-1}\phi^n \quad \text{as } n \to \infty$$ (4.4)

for some constant c. If (4.4) were to hold, though, one would expect that

$$f(x) \sim -c \log(\phi^{-1} - z) \quad \text{as } z \to \phi^{-1}.$$ (4.5)

It was not too difficult to prove (4.5), but I was unable to obtain a good enough estimate for the difference of the two sides in (4.5) to prove (4.4). The reason for that was that (4.4) is false, and (4.1) is the correct result. The crucial analytic result that led to the proof of (4.1) was the relation

$$f(z) = -c \, \log \, (\phi^{-1} - z) + w \, (\log \, (\phi^{-1} - z)) + O(|\phi^{-1} - z|) \tag{4.6}$$

as $z \to \phi^{-1}$, valid in a sector of the form $|\text{Arg} \, (\phi^{-1} - z)| < \pi/2 + \delta$ for some $\delta > 0$, where $w(z)$ is a nonconstant analytic function which is periodic with period $\log(4 - \phi)$. The proof of (4.6) involves intensive study of the polynomial iteration $T(z) = z^2 + z^3$, since $f(z)$ satisfies the functional equation

$$f(z) = z + f(z^2 + z^3) \, . \tag{4.7}$$

The conjecture that the behavior of $f(z)$ is given by (4.6) was made partially on the basis of computations with MACSYMA. Having failed to prove (4.4), I used the high-precision floating point facility of that system to explore the behavior of $f(z)$ for z near ϕ^{-1} by using the functional equation (4.7). If we let $T_0(z) = z$, $T_1(z) = T(z)$, and $T_{n+1}(z) = T(T_n(z))$ for $n \geqslant 1$, then (4.7) shows that

$$\begin{aligned} f(z) &= T_0(z) + f(T_1(z)) \\ &= T_0(z) + T_1(z) + f(T_2(z)) \end{aligned} \tag{4.8}$$

$$\dots$$

$$= \sum_{n=0}^{\infty} T_n(z) \, ,$$

provided $T_n(z) \to 0$ as $n \to \infty$. The expansion (4.8) (which gives an analytic continuation of $f(z)$ beyond the circle of convergence of its Taylor series, and has many interesting properties in its own right) can be used to study $f(z)$ numerically, provided one can do computations with enough precision to avoid loss of accuracy. Being able to use 50 or 100 decimal digit numbers was very helpful in this situation. By using the results of those computations and simultaneously investigating what phenomena could arise from iterating the map $T(z) = z^2 + z^3$, I was led to conjecture (4.6) and later to prove it. Also, by studying the polynomials $T_n(z)$, I was led to an understanding of how the behavior of $f(z)$ near its singularity at $z = \phi^{-1}$ influenced the behavior of the coefficients a_n, which led to the proof of (4.1).

The role of computer algebra in this case was purely to provide insight, not to help with the proof itself. The result certainly could have been obtained without the use of symbolic mathematics, and perhaps would have been, had I not had access to such systems, but the work would have been harder and would undoubtedly have taken longer.

5. STRING ENUMERATION

Another example where computer algebra had as its main function providing insight is given by the work on enumeration of strings [6,7]. As is shown in those papers and the references cited in them, a surprising variety of problems in combinatorics, probability theory, games, and algorithmic complexity can be reduced to the question of evaluating the number of strings of length n over some alphabet of q symbols, call it $f(n)$, which contain none of a given set, call them A, B ,..., T of words as subwords (i.e., blocks of consecutive characters). The basic result of [6], from which most of the others follow, is a formula for the generating function of $f(n)$,

$$F(z) = \sum_{n=0}^{\infty} f(n)z^{-n} .$$ (5.1)

(We take $f(0) = 1$. $F(z)$ is written as a power series in z^{-1} in order to obtain nicer formulas.) To explain the formula for $F(z)$, we define the correlation polynomial XY_z of the words $X = x_1...x_k$ and $Y = y_1...y_m$ to be

$$XY_z = \sum_{j=1}^{k} c_j z^{j-1}, \tag{5.2}$$

where $c_j = 1$ if the suffix of X of length j (i.e., $x_{k-j+1}...x_k$) equals the prefix of Y of length j (i.e., $y_1...y_j$), and $c_j = 0$ otherwise. (There are special rules for the case $m < k$.) The first result of [6] is that if the set of excluded words consists of a single word, call it A, then

$$F(z) = \frac{zAA_z}{1 + (z - q)AA_z}. \tag{5.3}$$

The proof of (5.3) that we gave is very simple, takes only a couple of lines, and certainly does not require, nor is it even helped by, the use of computer algebra. (Moreover, it turned out that this particular result had been obtained earlier by Solov'ev [16].) However, computer algebra was extremely helpful in obtaining that proof in the first place. Guibas and I started out by constructing finite state automata, reducing the number of states in them, and obtaining generating functions from them which determined $F(z)$. It was a very laborious process, and we used MACSYMA to carry out the symbolic manipulations that were involved. The formulas for $F(z)$ that came out were surprisingly compact, and inspection of a number of them showed that they were all of the simple form (5.3). Once we saw this elegant formula, we concluded that a simple proof had to exist, and it did not take very long to find it. Thus in this case the role of computer algebra was to bring out the simplicity that was hidden in our complicated initial approaches. We might have found the result (5.3) by hand, but it would have taken us a lot longer.

Once a simple proof of (5.3) was found, it was a simple matter to generalize it. If the set of excluded words is $\{A,B, ..., T\}$, and no word in this set contains another (as we may assume without loss of generality), then $F(z)$ is determined by the following system of linear equations:

$$(z - q)F(z) + zF_A(z) + zF_B(z) + ... + zF_T(z) = z$$

$$F(z) - zAA_zF_A(z) - zBA_zF_B(z) - ... - zTA_zF_T(z) = 0 \qquad (5.4)$$

$$\vdots$$

$$F(z) - zAT_zF_A(z) - zBT_zF_B(z) - ... - zTT_zF_T(z) = 0$$

Here $F_A(z)$, ..., $F_T(z)$ are generating functions related to $F(z)$. The system (5.4) can easily be shown to be nonsingular [6], and so $F(z)$ is seen to be a rational function. Computer algebra is not needed in the proof of (5.4), and was not even used in deriving the proof, as it was fairly easy to generalize the proof of (5.3).

Computer algebra was also useful in deriving other string enumeration results, but again at the level of providing insight, rather than providing proof. As an example, we might consider the asymptotic estimation of $f(n)$. If the set of excluded words consists of A only, then (5.3) holds, and it can be shown that the behavior of $f(n)$ is determined almost completely by a first order pole of $F(z)$ near $z = q$ [6,7]. No computer algebra is involved there, as only basic complex analysis is used. However, when several words are excluded, the situation can become much more complicated. For example, it can happen that $f(n) = 0$ for all sufficiently large n, as occurs for the set $\{000, 111, 10\}$ if $q = 2$, in which case $F(z)$ is a polynomial on z^{-1}. Thus in general we cannot say too much about the generating function $F(z)$ determined by (5.4). However, something can be done when we consider repetitive patterns. Suppose that B is a nonperiodic word of length m [7], so it cannot be written as $B = CC...C$ for any word C shorter than B. A B^*-run of length k is a word $A = B''BB...BB'$ of length k where B'' is a suffix of B and B' is a prefix of B. Thus, if $B = 010$, then $A = 001001$ is a B^*-run of length 6. One can ask about the distribution of lengths of maximal B^*-runs in random sequences. Now the number of strings of length n that contains no B^*-run of length k is exactly $f(n)$, the number of strings of length n that contain none of $A(1)$, ..., $A(m)$, where

$$A(r) = b_{m-r+1} \, b_{m-r+2} ...b_m \, BB \, ... \, BB' \, ,$$

and $B' = B'(r)$ is a prefix of B of the appropriate length. It can then be shown [7] that (for $q = 2$, say)

$$F(z) = \frac{z^{(k+1)m+1} \, (z^m - 1)^{m-1} + g_1(z)}{(z - 2)z^{(k+1)m} \, (z^m - 1)^{m-1} + g_2(z)} \, , \tag{5.5}$$

where $g_1(z)$ and $g_2(z)$ are polynomials of small degrees and small coefficients that satisfy some additional conditions (see [7] for details). From (5.5) one can then show that a good approximation to $f(n)$ is given by

$$f(n) \approx 2^n \, \exp(-nm\,2^{-k-1}) \, . \tag{5.6}$$

Also, $E_B^*(n)$, the expected length of the maximal B^*-run in a random binary sequence can be shown to satisfy

$$E_B^*(n) = lg\ n - 3/2 + \gamma(\log\ 2)^{-1} + lg\ m \qquad (5.7)$$

$$+ v\,(lg\ n + lg\ m) + o\,(1)\ \text{as}\ n \rightarrow \infty\ ,$$

where $lg\ x$ is the logarithm of x to base 2, γ is Euler's constant (0.577...), and $v\,(x)$ is a certain nonconstant continuous periodic function of period 1 and mean value 0.

The proofs of (5.6) and (5.7) did not rely in any way on computer algebra. Even (5.5), which is a very algebraic result, and was proved by some very messy algebraic manipulations, was proved without the help of symbolic manipulation programs, since the algebra involves exponents such as k and m as variables, as well as z. However, computer algebra was very helpful in guessing that something like (5.5) might hold. The functions $F(z)$ were computed for various B^*-runs, and from their algebraic forms and their singularities, the correct conjectures were derived.

6. MISCELLANEOUS APPLICATIONS AND CONCLUSIONS

Many more examples of the application of computer algebra to mathematics could be presented. I will note a few minor applications. In recent work on discrete logarithms [14], it was desirable to verify that for various degrees n, there are polynomials of the form $x^n + f(x)$ which are irreducible over $GF(2)$ and such that the degree of $f(x)$ is small. Since about one out of n polynomials of degree n is irreducible over $GF(2)$, it was expected and confirmed by computation that one can find such $f(x)$ with deg $f(x) \leq \log_2 n$. This was a straightforward problem perfectly suited for a symbolic manipulation program but too tedious to do by hand. In the work on cellular automata [9], computer algebra was used to explore the behavior of automata with so-called linear evolution rules. Once the data was collected, conjectures were made and proved by ordinary methods. In the work on knapsack cryptosystems described in [13], the theoretical attacks that were proposed were tested on a symbolic manipulation system mainly because it was easy to program all the required operations (random number generation, modular multiplication of large integers, reduction of lattice bases, etc.) in it. In the work on new bounds for "kissing numbers" [16], computer algebra was used to generate the required Jacobi polynomials. In all of the above applications, computer algebra was not essential, but it was very useful in taking care of most of the computational drudgery.

One of the important functions of computer algebra is to carry out little computational tasks which can be done by hand, but are tedious and not interesting. In these cases computer algebra can make mathematical research somewhat more efficient and much more pleasant. In some cases, such as the computation described on Section 2, computer algebra can carry out giant computations that appear to be unavoidable, but are impractical to do by hand. The most important applications of computer algebra, though, are to situations such as those in sections 4 and 5, where the symbolic computations serve to provide insight and lead to standard proofs being constructed.

In summary, I feel that symbolic manipulation programs are already very useful, and will become even more useful in mathematical research. Yet I see their role as essential that of a "bigger scratchpad," to quote P. Halmos. I feel safe in predicting that they will not change the basic nature of mathematical research.

REFERENCES

[1] E. Bombieri, Thompson's problem ($\sigma^2 = 3$), (with appendices by A. Odlyzko and D. Hunt), *Inventiones math.* **58** (1980), 77-100.

[2] W. S. Brown, *ALTRAN User's Manual*, 4th ed., Bell Laboratories 1977.

[3] J. Calmet and J. A. van Hulzen, Computer algebra applications, pp. 245-258 in *Computer Algebra*, B. Buchberger, G. E. Collins, and R. Loos, eds., *Computing Suppl.* **4** (1982), Springer.

[4] C. A. Cole, S. Wolfram, et al., *SMP-Handbook*, Caltech 1981.

[5] K. O. Geddes, G. H. Gonnet, and B. W. Char, *MAPLE User's Manual*, Univ. of Waterloo, 1983.

[6] L. J. Guibas and A. M. Odlyzko, String overlaps, pattern matching, and nontransitive games, *J. Comb. Theory (A)* **30** (1981), 183-208.

[7] L. J. Guibas and A. M. Odlyzko, Long repetitive patterns in random sequences, *Z. Wahrscheinlichkeits v. Geb.* **53** (1980), 241-262.

[8] R. W. Hamming, *Numerical Methods for Scientists and Engineers*, 2nd ed., McGraw Hill 1973.

[9] O. Martin, A. M. Odlyzko, and S. Wolfram, Algebraic properties of cellular automata, *Commun. Math. Physics* **93** (1984), 219-258.

[10] MATHLAB Group; *MACSYMA Reference Manual*, MIT Laboratory for Computer Service 1977.

[11] A. M. Odlyzko, Lower bounds for discriminants of number fields, *Acta Arith.* **29** (1976), 275-297.

[12] A. M. Odlyzko, Periodic oscillations of coefficients of power series that satisfy functional equations, *Advances in Math.* **44** (1982), 180-205.

[13] A. M. Odlyzko, Cryptanalytic attacks on the multiplicative knapsack cryptosystem and on Shamir's signature scheme, *IEEE Trans. Inform. Theory IT-30* (1984), 594-600.

[14] A. M. Odlyzko, Discrete logarithms in finite fields and their cryptographic significance, *Proc. Eurocrypt 84*, Springer Lecture Notes in Computer Science, to appear.

[15] A. M. Odlyzko and N. J. A. Sloane, New bounds on the number of unit spheres that can touch a unit sphere in n dimensions, *J. Comb. Theory (A)* **26** (1979), 210-214.

[16] A. D. Solov'ev, A combinatorial identity and its application to the problem concerning the first occurrence of a rare event, *Theory Prob. Appl.* **11** (1966), 276-282.

5

PAST, PRESENT, AND FUTURE APPLICATIONS OF COMPUTER ALGEBRA IN CHEMISTRY

T. E. RAIDY

Department of Chemistry, University of South Carolina, Columbia, South Carolina 29208

ABSTRACT

Although one of the first applications of computer algebra occurred in chemistry, chemists have only recently begun to use computer algebra as a research tool. This paper will begin with a survey of past (largely uncited) and present applications of computer algebra in chemistry. Following this introduction, applications in chemical research and chemical education will be discussed in terms of the dramatic increase in availability and in power of computer algebra languages.

INTRODUCTION

This paper is intended to provide an historical perspective or review of the application of computer algebra (CA) in chemical research. There is some difficulty in attempting to document past uses of CA by chemists as many chemists have either failed to note the role of CA in their research or have failed to cite relevant CA literature in their publications. In addition to the review of past and present applications of CA in chemistry, speculations on future applications of CA in chemical education and in chemical research will be given.

As indicated by the title, this presentation is divided into three sections; past, present, and future. Further, the section on past applications of CA divides naturally into three distinct eras. The use of arithmetic computer languages (e.g., FORTRAN) to perform algebraic manipulations characterizes the earliest period. Considerable effort was required for these early applications which were often unsuccessful. The second era saw the development of the LISP programming language and several primitive computer algebra languages. The advances during this period, however, were all but ignored by the chemical community. The third and present period has been and is noted by the spread of existing CA languages, by the development of powerful general purpose CA languages, and by the availability of supermicro- and mini-computers with sufficient memory to perform algebraic manipulations. It is also an era of increasing interest in and use of CA languages in all areas of scientific research.

Past Applications

In 1956, Boys et al. (I) reported the first chemical application of CA with the sentence, "The machine has been caused to perform most of the organization of the calculations and a considerable portion of the mathematical analysis, as well as the arithmetic" Boys' following sentence, "It is possible, though not certain, that more effort has been required for the development of these procedures than might have been needed for the calculations without automatic computation, but this initial effort is largely non-recurrent", clearly and succinctly summarizes the first decade of computer algebra. For example, in 1959 Gimarc used FORTRAN in an attempt to collect terms in an expansion of products of polynomials. Gimarc states, "The resulting program, even if completed, would not have been cited or noted in any subsequent publication"(2). Similar efforts by other chemists during this period were either equally unsuccessful or unnoted. An exception to this trend was the treatment of a problem in theoretical chemistry by Wactlar and Barnett(3). However, their paper was actually one of a computer science series by Barnett illustrating the use of FORTRAN to perform algebraic manipulations.

Perhaps the most significant advance in symbolic computing was the development of the LISP programming language in I962(4). While LISP serves as the base for many computer algebra languages, LISP itself has been used directly in at least two chemical applications. G.S. Chandler and T. Thirunamachandran with the aid of J. A. Campbell used LISP to sum and substitute terms in a treatment of phosphorus orbital configurations(5). Later, Laurenzi et al.(6) wrote a LISP program to perform differentiations and Laplace transform subsitutions.

In a parallel development, the first languages or programs specifically designed to perform algebraic manipulations were developed in North America and Europe. These included GRAD ASSISTANT(7), ALTRAN(8), LAM(9), FORMAC (I0), SYMBAL(II), CAMAL(I2), and SAC-I(I3). Those based in FORTRAN or PL/I were clearly an outgrowth of earlier CA efforts. During this period the author(I4) is not aware of any uses of CA by chemists other than the examples cited above(3,5).

As these early languages, specifically ALTRAN and FORMAC, and the newer LISP based languages MACSYMA(I5) and REDUCE(I6) spread through research communities, chemists began to apply these tools to problems in chemistry. Benesch(I7) employed FORMAC to sum linear combinations of products in treating Hylleraas type wave functions. R. Bogen, one of the developers of MACSYMA, combined with chemists D.W. Underhill and J.A. Reeds(I8) to apply MACSYMA to the analysis of spatial moments of sorbates. S. Aubry(I9) used REDUCE to perform the repeated differentations necessary to obtain a formal expression for a Liouville operator. Matrix elements of a perturbed harmonic oscillator were generated by a FORMAC program written by C. W. David(20).

C. S. Johnson, Jr.(21) notes that the use of CA by chemists is probably considered a secondary aspect of most research efforts. As such, it is not surprising that there are so few direct citations of the use of CA in chemical applications. The author is aware of several examples of this neglect. Beginning in 1972, several members of the Theoretical Chemistry Group at the University of Waterloo applied the ALTRAN language to research problems involving polynomial expansions. Although different applications played a role in 5-8 publications, the only clear citation of CA appears in an unpublished thesis(22). Kreek and Le Roy(23) acknowledged their part in these efforts with, "In spite of the complexity of Eq.(9), values of coefficients of $t_j V^{(\gamma)}$ may be readily evaluated on a computer". Similar applications at other institutions also went uncited. For example, in a series of papers culminating with Santry and Raidy(24), no reference was made to partial differentiation performed with REDUCE. Maricq and Waugh(25) comment that, "The higher moments, which involve some considerable algebraic manipulations, were evaluated with the aid of the symbolic computer program MACSYMA", however no citation of MACSYMA appears in their references.

More recently, there has been an increase in the direct citation of CA language applications in chemistry. Grimmelmann et al.(26) differentiated polynomial terms with ALTRAN. The use of the REDUCE language was reported in six publications in 1982-83. Evans et al.(27) obtained exact numerical solutions of a polynomial summation. Chenon et al.(28) differentiated magnetization mode variables with respect to spectral densities. Kaguei and Wakao(29) in a chemical engineering application, studied column flow rate parameters in adsorption packed beds and in chemical reactors, performing both differentiation and matrix manipulations with REDUCE. Bessel functions and continued expansions were employed by F. E. Harris(30)in the evaluation of GTO integrals. Shmueli and Wilson(31) performed summations while Shmueli and Kaldor(32) used REDUCE to perform trigonometric summations.

An encouraging trend in the chemical application of CA is the appearance of articles by chemists demonstrating the value of CA as a chemical research tool. In 1982, both J. F. Ogilvie(33) and T. E. Raidy(34) presented brief reviews of CA languages and of their application to problems in chemistry. Two recent, heuristically oriented papers(21,35) compare CA languages and give programming examples from a chemist's point of view. In a follow up paper by T. E. Raidy(36), the complete mathematical workup and REDUCE program for the calculation of closed expressions for a simple set of gaussian molecular integrals is presented as an example application of CA to problems arising in computational chemistry. Furthermore, a referee suggested that Shmueli and Kaldor(32) include some of the details of their REDUCE program. The American Chemical Society's Symposium on Symbolic Algebraic Manipulations at which this paper was presented is further evidence of this trend in chemistry.

Present Applications

Recent published applications of CA in chemistry cite the use of REDUCE to perform density matrix calculations(37) and the use of MACSYMA to perform partial differentiation(38). The research group at the University of South Carolina NSF Nuclear Magnetic Resonance Facility now uses REDUCE and muMATH (39), while MACSYMA, SMP(40), or Maple(41) will be available soon. Current applications at the facility include programs that perform expansions of Wigner rotation matrix terms. This has been helpful in examining magic angle spinning expressions, in obtaining expressions for the angular dependence of slow molecular motion in solids, and in obtaining expressions for the antisymmetric part of the chemical shift tensor. The general matrix handling capabilities of CA are being exploited to obtain closed expressions for NMR lineshapes. At the University of Waterloo, C. Schwartz (42) is analyzing matrix terms with MACSYMA. These terms appear as elements of a problem in scattering theory. A number of other chemists have recently reported increased access to CA languages(21). Moreover, the papers presented in this volume serve as examples of current applications of CA in chemistry and other fields.

Future Applications

Regardless of how the chemical community views CA, it will have an impact on chemistry. Whether directly as part of a computer science/mathematics component of a chemistry curriculum or indirectly as a skill taught in mathematics courses, CA will be a tool applied to both coursework and research problems in chemistry.

The development of supermicrocomputers with sufficient memory to handle CA systems such as MACSYMA, SMP, or Maple are readily available. The muMATH system which was designed in part to be used as an instructional tool, runs on several micrcomputers and at least one hand-held calculator version - picoMATH(43) - has been demonstrated. D. R. Stoutemyer(44) and C. S. Johnson, Jr.(21) have suggested several applications of CA in scientific education. K. O. Geddes(35), one of the developers of Maple, will teach introductory calculus with Maple in the Fall 1984 semester. Maple is available to students on a combination of mainframes, minis, and supermicrocomputers.

To illustrate the power of CA in chemical education consider the solution of a straightforward stoichiometry problem taken from a freshman chemistry test.

> Calculate the mass of ZnS consumed in producing 10.0 grams of ZnO according to the reaction
> $$2 \ ZnS + 3 \ O_2 \to 2 \ ZnO + 2 \ SO_2$$

Using the "label-factor method", the solution would be obtained with muMath as:

    ? POINT: 2;
    ? (10.0 g ZnO) (1 mol ZnO/(81.4 g ZnO)) (2 mol ZnS/(2 mol ZnO)) (97.4 g ZnS/(1 mol ZnS));
    @: 11.9 g ZnS

The numeric result (11.9) is displayed with the correct units (g ZnS).

In the research field the impact of CA is already apparent. Chemists are late arrivals in an active field. Powerful computer algebra languages are being marketed as commercial products or being sold as prime components of computer workstations. While chemists have used the summation, the substitution, the differentiation, and the general algebraic capabilities of CA languages, other aspects of CA such as integration, matrix handling, differential equations, and tensor analysis have yet to be fully employed in any published chemical application. As well, physical chemists have dominated the early use of CA (33), but surely other areas of chemistry and chemical engineering would benefit from the application of CA to mathematical problems common to these fields.

Possibly most promising of all is the hybrid combination of computer algebra and numeric computation. Brown and Hearn(45) review some of the early successes of this hybrid approach which combines the clear and exact solutions of the algebraic systems with well understood numeric methods. Problems considered intractable for CA or numeric solution alone have proven amenable to solution with the hybrid approach. Insights provided by CA have led to the simplification of numerical computations. Similarly, in the treatment of complex problems in physics, straightforward numerical computations have eliminated algebraic bottlenecks that normally would have prevented further analysis of the problems.

Computer algebra provides chemists with another tool in their investigation of nature. It should prove to be an invaluable aspect of chemical education and chemical research.

REFERENCES

1. Boys, S. F., Cook,G. B., Reeves,C. M. and Shavitt,I. Nature (London) 178: 1207, 1956.
2. Gimarc, B. M. private communication, 1984.
3. Wactlar, H. D. and Barnett,M. P. Comm. ACM 7: 704 1964.
4. McCarthy, J.D. LISP 1.5 Manual, MIT Press, Cambridge, MA, 1962.
5. Chandler,G. S., Thirunamachandrum,T. and Campbell, J. A. J. Chem. Phys. 49: 3640, 1968.
6. Laurenzi, B. J., Williams, D. G. and Bhatia,G. S. J. Chem. Phys. 61: 2077, 1974.
7. Fletcher, J. G., Clemens,.R. W., Matzner,R. A., Thorne, K. S. and Zimmerman, B.A. Ap. J. Lett. 148: 91, 1967.
8. Brown, W. S. ALTRAN User's Manual, 4th ed., Bell Laboratories, Murray Hill, NJ, 1977.
9. D'Inverno, R. A. Comput. J. 12: 124, 1969.
10. Fike, C. PL/1 for Scientific Programmer's, Prentice-Hall, Englewood Cliffs, NJ, Chap. 12, 1970.
11. Engeli, M. ACM SIGSAM Bull: 21, 1975.
12. Fitch, J. P. CAMAL User's Manual, Comput. Lab., Cambridge, UK, 1975.
13. Collins, G. E. Proc 2nd Symp. on Symbolic and Algebraic Manipulation, ACM, New York, p. 144, 1971.
14. The author would appreciate any communications concerning early applications of computer algebra in chemistry not mentioned in this article.
15. Bogen, R. et al. MACSYMA Reference Manual, Project MAC, M. I.T., Cambridge, MA, 1974.
16. Hearn, A. C. REDUCE User's Manual, 2nd ed., University of Utah, 1973.
17. Benesch, R. J. Phys. B 4: 1403, 1971; Phys. Rev. A 6: 573, 1972.
18. Underhill, D. W., Reeds, J. A. and Bogen, R. Anal. Chem. 45: 2314, 1973.
19. Aubry, S. J. Chem. Phys. 62: 3217, 1975.
20. David, C. W. Comput. Chem. 1: 93, 1976.
21. Johnson, Jr.,C. S. J. Chem. Inf. Comput. Sci. 23: 151, 1983.
22. Raidy, T. E. Ph. D. Thesis, University of Waterloo, 1976.
23. Kreek, H. and LeRoy, R. J. J. Chem. Phys. 63: 338, 1975.
24. Santry, D. P. and Raidy, T. E. Theoret. Chim. Acta 53: 121, 1979.
25. Maricq, M. M. and Waugh, J. S. J. Chem. Phys. 70: 3300, 1979.
26. Grimmelmann, E. K., Tully, J. C. and Helfand, E. J. Chem. Phys. 74: 5300, 1981.
27. Evans, W. J., McCourtney, E. J. and Shrager, R. I. J. Amer. Oil Chem. Soc. 59: 189,1982.
28. Chenon, M. T., Bernassau, J. M., Mayne, C. L. and Grant, D. M. J. Phys. Chem. 86: 2733, 1982.
29. Kaguei, S. and Wakao, N. J. Chem. Eng. Jap. 16: 78, 1983.
30. Harris, F. E. Int. J. Quant. Chem. 23: 1469, 1983.
31. Shmueli, U. and Wilson, A. J. C. Acta Cryst. A39: 225, 1983.
32. Shmueli, U. and Kaldor, U. Acta Cryst. A39: 615, 1983.
33. Ogilvie, J. F. Comput. Chem. 6: 169, 1982.
34. Raidy, T. E. 34th Southeast Regional ACS Meeting, 1982.
35. Raidy, T. E. J. Chem. Ed. 61: 629, 1984.
36. Raidy, T. E. ,submitted to J. Chem. Inf. Comput. Sci.

118

37. Nakashima, T. T., McClung, R. E. D. and John, B. K . J. Magn. Res. 56: 262, 1984.
38. Waterland, R. L. and Delos, J. B. J. Chem. Phys. 80: 2034, 1984.
39. muMATH, Microsoft Consumer Products, Bellevue, WA.
40. Inference Corp., Computer Mathematics Group, 5300 West Century Blvd., Los Angeles, CA 90045.
41. Char, B. W., Geddes, K. O., Gentleman, W. M. and Gonnet, G. H. Lecture Notes in Comput. Sci. 162: 101, 1983.
42. Schwartz,C. private communication, 1984.
43. Stoutemyer, D. R. SIGSAM Bull. 14: 5, 1980.
44. Stoutemyer, D. R. Amer. J. Phys. 49: 85, 1981.
45. Brown, W. S. and Hearn, A. C. Comput. Phys. Comm. 17: 207, 1979.

6

SYMBOLIC COMPUTATION IN CHEMICAL EDUCATION

Don McLaughlin

Chemistry Department, The University of New Mexico

1. INTRODUCTION

The past two decades have witnessed revolutions in computing techniques used by students at all levels of science education. The slide rule has taken its place beside the abacus as a historical curiosity, and the electronic pocket calculator is now edging past the computing power of the first "giant electronic brains". The end of the computer age is not yet in sight, and we are about to witness another equally impressive decade of developments which will most likely render today's calculational methods obsolete.

One area of intense interest is artificial intelligence, which should probably be considered a misnomer in light of its successful application in "discovering" such subtle chemical and physical relationships as the Periodic Table[1] and Boyle's Law[2]. In this article we will explore an area of artificial intelligence which deals with employing computers to manipulate symbolic mathematical expressions. Just as the hand-held calculator replaced pencil and paper in doing arithmetic, it is not inconceivable that the next generation of hand-held computers will replace pencil and paper in solving problems in higher mathematics such as algebra, calculus and differential equations.

It was soon recognized in the early days of the development of high-level computer languages that general purpose digital electronic computers designed to manipulate bits of numerical information could just as easily manipulate bits of more general symbolic information[3]. Computer programming languages quickly separated into two divisions, emphasising either numerical or symbolic manipulation. The result of the first division is exemplified in the various dialects of the FORTRAN (FORmula TRANslation) programming languages, and representative of the second are the LISP (LISt Processing) programming languages. Because so many practical scientific applications have required numerical solutions to problems as opposed to analytical solutions, generally greater emphasis has been placed on the numerically oriented computer languages. However, in recent years both approaches have begun to merge as each language has evolved, expanded, or been implemented to include desirable features of the other. This process has been facilitated by the production of faster computers and more versatile operating systems.

Our intention here is to explore the current status of symbolic computing as a possible tool in the teaching and learning of chemistry. Symbolic computation will be explored for its pedagogical value through several exercises encountered in the chemistry curriculum.

2. BASICS OF SYMBOLIC COMPUTATION

Computer programs written to perform symbolic (or algebraic) manipulation attempt to communicate in familiar mathematical terminology, which suggests that learning to compute symbolically should be a relatively easy process. For demonstration purposes one of the rather well-developed symbol manipulation programs, MACSYMA[4] will be used. Although MACSYMA is written in LISP, it is possible to solve many simple problems without knowing much about LISP. This is because MACSYMA accepts input in the form of high level "commands" which are translated to LISP and executed by calling the appropriate routines. Since the commands accept arguments and produce output, MACSYMA is referred to as a "functional" language. Because the operands and output include functions as well as variable quantities, however, it may be helpful to think of it as an "operational" language, that is, one which performs various operations on rather large classes of objects. Advanced problem solving eventually requires familiarity with

119

block-structured programming techniques used in, say FORTRAN, Pascal or C.

One important feature of symbol manipulation programs is their interactive design (i.e. immediate response to a single command or set of commands. This not only allows for instantaneous editing and correcting but provides freedom to explore various pathways leading to the desired result expressed in a designated format. Although the ability to control the progress of the manipulations can become an indispensable advantage for complex problems, this interactive feature will be minimized in the examples to be presented below; only transcripts of final versions of tested sessions will be shown. However, it should be emphasized that perfected demonstrations are a a poor substitute for experiential learning.

2.1. Commands

The transcript of the first interactive session below introduces some notation and syntax of simple symbolic manipulation commands. Each input line is automatically referenced and indexed for possible future reference and causes a response by the system. Comments (delimited by slash and star notation borrowed from the PL1 language) are not processed. Numbers, names, simple mathematical operations and library function calls are syntactically similar to other languages. The display of the output using typeset notation is a novel feature however. One significant difference is the symbol used for assignment. Since the equals sign ($=$) must be reserved for expressing equations in symbolic form, another symbol is needed for "replacement of" or "assignment to" variables. In MACSYMA the assignment symbol is the colon (:).

(c1) /* Arithemetic */

/* Note each command (or set of commands) is terminated with a semi-colon. */

/* (Note also that expressions are automatically evaluated, reduced and expressed in a "typeset" form.) */

2 + 3*4 -5/6;

$$\frac{79}{6} \tag{d1}$$

(c2) /* Powers */

2^10;

$$1024 \tag{d2}$$

(c3) /* Variables */

x^-y;

$$\frac{1}{x^y} \tag{d3}$$

(c4) /* Variable assignment */

x:2;

$$2 \tag{d4}$$

(c5) y:10;

$$10 \hspace{10em} \text{(d5)}$$

(c6) /* Evlauate the expression given on line d3 with x and y "bound" to numbers. */

ev(d3);

$$\frac{1}{1024} \hspace{10em} \text{(d6)}$$

(c7) /* % is a special symbol used to refer to the result of the previous computation. */

ev(%,numer);

$$0.0009765625 \hspace{8em} \text{(d7)}$$

(c8) /* % is also used to name the following three mathematical constants: */

%e^(%i*%pi);

$$-1 \hspace{10em} \text{(d8)}$$

(c9) /* Factorial notation is also understood */

100!;

$$93326215443944152681699238856266700490715968264381621$$
$$46859296389521759999322991560894146397615651828 62536$$
$$979208272237582511852109168640000000000000000000000000 \hspace{4em} \text{(d9)}$$

(c10) /* Note that the value assigned to x above has not been forgotten. */

log(x)^3;

$$\log(2)^3 \hspace{10em} \text{(d10)}$$

(c11) /* This could be easily checked by referencing the symbol x. */

x;

$$2 \hspace{10em} \text{(d11)}$$

(c12) /* In general, assignment using the symbol : defines a "shorthand" label for referencing the expression on the right of the colon. */

eq1:u+v-4=0;

$$-4 + u + v = 0 \tag{d12}$$

(c13) eq2:u-v+2=0;

$$2 + u - v = 0 \tag{d13}$$

(c14) eq3:eq1+eq2;

$$-2 + 2u = 0 \tag{d14}$$

(c15) /* Solve the simple equation labeled "eq3": */

solve(eq3,u);

$$[u = 1] \tag{d15}$$

(c16) /* Solve a harder equation: */

solve(log(z)=1,z);

$$[z = e] \tag{d16}$$

(c17) /* If an attempt had been made to solve for the variable x, an error would have occurred because x is still assigned the number 2. The "binding" of x to 2 can be removed however with the "remvalue" (or the "kill") command: */

remvalue(x);

$$[x] \tag{d17}$$

(c18) x;

$$x \tag{d18}$$

(c19) /* (Expressions which do not evaluate to anything are merely echoed.) */

/* Now its ok to solve for x as an unknown. */

solve(log(x)=1,x) ;

$$[x = e] \tag{d19}$$

(c20) /* Finally, note that if the input is terminated with a $ instead of a ; displaying (but not evaluating) the output is surpressed. */

eq1:u+(4/2)*v+5=0$

(c21) eq1;

$$5 + u + 2v = 0 \tag{d21}$$

(c22) /* End the session */

exit();

2.2. Functions

The real power of symbol manipulation begins to appear with operational "functions", which invoke transparent routines for performing sophisticated mathematical algorithmic procedures. A standard demonstration is the factorization of of a polynomial.

(c1) /* Construct a large polynomial with the command expand: */

(x+1)^10;

$$(1 + x)^{10} \tag{d1}$$

(c2) expand(%);

$$1 + 10\,x + 45\,x^2 + 120\,x^3 + 210\,x^4 + 252\,x^5 + 210\,x^6 + 120\,x^7 + 45\,x^8 + 10\,x^9 + x^{10} \tag{d2}$$

(c3) /* Now factor the expanded polynomial and recover the original expression: */

factor(%);

$$(1 + x)^{10} \tag{d3}$$

(c4) exit();

Clearly a new realm of computing capability has been demonstrated here with the introduction of callable routines having a comprehension of mathematical manipulation. This example illustrates the ability to decode symbolic expressions and manipulate them according to predetermined rules, in this case the rules of algebra.

One drawback of symbolic programming compared to numerical programming can be illustrated by the preference for integer arithmetic over floating-point arithmetic. In the next example, unnecessary complexity is introduced by attempting to force floating-point computation.

(c1) /* Solve a polynomial equation in one variable. */

solve(x^2-1=0,x); Solution:

$$x = -1 \tag{e1}$$

$$x = 1 \tag{e2}$$

$$[e1, e2] \qquad (d2)$$

(c3) /* (The brackets indicate that the answer is displayed as a "list" of two solutions) */

/* Now turn on a MACSYMA flag which causes floating point conversion. */

numer:true;

$$true \qquad (d3)$$

(c4) /* Solve the same equation as before. */

solve(x^2-1=0,x);

RAT replaced 6.283185307179586 by 732133415/116522652 = 6.283185307179586

RAT replaced 3.141592653589793 by 732133415/233045304 = 3.141592653589793 Solution:

$$x = e^{3.141592653589793\,i} \qquad (e4)$$

$$x = e^{6.283185307179586\,i} \qquad (e5)$$

$$[e4, e5] \qquad (d5)$$

(c6) /* RAT is a library function; it was called automatically and used to convert the 16 signficant-figure decimal approximation of pi to an equivalent rational fraction. RAT can also be called directly: */

rat(%pi);

RAT replaced 3.141592653589793 by 732133415/233045304 = 3.141592653589793

$$\frac{732133415}{233045304} \qquad (d6)$$

(c7) exit();

Since most first-year General, Analytical and Organic Chemistry courses do not require mathematical skills much beyond algebra, further examples illustrating the algebraic power of symbol computing can be conveniently placed in the context of these elementary courses.

3. APPLICATIONS IN LOWER DIVISION CHEMISTRY COURSES

3.1. General Chemistry

Symbolic algebra computation does not offer much advantage over digital computation in solving ratio-proportion problems. These include single-step and multiple-step units conversions, stoichiometry, and many ideal-gas-law problems. It would be possible to include the units

125

explicitly using symbolic computation thereby avoiding errors in setting up problems, but the actual computation ordinarily involves floating-point multiplications and divisions, for which hand-held calculators are adequate.

3.1.1. Temperature Conversions

The situation is quite different when it comes to applications requiring algebra for solution. For example, most students can convert between temperature scales when an explicit formula is given for the output temperature in terms of the input temperature, but some have a more difficult time when the formula must be rearranged, and many faulter in solving two linear equations simultaneously. Assuming that the algebraic equations can be extracted from the statement of the problem in words, see how computer algebra reduces the remaining algebraic hurdles.

(c1) /* Convert 50° Fahrenheit to Celsius. */

/* The implicit conversion formula is */

eq:F=(9/5)*C+32;

$$F = 32 + \frac{9\,C}{5} \tag{d1}$$

(c2) /* The usual solution is to substitute the given temperature into the conversion formula and solve the resulting equation in one unknown. */

subst(40,F,eq);

$$40 = 32 + \frac{9\,C}{5} \tag{d2}$$

(c3) solve(%,C); Solution:

$$C = \frac{40}{9} \tag{e3}$$

$$[e\,3] \tag{d3}$$

(c4) /* Another way is to solve two simultaneous equations. */

solve([F=50,eq]); Solution

$$F = 50 \tag{e4}$$

$$C = 10 \tag{e5}$$

$$[[e\,4,e\,5]] \tag{d5}$$

(c6) /* Finding the temperature at which F=2C can also be done either way but solving two simultaneous equations may be mathematically more illuminating. */

solve([F=2*C,eq]); Solution

$$C = 160 \qquad (e6)$$

$$F = 320 \qquad (e7)$$

$$[[e6, e7]] \qquad (d7)$$

(c8) exit();

3.1.2. Balancing Chemical Reactions

Chemical reactions are balanced conveniently using symbolic algebra. Mathematically the process can be achieved by solving systems of linear algebraic equations. The atomic theory insures that the variables involved will either be integers or rational fractions of integers. In this case symbolic manipulation shows a decided advantage over algorithms which are restricted to floating-point arithmetic.

(c1) /*   Balance: $aCu_2S + bH^+ + cNO_3^- \rightarrow dCu^{2+} + eNO + fS_8 + gH_2O$ */

/* The conservation equation for each element can be assigned a short-hand label using a name that reflects the element being conserved (Q stands for charge). */

Cu:2*a=d\$

(c2) S:a=8*f\$

(c3) H:b=2*g\$

(c4) N:c=e\$

(c5) O:3*c=e+g\$

(c6) Q:b-c=2*d\$

(c7) /* The linear equations are to be solved simultaneously. The library routine "solve" expects a [list of equations] and a [list of unknowns], each contained in square brackets. There is one more unknown than equations, which reflects the ambiguity in balanced chemical equations. */

coefs:solve([Cu,S,H,N,O,Q],[b,c,d,e,f,g]);

$$[[b = \frac{16a}{3}, c = \frac{4a}{3}, d = 2a, e = \frac{4a}{3}, f = \frac{a}{8}, g = \frac{8a}{3}]] \qquad (d7)$$

(c8) /* Fractions may cleared by finding the lowest common multiple of the denominators. One way to accomplish this is to define a routine which takes a list of numbers and computes the l.c.m. stepwise with the MACSYMA function for finding the greatest common divisor, "ezgcd" */

l:makelist(denom(rhs(part(coefs[1],i))),i,1,6);

$$[3,3,1,3,8,3] \qquad (d8)$$

(c9) (lcm:l[1], for n:2 thru length(l) do lcm:lcm*l[n]/first(ezgcd(lcm,l[n])));

$$done \qquad (d9)$$

(c10) coefs:multthru(lcm,makelist(rhs(part(coefs[1],i)),i,1,6));

$$[128\,a,32\,a,48\,a,32\,a,3\,a,64\,a] \qquad (d10)$$

(c11) exit();

It is possible using a judicious choice of MACSYMA commands to construct a more general chemical equation balancing algorithm. In order to so this a bridge must be constructed between the non-mathematical notation used by chemists to represent chemical reactions and the propensity for MACSYMA to treat all quantities as mathematical objects. One way to communicate chemical formulas to MACSYMA is through lists which can be decoded to extract the numerical and symbolic information needed to balance the reaction. In order to display the output in familiar chemical notation MACSYMA can be coaxed into treating chemical formulas as subscripted variables raised to powers. A routine which uses these techniques is presented below. It was stored as a text file called "balancefile". After MACSYMA is initiated balancefile is "loaded" and executed (compiled); A call to the compiled function named balance() executes the routine. The first example executed is the formation of water; the second repeats the previous example.

(c1) load(balancefile)$

(c2) balance():=(remvalue(all), remarray(all),
        num:read("Enter the number of reactant and product molecules as a list of the form
[nr,np]:"),
            nr:part(num,1), np:part(num,2), nm:nr+np,
        naqs:read("Enter the numbers of atoms and charges on all the molecules as a double list
of the form   [[na1,q1],[na2,q2],...]:"),
        molecs:read("Enter all the molecular formulas as a double list of the form
[[A1[a1],B1[b1],...],[A2[a2],B2[b2],...],...]:"),
    for i:1 thru nm do (
        na:part(naqs,i,1), charge[i]:part(naqs,i,2), molec:part(molecs,i),
            for j:1 thru na do (
            atom[i,j]:part(molec,j,0), subscript[i,j]:part(molec,j,1),
            if subscript[i,j]=1 then sym[j]:atom[i,j] else sym[j]:part(molec,j),
                if i=1 and j=1 then elements:[atom[i,j]],
                if member(atom[i,j],elements)
                then for k:1 thru length(elements)
                    do (if atom[i,j]=elements[k] then
                        if i <= nr then a[i,k]:subscript[i,j]
                            else a[i,k]:-subscript[i,j])
                else (elements:append(elements,[atom[i,j]]),
                    for k:1 thru length(elements) do

```
                    (if atom[i,j]=elements[k] then a[i,k]:subscript[i,j]) )
                    ),
        molecule[i]:product(sym[j],j,1,na)`string(charge[i]),
        if charge[i]=1 then molecule[i]:product(sym[j],j,1,na)`"+",
        if charge[i]=0 then molecule[i]:product(sym[j],j,1,na),
        if i <= nr then eqncharge[i]:charge[i] else eqncharge[i]:-charge[i]
                    ),
   neqns:length(elements), nunks:nm,
   charges:makelist(eqncharge[i],i,1,nm),
   rmolecules:makelist(molecule[i],i,1,nr),
   pmolecules:makelist(molecule[i],i,nr+1,nm),
   unknowns:makelist(c[i],i,1,nunks),
        for l:1 thru nunks do for m:1 thru neqns do
            if not numberp(a[l,m]) then a[l,m]:0,
     a:transpose(genmatrix(a,nunks,neqns)),
     a:addrow(a,charges),
        solutions:linsolve(transpose(a.unknowns)[1],unknowns),
        tmp:makelist(rhs(solutions[i]),i,1,nm)/%rnum_list[1],
        den:makelist(denom(rhs(solutions[i])),i,1,nm),
           lcm:den[1],
           for n:2 thru length(den) do lcm:lcm*den[n]/first(ezgcd(lcm,den[n])),
           rcoeffs:makelist(lcm*tmp[i],i,1,nr),
           pcoeffs:makelist(lcm*tmp[i],i,nr+1,nm),
        print("The balanced reaction is:"),
        reaction:rcoeffs.rmolecules=pcoeffs.pmolecules
        )$
```

Batching done.

$$BATCH\ DONE \tag{d4}$$

(c5) balance();

        Enter the number of reactant and product molecules as a list of the form [nr,np]:
[2,1];

Enter the numbers of atoms and charges on all the molecules as a double list of the form [[na1,q1],[na2,q2],...]:
[[1,0],[1,0],[2,0]];

   Enter all the molecular formulas as a double list of the form [[A1[a1],B1[b1],...],[A2[a2],B2[b2],...],...]:
[[H[2]],[O[2]],[H[2],O[1]]];

Dependent equations eliminated: (3)

               The balanced reaction is:

$$2H_2 + O_2 = 2H_2O \tag{d5}$$

(c6) balance();

> *Enter the number of reactant and product molecules as a list of the form   [nr,np]:*

[3,4];

Enter the numbers of atoms and charges on all the molecules as a double list of the form   [[na1,q1],[na2,q2],...]:

[[2,0],[1,+1],[2,-1],[1,+2],[2,0],[1,0],[2,0]];

> *Enter all the molecular formulas as a double list of the form   [[A1[a1],B1[b1],...],[A2[a2],B2[b2],...],...]:*

[[Cu[2],S[1]],[H[1]],[N[1],O[3]],[Cu[1]],[N[1],O[1]],[S[8]],[H[2],O[1]]];

> *The balanced reaction is:*

$$128 H^+ + 32(O_3 N)^{-1} + 24 Cu_2 S = 3 S_8 + 48 Cu^2 + 64 H_2 O + 32 N O \qquad (d6)$$

(c7) exit();

The second reaction above illustrates the limitation of the routine in its rudimentary from to handle radicals as groups. Also it is seen that MACSYMA rearranges symbols according to a preconceived hierarchy (reverse alphabetical order here).

3.1.3. Empirical Formulas

One of the calculations in beginning chemistry has suffered from improvements in calculational tools. Empirical chemical formulas are derived from experimental data by converting weight ratios to number ratios with atomic weights. It is relatively easy to convert the decimal ratios obtained during the conversion process to integer ratios with a slide rule by scanning for a pair of integers lined up on the C and D scales. The pocket calculator produces a floating point decimal number for the number ratio, but the student must deduce the equivalent integer ratio. This is fairly easy for ratios like 2/3 and 1/4; but not many people can recognize 0.2857... as equivalent to 2/7. Unfortunately integer-oriented programs like MACSYMA only make matters worse. If the data are exact (i.e. fabricated) then exact decimal equivalents produce integer fractions. However the introduction of even a small amount of experimental error produces unrecognizable results.

(c1) /* Simple empirical formula */

/* (Assume integer atomic weights for simplicity) */

%N:28/108;

$$\frac{7}{27} \qquad (d1)$$

(c2) %O:80/108;

$$\frac{20}{27} \qquad (d2)$$

(c3) atomratio:(%N/14)/(%O/16);

$$\frac{2}{5}$$ (d3)

(c4) /* Now introduce a small amount of experimental error: */

%N:0.258;

$$0.258$$ (d4)

(c5) %O:0.741;

$$0.741$$ (d5)

(c6) atomratio:(%N/14)/(%O/16);

$$0.3979178716020821$$ (d6)

(c7) /* An attempt to use the rat function to convert to a rational fraction only makes matters worse: */

rat(atomratio);

RAT replaced 0.3979178716020821 by 688/1729 = 0.3979178716020821

$$\frac{688}{1729}$$ (d7)

(c8) /* Rat can even get into trouble when the apparent "exact" ratio is used: */

%N:28/108$

(c9) %O:80/108$

(c10) atomratio:(%N/14.0)/(%O/16.0);

$$0.4$$ (d10)

(c11) rat(atomratio);

RAT replaced 0.4 by 9007199254740993/22517998136852483 = 0.4

$$\frac{9007199254740993}{22517998136852483}$$ (d11)

(c12) exit();

131

3.2. Analytical Chemistry

Solution equilibria are treated in analytical chemistry. Results for complex multicomponent equilibria are usually obtained by making simplifying approximations to the relevant equations. Some teachers may feel that when exact solutions become available via computer algebra, physical insight into the nature of equilibria will be lost.[5] The pedagogical loss is related to the ability of the computer to solve complex problems. If problems of all levels of complexity could be handled with equal facility, then the pedagogical value would be reduced to something comparable to using logarithm tables to solve pH problems instead of using a calculator.

3.2.1. Titration Curves

The calculation of a titration curve illustrates the procedures involved with employing a computer algebra program. A weak diprotic acid titrated with strong base provides an intermediate-level case. Since the hydrogen ion concentration can be obtained in exact polynomial form, there is no conceptual difference between acids with quite close equilibrium constants and those whose equilibrium constants differ greatly. Approximate treatments would apply various approximations depending on the circumstances.

Equilibria and (mass and charge) conservation equations are solved simultaneously to produce an polynomial expression involving the hydrogen ion. The standard procedure is to solve the quartic in hydrogen ion concentration in terms of the volume of base added. However, this produces multiple solutions, only one of which is physically acceptable. Finding the one real solution is not trivial with MACSYMA. An alternative procedure solves the inverse problem, i.e. the added base volume as a function of pH, a much simpler problem both mathematically and computationally.[6] The titration curve obtained in this case is simply a rotation of the familiar pH vs mL(base) curve. The algebra problem is solved once in the first block of code, resulting in the volume of base as a function of pH. The parameters of the problem are read in in the second block and a library plotting routine calls the function for a number of values of pH to generate the titration curve.

(c1) load(titdpa)$

(c2) print("Titration curve for a diprotic weak acid titrated with strong base")$

Titration curve for a diprotic weak acid titrated with strong base

(c3) print("To initialize pH functions, type titration()")$

To initialize pH functions, type titration()

(c4) print("To plot each titration curve, type curve()")$

To plot each titration curve, type curve()

```
(c5) titration():=(
        K1:K[1]=H*HA/H2A,
        K2:K[2]=H*A/HA,
        Kw:K[w]=H*OH,
        Mass:Ca*Va/V=H2A+HA+A,
        Charge:Cb*Vb/V+H=HA+2*A+OH,
          Heqn:eliminate([K1,K2,Kw,Charge,Mass],[H2A,HA,A,OH])[1],
             Vbeqn:ratsubst(Va+Vb,V,Heqn),
             Vb:solve(Vbeqn,Vb),              Vb:-factor(subst(10^-pH,H,rhs(Vb[1])))   )$
```

```
(c6) curve():=(
        title:read("Enter the name of the acid to be titrated:"),
         Keqs:read("Enter a list of two equilibrium constants:"),
             K[1]:part(Keqs,1), K[2]:part(Keqs,2),
         incons:read("Enter values for acid and base concentrations as a list:"),
             Ca:part(incons,1),Cb:part(incons,2),
         Va:read("Enter a value for the volume of acid:"),
         pHs:read("Enter values for the initial pH and final pH:"),
             initialpH:part(pHs,1),finalpH:part(pHs,2),
        K[w]:1.e-14,
    plot(Vb,pH,initialpH,finalpH,50),
  plotout(vp)   )$
```

Batching done.

$$BATCH\ DONE \tag{d8}$$

(c9) titration();

$$\frac{Va\left(-1-K_1 10^{pH}-K_1 K_2 10^{2pH}+K_1 Ca\, 10^{2pH}+2K_1 K_2 Ca\, 10^{3pH}+10^{2pH}K_w+K_1 10^{3pH}K_w+K_1 K_2 10^{4pH}K_w\right)}{(1+K_1 10^{pH}+K_1 K_2 10^{2pH})(-1-Cb\, 10^{pH}+10^{2pH}K_w)} \tag{d9}$$

(c10) curve();

$$Enter\ the\ name\ of\ the\ acid\ to\ be\ titrated:$$

Carbonic;

$$Enter\ a\ list\ of\ two\ equilibrium\ constants:$$

[4.2e-7,4.8e-11];

$$Enter\ values\ for\ acid\ and\ base\ concentrations\ as\ a\ list:$$

[0.1,0.1];

$$Enter\ a\ value\ for\ the\ volume\ of\ acid:$$

100.;

$$Enter\ values\ for\ the\ initial\ pH\ and\ final\ pH:$$

[3.0,12.5];

$$done \tag{d10}$$

(c11) curve();

$$Enter\ the\ name\ of\ the\ acid\ to\ be\ titrated:$$

Oxalic;

Enter a list of two equilibrium constants:

[6.5e-2,6.1e-5];

Enter values for acid and base concentrations as a list:

[0.1,0.1];

Enter a value for the volume of acid:

100.;

Enter values for the initial pH and final pH:

[1.0,12.0];

                                 done   (d11)

(c12) exit();

Carbonic Acid Titration Curve
$pK_1 = 6.36$ $pK_2 = 10.32$

Oxalic Acid Titration Curve
$pK_1 = 1.19$ $pK_2 = 4.21$

(pH increases to the right along the horizontal axis; mL base increases *down* the vertical axis.)

3.3. Organic Chemistry

Not many mathematical demands are placed on introductory organic chemistry students at the present time. Nevertheless certain areas in the organic curriculum could be enhanced by more calculational power. Here we will treat two; simple Hückel molecular orbital (HMO) theory and kinetics.

3.3.1. Hückel Molecular Orbitals

HMO theory is a simple approximation to accurate electronic structure theory, which in turn forms the basis of understanding chemical reactivities and states. Mathematically HMO theory employs linear algebra techniques. The matrix operations involved quickly become so complex as to exclude all but the simplest examples from the textbooks and exercises. The statement of the problem is transparently simpler in a symbolic language; however the solution of the problem is more efficient using a numerical program.

(c1) /* HMO for butadiene */

/* Construct the Hückel matrix. */

mat:matrix([a,b,0,0],[b,a,b,0],[0,b,a,b],[0,0,b,a]);

$$\begin{bmatrix} a & b & 0 & 0 \\ b & a & b & 0 \\ 0 & b & a & b \\ 0 & 0 & b & a \end{bmatrix} \tag{d1}$$

(c2) /* The eigen package is not on-line and needs to be loaded from the library. Eigen returns a list of eigenvalues, followed by a list of their degeneracies. */

load(eigen); [load /usr/mac/share/eigen.l]

$$/usr/mac/share/eigen.l \tag{d1}$$

(c2) eigenvalues(mat);

$$[[\frac{2a+(1-\sqrt{5})b}{2}, \frac{2a+(1+\sqrt{5})b}{2}, \frac{2a+(-1-\sqrt{5})b}{2}, \frac{2a+(-1+\sqrt{5})b}{2}],[1,1,1,1]] \tag{d2}$$

(c3) exit();

Larger molecules may be treated in a similar fashion. Note how the operations on the Hückel matrix are telescoped.

(c1) /* HMO for naphthalene */

load(eigen)$ [load /usr/mac/share/eigen.l]

(c2) ev(eigenvalues(matrix([a,0,0,0,0,0,0,b,b,0],
                [0,a,b,0,0,0,0,0,b,0],
                [0,b,a,b,0,0,0,0,0,0],
                [0,0,b,a,b,0,0,0,0,0],
                [0,0,0,b,a,0,0,0,0,b],

[0,0,0,0,0,a,b,0,0,b],
[0,0,0,0,0,b,a,b,0,0],
[b,0,0,0,0,0,b,a,0,0],
[b,b,0,0,0,0,0,0,a,b],
[0,0,0,0,b,b,0,0,b,a]]),expand);

$$[[a - \frac{b}{2} - \frac{\sqrt{5}\,b}{2}, a - \frac{b}{2} + \frac{\sqrt{5}\,b}{2}, a + \frac{b}{2} - \frac{\sqrt{13}\,b}{2}, a + \frac{b}{2} + \frac{\sqrt{13}\,b}{2},$$

$$a - \frac{b}{2} - \frac{\sqrt{13}\,b}{2}, a - \frac{b}{2} + \frac{\sqrt{13}\,b}{2}, a - b, a + \frac{b}{2} - \frac{\sqrt{5}\,b}{2}, a + \frac{b}{2} + \frac{\sqrt{5}\,b}{2}, a + b],$$

$$[1,1,1,1,1,1,1,1,1,1]]$$

(d2)

(c3) exit();

It would not be out of the question to consider an extension of this program which would receive chemical composition and structural information as input and produce HMO energies and molecular properties as output.

3.3.2. Integrated Rate Expressions

Chemical kinetics naturally involves differential equations. We will take for our example the steady-state approximation, which is used to reduce systems of differential equations associated with complex reaction mechanisms to tractable form. It is pedagogically useful to compare steady-state solutions with accurate solutions under various conditions to assess the validity of the steady-state approximation. Consider the simplest case of reactant producing product through a single intermediate in first-order steps[7]:

$$A \rightarrow B \rightarrow C.$$

A direct approach to this problem can be made using symbolic differential equation solvers.

(c1) /* Consecutive first-order kinetics demonstrates the steady-state approximation */

/* Initial values of concentrations */

atvalue(a(t),t=0,a[0])$

(c2) atvalue(b(t),t=0,0)$

(c3) atvalue(c(t),t=0,0)$

(c4) /* The kinetics equations are:*/

eq1:diff(a(t),t) = -k[1]*a(t);

$$\frac{d}{dt}a(t) = -k_1 a(t)$$

(d4)

(c5) eq2:diff(b(t),t) = k[1]*a(t)-k[2]*b(t);

$$\frac{d}{dt}b(t) = k_1 a(t) - k_2 b(t)$$

(d5)

(c6) eq3:diff(c(t),t) = k[2]*b(t);

$$\frac{d}{dt}c(t) = k_2 b(t) \tag{d6}$$

(c7) /* load the diff. eq. solver and solve */

load(desoln);

desoln.l being loaded. [load desoln.l]

$$desoln.l \tag{d7}$$

(c8) desolve([eq1,eq2,eq3],[a(t),b(t),c(t)]);

$$[a(t) = a_0 e^{-k_1 t}, b(t) = \frac{a_0 k_1 e^{-k_1 t}}{-k_1 + k_2} - \frac{a_0 k_1 e^{-k_2 t}}{-k_1 + k_2}, c(t) = a_0 - \frac{a_0 k_2 e^{-k_1 t}}{-k_1 + k_2} + \frac{a_0 k_1 e^{-k_2 t}}{-k_1 + k_2}] \tag{d8}$$

(c9) /* (Save these solutions for future reference) */

truesoln:%$

(c10) /* Now solve for the steady-state approximate solution: */

atvalue(bs(t),t=0,0)$

(c11) atvalue(cs(t),t=0,0)$

(c12) eq4:diff(as(t),t) = -k[1]*as(t);

$$\frac{d}{dt}as(t) = -k_1 as(t) \tag{d12}$$

(c13) eq5:0 = k[1]*as(t)-k[2]*bs(t);

$$0 = k_1 as(t) - k_2 bs(t) \tag{d13}$$

(c14) eq6:diff(cs(t),t) = k[2]*bs(t);

$$\frac{d}{dt}cs(t) = k_2 bs(t) \tag{d14}$$

(c15) /* The differential equation solver can also handle this system of mixed differential and algebraic equations: */

sssoln:desolve([eq4,eq5,eq6],[as(t),bs(t),cs(t)]);

$$[as(t) = as(0)e^{-k_1 t}, bs(t) = \frac{as(0)k_1 e^{-k_1 t}}{k_2}, cs(t) = as(0) - as(0)e^{-k_1 t}] \tag{d15}$$

(c16) /* Now lets compare the steady-state concentration of species B with the actual B concentration at various times, for the special case of $k_2 = 10k_1$, and $a_0 = 1$ */

/* Calculate the ratio of B(steady-state) to B(actual) for the above conditions */

ratio(k2t):=(subst(k2t/k[2],t,(subst(.1*k[2],k[1],subst(1,as(0),rhs(sssoln[2])))/
subst(.1*k[2],k[1],subst(1,a[0],rhs(truesoln[2]))))));

$$ratio(k2t) := substitute(\frac{k2t}{k_2}, t, \frac{substitute(0.1\,k_2, k_1, substitute(1, as(0), rhs(sssoln_2)))}{substitute(0.1\,k_2, k_1, substitute(1, a_0, rhs(truesoln_2)))}) \quad (d16)$$

(c17) /* This ratio should approach (9/10) at large values of k_2t: */

value:0.1;

$$0.1 \qquad (d17)$$

(c18) for i:1 thru 4 do (display(value,ev(ratio(value),numer)),value:10*value);

$$value = 0.1$$

$$ev(ratio(value), numer) = 10.4567490889257$$

$$value = 1.0$$

$$ev(ratio(value), numer) = 1.516605975364507$$

$$value = 10.0$$

$$ev(ratio(value), numer) = 0.9001110825323516$$

$$value = 100.0$$

$$ev(ratio(value), numer) = 0.9$$

$$done \qquad (d18)$$

(c19) exit();

Systems of first order differential equations can also be solved by Laplace transform techniques and by matrix algebra.[8] Since most chemists are probably more familiar with matrix algebra than with Laplace transforms, we will illustrate the solution to the consecutive kinetics mechanism using a matrix method.

(c1) /* Three consecutive kinetic equations, using matrix methods. */

/* Set up the rate coefficient matrix K such that -dC/dt = K.C */

Kmat:matrix([k[1],0,0],[-k[1],k[2],0],[0,-k[2],0]);

$$\begin{bmatrix} k_1 & 0 & 0 \\ -k_1 & k_2 & 0 \\ 0 & -k_2 & 0 \end{bmatrix} \tag{d1}$$

(c2) /* Assign a column matrix of the initial concentrations: */

C[0]:matrix([a[0]],[0],[0]);

$$\begin{bmatrix} a_0 \\ 0 \\ 0 \end{bmatrix} \tag{d2}$$

(c3) /* simtran returns two lists; the first has sublists of eigenvalues and corresponding degeneracies, and the second has eigenvectors. It also produces matricies of eigenvectors such that leftmatrix.kmat.rightmatrix = a diagonal eigenvalue matrix. */

eigvals:first(first(simtran(Kmat)));

$$[k_2, k_1, 0] \tag{d3}$$

(c4) /* Simplify the eigenvector matrices */

Xinv:radcan(leftmatrix)$

(c5) X:radcan(rightmatrix)$

(c6) /* Construct the matrix function exp(-t*eigvals) */

f[i,j]:=if i=j then exp(-t*eigvals[i]) else 0$

(c7) fmat:genmatrix(f,3,3);

$$\begin{bmatrix} e^{-k_2 t} & 0 & 0 \\ 0 & e^{-k_1 t} & 0 \\ 0 & 0 & 1 \end{bmatrix} \tag{d7}$$

(c8) /* coefmat transforms initial concentrations into concentrations at time t (combine simplifies the result): */

coefmat:combine(X.fmat.Xinv)$

(c9) /* The matrix of concentrations as functions of time is: */

C:coefmat.C[0];

$$
\begin{bmatrix}
a_0 e^{-k_1 t} \\
\dfrac{a_0(k_1 e^{-k_1 t} - k_1 e^{-k_2 t})}{-k_1 + k_2} \\
a_0(1 + \dfrac{-k_2 e^{-k_1 t} + k_1 e^{-k_2 t}}{-k_1 + k_2})
\end{bmatrix}
\qquad (d9)
$$

(c10) /* The steady-state solutions correspond to k2>>k1 and t>>1/k2: */

assume(a[0]>0,k[1]>0,k[2]>0)$

(c11) css:factor(limit(limit(C,k[2]-k[1],k[2]),k[2]*t,inf));

$$
\begin{bmatrix}
a_0 e^{-k_1 t} \\
\dfrac{a_0 k_1 e^{-k_1 t}}{k_2} \\
a_0 e^{-k_1 t}(-1 + e^{k_1 t})
\end{bmatrix}
\qquad (d11)
$$

(c12) exit();

4. APPLICATIONS IN PHYSICAL CHEMISTRY

Three areas in introductory physical chemistry which employ non-trivial amounts of mathematics are classical thermodynamics, quantum mechanics, and statistical thermodynamics. Representative examples from each of these areas will be discussed in turn. Further discussion is given by Ogilvie.[9]

4.1. Classical Thermodynamics

Classical thermodynamics by its nature employs multivariate calculus. Rather than demonstrate elementary thermodynamic operations involving multivariate mathematical quantities such as partial derivatives, we will go straight to a sophisticated technique for transforming variables. A problem of wide practical application deals with relating thermodynamic variables and their derivatives to measurable quantities. A very general method for transforming partial derivatives which takes advantage of the Lie algebra structure of the derivatives employs Jacobian determinant notation. Our example uses an algorithm of E. T. Jaynes which has been incorporated into a textbook by Tribus.[10] The object is to eliminate from a given partial derivative any thermodynamic potentials and express the derivative as a function of pressure, volume, temperature, coefficients of compressibility and thermal expansion, and heat capacities. Maxwell relations and differential forms of the thermodynamic potentials expressed in their natural variables are used to perform the transformation of independent variables. In order to carry out this process with MACSYMA a new functional notation is defined using braces to represent the Jacobian determinants. In this notation, $(\frac{\partial X}{\partial Y})_Z = \frac{\{X,Z\}}{\{Y,Z\}}$. After the properties of Jacobians are defined and the thermodynamic equations are introduced in Jacobian notation, quite arbitrary partial derivatives can be reduced automatically. (Some attention should be paid to the order of terms in the Jacobians; if a rule such as $\{X,Y\} = -\{Y,X\}$ for general X and Y is defined, the simplification process will run indefinitely for an expression involving $\{X,Y\}/\{Y,X\}$.) As an example, the relation between the two heat capacities C_v and C_p will be obtained in terms of convenient experimental parameters.[11]

4.1.1. Transformation of Variables

(c1) /* Transformation of variables in first derivatives by Jacobians. */

matchfix(" { ", " } ")$

(c2) /* X is a generic variable */

matchdeclare(X,true)$

(c3) /* Define a rule for Jacobians */

let({X,X},0)$

(c4) /* Entropy derivatives in terms of Jacobians */

let({S,V},{S,T}*{V,P}/{T,P}-{S,P}*{V,T}/{T,P})$

(c5) /* Maxwell's relation in terms of Jacobians */

let({S,T},{V,P})$

(c6) /* Experimental thermic and caloric derivatives in terms of Jacobians */

let({S,P},Cp*{T,P}/T)$

(c7) let({V,P},alpha*V*{T,P})$

(c8) let({V,T},+beta*V*{T,P})$

(c9) /* Heat capacity relation */

Cv=ratexpand(letsimp(-T*{S,V}/{V,T}));

$$Cv = Cp - \frac{T\,V\,\alpha^2}{\beta} \qquad \text{(d9)}$$

(c10) exit();

4.2. Quantum Theory

The old quantum theory is a form of classical mechanics modified to provide quantized orbits. Many problems are solved using elementary algebra. A simple example obtains the quantized orbit properties of two attractively charged particles (the Bohr atom).

4.2.1. Bohr Atom

(c1) /* Simple Bohr atom */

momentum:m*v*r=n*hbar;

$$m\, r\, v = hbar\, n \tag{d1}$$

(c2) force:m*v^2/r=z*e^2/r^2;

$$\frac{m\, v^2}{r} = \frac{e^2 z}{r^2} \tag{d2}$$

(c3) energy:E=(1/2)*m*v^2-z*e^2/r;

$$E = \frac{m\, v^2}{2} - \frac{e^2 z}{r} \tag{d3}$$

(c4) solve([momentum,force,energy],[r,v,E]);

$$[[r = \frac{hbar^2 n^2}{e^2 m\, z}, v = \frac{e^2 z}{hbar\, n}, E = -\frac{e^4 m\, z^2}{2\, hbar^2 n^2}]] \tag{d4}$$

(c5) exit();

New quantum mechanics solves eigenvalue problems realized as differential or matrix equations. However one dimensional problems at least can be reduced to elementary algebra using ladder techniques. It is speculated that all common textbook examples can be treated quite simply with these methods.[12] Traditional textbooks treat eigenvalue problems as a branch of differential equations. However the rather artificial constraints placed on the eigenvalue problem in differential equation form required to obtain physically meaningful solutions (e.g. boundary behavior) do not lend themselves to automatic mathematical manipulation. Symbolic programming can be useful in obtaining intermediate complicated formulas used in electronic structure theory. An example of ordinary differential equation solving was given above in the subsection on organic chemistry. Examples of matrix algebra relevant to quantum chemistry were discussed there also.

4.3. Statistical Mechanics

To illustrate applications to classical statistical mechanics, we will obtain the most probable velocity of a Maxwellian distribution function as an extremum.

4.3.1. Maxwell Gas

(c1) /* Most probable velocity of a Maxwell distribution */

n(v):=4*%pi*n[0]*[m/(2*%pi*k*T)]^(3/2)*exp(-m*v^2/(2*k*T))*v^2;

$$n(v) := 4\pi n_0 [\frac{m}{2\pi k\, T}]^{\frac{3}{2}} \exp(-\frac{m\, v^2}{2\, k\, T}) v^2 \tag{d1}$$

(c2) ratsimp(diff(n(v),v));

$$[-\frac{(-2\sqrt{2}\,n_0\,T\,k\,\dfrac{m}{T\,k^{\frac{3}{2}}}v+\sqrt{2}\,n_0\,m\,\dfrac{m}{T\,k^{\frac{3}{2}}}v^3)\,e^{-\frac{m\,v^2}{2\,T\,k}}}{\sqrt{\pi}\,T\,k}] \tag{d2}$$

(c3) solve(%,v);

$$[v=-\sqrt{2}\,\sqrt{\frac{T\,k}{m}},v=\sqrt{2}\,\sqrt{\frac{T\,k}{m}},v=0] \tag{d3}$$

(c4) exit();

4.3.2. Planck Radiation Formula

A second example derives the Stefan-Boltzmann and Wien radiation laws from the Planck radiation distribution function.

(c1) /* Stefan-Boltzmann law from the Planck radiation distribution function */

assume(c>0,h>0,k>0,T>0);

$$[c>0,h>0,k>0,T>0] \tag{d1}$$

(c2) rho(nu):=(8*%pi*h/c^3)*nu^3/(exp(h*nu/(k*T))-1);

$$\rho(\nu):=\frac{\dfrac{8\pi h}{c^3}\nu^3}{\exp(\dfrac{h\nu}{k\,T})-1} \tag{d2}$$

(c3) integrate(rho(nu),nu);

$$\frac{8\pi h\,(-\dfrac{\nu^4}{4}+\dfrac{6\,T^4 k^4(\dfrac{h^3\nu^3\log(1-e^{\frac{h\nu}{T\,k}})}{6\,T^3 k^3}+\dfrac{h^2\nu^2 li_2[e^{\frac{h\nu}{T\,k}}]}{2\,T^2 k^2}-\dfrac{h\nu li_3[e^{\frac{h\nu}{T\,k}}]}{T\,k}+li_4[e^{\frac{h\nu}{T\,k}}])}{h^4})}{c^3} \tag{d3}$$

(c4) E=limit(%,nu,0);

$$E=\frac{8\pi^5\,T^4 k^4}{15\,c^3 h^3} \tag{d4}$$

(c5) /* (limit doesn't evaluate to zero as nu goes to infinity) */

/* Wien displacement law */

solve(diff(rho(nu),nu),h*nu/(k*T));

$$[\frac{h\,\nu}{T\,k} = \log(\frac{3\,T\,k}{3\,T\,k - h\,\nu})] \tag{d5}$$

(c6) /* (Thus frequency is inversely proportional to temperature.) */

exit();

     Obtaining and using statistical mechanical partition functions employs mathematical manipulations which should be straightforward for symbolic algebra programs. The following examples obtain limiting values for the heat capacity of an Einstein solid, and analytical and numerical forms for the entropy of a monatomic ideal gas, starting with the quantized energy expressions.

4.3.3. Einstein Solid

(c2) /* Heat capacity of an Einstein solid. */

assume(h>0,v>0,k>0,T>0,N>0);

$$[h>0,v>0,k>0,T>0,N>0] \tag{d2}$$

(c3) depends([q,E],T);

$$[q(T),E(T)] \tag{d3}$$

(c4) q:sum(exp(-(n+1/2)*h*v/(k*T)),n,0,inf);

$$\frac{e^{-\frac{h\,v}{2\,T\,k}}}{1 - e^{-\frac{h\,v}{T\,k}}} \tag{d4}$$

(c5) E:factor(k*T^2*diff(log(q^(3*N)),T));

$$\frac{3\,N\,h\,v\,(1 + e^{\frac{h\,v}{T\,k}})}{2\,(-1 + e^{\frac{h\,v}{T\,k}})} \tag{d5}$$

(c6) Cv:factor(diff(E,T));

$$\frac{3\,N\,h^2\,v^2\,e^{\frac{h\,v}{T\,k}}}{T^2\,k\,(-1 + e^{\frac{h\,v}{T\,k}})^2} \tag{d6}$$

(c7) limit(Cv,T,0);

$$0 \tag{d7}$$

(c8) /* (Use Taylor series expansions to get upper limit.) */

tlimit(Cv,T,inf);

$$3 N k \tag{d8}$$

(c9) exit();

4.3.4. Sackur-Tetrode Gas

(c1) /* Entropy of a monatomic ideal gas (Sackur-Tetrode equation). */

/* One degree of translation */

E(n):=n^2*h^2/(8*m*a^2);

$$E(n) := \frac{n^2 h^2}{8 \, m \, a^2} \tag{d1}$$

(c2) /* Define the classical mechanical partition function as an integral. */

/* Let the integrator know that the parameters have positive values: */

assume(T>0,h>0,k>0,m>0,a>0,g[e]>0)\$

(c3) q[tr1]:rootscontract(integrate(exp(-E(n)/(k*T)),n,0,inf));

$$\frac{a \sqrt{(2 \pi T k m)}}{h} \tag{d3}$$

(c4) /* Cube and substitute V for a-cubed (g[e] is the electronic degeneracy). */

q[tr3]:g[e]*subst(V,a^3,q[tr1]^3);

$$\frac{2\sqrt{2} \, \pi^{\frac{3}{2}} T^{\frac{3}{2}} V \, g_e \, k^{\frac{3}{2}} m^{\frac{3}{2}}}{h^3} \tag{d4}$$

(c5) /* Q is the molar partition function. */

Q:(q[tr3]*%e/N)^N\$

(c6) logexpand:super\$

(c7) S:k*log(Q)+k*T*diff(log(Q),T)\$

(c8) S:subst(N*k*T/P,V,subst(M/N,m,S));

$$\frac{3 N k}{2} + k \, (N + \frac{3 \log(2) N}{2} + \frac{3 \log(\pi) N}{2} + \frac{3 N \, (\log(M) - \log(N))}{2} \tag{d8}$$
$$+ \frac{3 N \log(T)}{2} + N \log(g_e) - 3 N \log(h) + \frac{3 N \log(k)}{2}$$

$$+ N(-\log(P) + \log(T) + \log(k)))$$

(c9) S:ev(S,N=6.0220e23,h=6.626e-27,k=1.3807e-16)\$

(c10) /* Standard entropy in calories: */

S[0]:expand(ev(S,T=298.16,P=1.0132e6,numer)/4.184e7);

$$25.99181902933839 + 2.980846821223709\log(M) + 1.98723121414914\log(g_e) \tag{d10}$$

(c11) /* Entropy of argon in calories (observed value is 36.983 cal/mol-deg): */

S[Ar]=ev(S[0],g[e]=1,M=39.948);

$$S_{Ar} = 36.98392600214311 \tag{d11}$$

(c12) exit();

Had an attempt been made to substitute R/k for N in the analytical entropy expression, an incorrect numerical value would have resulted. This is due to the fact that N does not always occur multiplied by k, which causes superfluous R variables to be introduced into the entropy expression. This illustrates the care which must be given in treating algebraic variables and expressions when using computer algebra. In examples like this difficulties are avoided by paying strict attention to the expected functional dependency on independent variables; entropy can depend on k or R, but should not be expected to depend on both. On the other hand, replacing atomic mass m with molar mass M through M/N does not increase the number of independent variables.

5. DISCUSSION

An initial encounter with symbolic computer algebra might leave one with the anticipation that convenient *ab initio* molecular property calculations are right around the corner, or with the fear that there soon won't be any difficult problems to assign students. Actually, both views are premature. It is possible to perform tedious and complex mathematical operations with a variety of available programs. Yet it is still also quite true that clever solutions are obtained by clever programming, even with the highest-level symbol manipulators available. Here we have seen that MACSYMA doesn't understand the language of chemistry very well, which is only sensible since most of chemistry hasn't been reduced to pure mathematics yet.

Teachers who fear that the adoption of sophisticated artificial intelligence techniques will undermine intuitive thinking should take heart in the thought that chemistry, at least, is not supposed to be purely mathematical, and that as more labor can be passed to the computer, more time can be spent on other creative tasks. Of course it goes without saying that like all other useful intellectual tools, once the idea has been demonstrated, the adoption is inevitable. Along this same line of reasoning it seems clear that the more mathematically oriented the discipline, the greater the interest in symbolic computation. What chemists might find amusing, physicists and mechanical engineers may find essential.

It may appear that symbol manipulation is too far separated from digital computation to produce useful numerical results for complex problems. However, not only can symbolic algebra produce tedious, complex code in numerically-related languages,[13] but symbol manipulation codes have been produced in traditional numerically oriented languages, such as FORTRAN[14] and C[15], among others. Finally it should be remembered that even when the value of a simple transcendental function such as the exponential is required at a particular value of the argument, by definition it is necessary to evaluate the result by numerical approximation.

146

Many improvements in levels of communication, documentation, efficiency and application will be needed before any of the existing computer symbol programs will become generally useful to students of chemistry. These developments will occur over a period of time. Our tasks as chemists will be to encourage and contribute to the process by gaining experience, giving advise and promoting the use of symbolic computation.

Acknowledgements

I would like to thank Professor Stanly Steinberg of the University of New Mexico for introducing me to the world of symbolic computation. The MACSYMA project is supported, in part, by ERDA and NASA.

REFERENCES

1. Kowalski, B. R. and Bender, C. F., J. Amer. Chem. Soc. **94**, 5632 (1972).

2. Bradshaw, G. L., Langley, P., and Simon, H. A. *Proceedings of the Third National Conference of the Canadian Society for Computational Studies of Intelligence*, 1980, 19-25.

3. S. Rosen in *Programming Systems and Languages*, (McGraw-Hill, 1967), p.3.

4. *MACSYMA Reference Manual*, The Math Lab Group, Laboratory for Computer Science, MIT, Version 9, 1977.

5. Bruckenstein, S. and Kolthoff, I. M., in *Treatise on Analytical Chemistry, Part I, Volume 1*, (Wiley, 1959), sec 11.III.C.

6. Willis, C. J., J. Chem. Educ., **58**, 659 (1981).

7. R. E. Weston and H. A. Schwarz, *Chemical Kinetics*, (Prentice-Hall, 1972), Sec. 1.11.

8. C. L. Perrin, *Mathematics for Chemists*, (Wiley, 1970), Sec 8.15.

9. Olgivie, J. F., Computers and Chemistry, **6**, 169 (1982).

10. Tribus, M., *Thermostatics and Thermodynamics*, (Van Nostrand, 1961), Ch. 9.

11. Callen, H. B., *Thermodynamics*, (Wiley, 1960), Sec. 7.5.

12. Fernández, F. M. and Castro, E. A., Am. J. Phys. **52**, (4), 344 (1984).

13. Steinberg, S. and Roache, P., in *1984 MACSYMA User's Conference*, to be published.

14. Sakoda, J. M., *AFIPS Conference Proceedings*, (AFIPS Press, 1982), p. 827.

15. *SMP Primer and SUMMARY*, Inference Corporation, 1983.

7

A LISP System for Chemical Groups: Wigner - Eckart Coefficients
for Arbitrary Permutation Groups

Carl Trindle
Chemistry Department
University of Virginia
Charlottesville, VA 22901

Abstract:

Computer applications of group theory have almost invariably used
numerical representations of the fundamental quantities of the group. For
example, the combination laws in the crystallographic groups have been
described as matrix multiplications on tables of coordinates [1]. These
decimal values are approximate specifications of in principle exactly
equivalent points. The finite mathematics realized in computers does not
permit the numerical operations to represent faithfully the group theoretic
operations.

Symbolic manipulation systems make possible an exact representation
of the combination laws of a group, and the exact calculation of the characters, irreducible representations, and coupling coefficients which
symmetry-adapt primitive bases. We illustrate how the small MuMATH
(Registered trade mark of the Soft Warehouse) system can generate properties
of the permutation groups useful in the analysis of NMR spectra of flexible
or rearranging systems. The work rests on the properties of semi-direct
products, as described by Altmann [2].

Introduction:

No chemical spectroscopist or quantum chemist is without a nodding

acquaintance with group theory, the algebra of symmetry operators. This

is one of the few branches of mathematics useful in chemistry which has not

been thoroughly altered in practice and in tone by computing machinery.

The reasons for this are that: the group theory is essentially algebraic;

its elementary operations are not strictly numerical; and the computations

are primarily formal rather than numeric. At least in the symmetry theory

most immediately useful in spectroscopy, there is no need for intensive

148

quadrature, numerical manipulation of matrices, or any need to manipulate
large data sets. Application of (the results of) the simplest symmetry
theory is a small-scale formal exercise, eminently feasible without machine
assistance. There are applications of group theory, however, where computing
can become useful. This is the case when the permutation symmetry groups
of easily rearranged systems become important, as in the analysis of the
magnetic resonance spectra of flexible molecules. These groups are often
extremely large, as are their representations. Although several programs
accomplishing a numerical simulacrum of group theory have been described [3],
no strictly numerical representation of group properties can be satisfactory
for these huge groups. The work becomes bulky while still symbolic. It
is natural then to turn to machine aid, but only if the machine can deal
with the essentially symbolic character of the work.

Early Work in Symbol Manipulation

The very first software analyst, namely Ada, Countess of Lovelace [4],
was well aware that Babbage's analytical engine would be as capable of
algebraic manipulation as numerical manipulation. She wrote that the machine
could

> arrange and combine its numerical quantities exactly as
> if they were letters or any other general symbols; and
> in fact might bring out its results in algebraic notation,
> were provisions made accordingly.

Turing's generalization of the idea of a computing machine [5] also embraced
the possibility of manipulation of abstract symbols on an equal basis with
numerical work. Shortly after the von Neumann machine was realized, a
small group of computer scientists defined the discipline of artificial
intelligence (AI), which had as its expressed purpose the development of

automatic systems which would display behavior which would be considered intelligent if only a human being had acted similarly [6]. This behavioristic criterion is a weakened version of Turing's test of machine (or more generally, alien) intelligence [7], and evades troublesome questions about the internal mental states of the machine. While progress toward Turing's goal has been slower than the prophets of the 1950's expected, there is no doubt that the more modest goal has been met in remarkable and, more significant, extremely useful ways. One of the early dramatic successes was Slagle's program SAINT [8], which was able to recognize and evaluate most of the integrals presented to MIT freshmen. From this first step progress to programs which rival the best human abilities was remarkably rapid. An annotated summary is provided in Feigenbaum's Handbook of Artificial Intelligence [9].

LISP, Mu, and Data Structures

It is an oddity of AI that is spite of the unparalleled changes in hardware, organization, and theory of computing, the language and notation of the field have remained essentially unaltered for almost thirty years. The use of McCarthy's LISP [10], invented in the 1950's, as the lingua franca of AI in the 1980's is the modern equivalent of Newton's use of Latin and geometry in his Principia Mathematica. The well-tested flexibility and power of the tongue is such that ideas not so expressed are somehow suspect. This is partly a social phenomenon, but is also an eloquent tribute to the value of the invention. It defies the idea that rapid progress requires equally rapid obsolescence. Since the LISP idiom is the foundation of all symbol manipulation programs, and since it forms a part of this work in setting up problems for symbol manipulation programs, some brief discussion of LISP is now in order. All examples will be drawn from the working group theory programs, and the code is expressed in the MuSIMP dialect of LISP which accompanies the MuMATH algebra system. There is a direct parallel with conventional LISPs, so the experienced programmer can follow the dialect without difficulty. However, the choice of names for the functions and commands is much more memorable, making the MuSIMP code accessible to those unfamiliar with LISP. [11]

We begin with the fundamental (and at bottom, only) data structure in LISP, the list. We work with sets A and B which are represented as lists (a1, a2, ... ak) and (b1, ... bn). The quantities aj and bl may be the irreducible data element the ATOM, or they may themselves be lists. For the moment our sole operation on a list will be to detach the first entry, which we accomplish by an operator called FIRST(A) in Mu. This is called CAR(A) in usual LISP. The remainder of the list is called REST(A); traditionally, CDR(A). Not only do we take lists apart, we can combine fragments into lists by the operation ADJOIN (A,B), known as CONS (A,B) in LISP. This is almost all that happens in our programs.

It is frequently the case that a particular atom in a list can be attributed properties. In LISP dialects it is possible to assign and recall properties by operations on the atom's property list. The commands in Mu are PUTPROP (ATOM PROPNAME PROPVALUE) and GETPROP (ATOM PROPNAME).

Some Essential and Hardworking Procedures

The early stages of permutation group theory make heavy use of set-theoretic unions, intersections, and set partitions. These are so easily accomplished in LISP that they are often presented as exercises in the first steps of learning the language. They are frequently provided as LISP primitives in particular installations. To simplify the discussion,

we will assume that our list is composed entirely of atoms. If A is to be
a subset of B, it must surely be the case that al or FIRST(A) is a member
of B. (This is not sufficient, but is necessary.) We need a way to determine
whether an atom is found in a list; we will define a function MEMBER(FIRST(A),
B) which may have Boolean values T (true) or F (false). If B is empty, the
result must be F.

> WHEN EMPTY(B) F EXIT,

But if FIRST(A) is identical with FIRST(B) then the result must be T.

> WHEN EQ(FIRST(A) FIRST(B)) T EXIT,

otherwise, we can get a response only if FIRST(A) is in the rest of list B

> MEMBER (FIRST(A) REST(B)),

If we don't find that FIRST(A) is the first element of the list B, we solve
a similar problem with the shorter list REST(B). The problem is simplified
step by step until it becomes trivial to solve. Either we find a short list
with FIRST(A) as the first element, or we exhaust the list B. LISP rests
primarily on this strategem of reducing problems by referring to a similar
smaller problem. Programming in LISP requires facility in this recursive
expression of algorithms, which makes the language rather opaque for the
FORTRAN-bred master of linear strategems. Now to answer the question
whether A is a subset of B we call on the MEMBER operation, and apply the
MEMBER test to each member of A in sequence.

```
DEFINE FUNCTION SUBSET (A B),
WHEN EMPTY(B) F EXIT,
EMPTY(A) T EXIT,
AND (MEMBER (FIRST(A)B)), SUBSET(REST(A) B),
ENDFUN:
```

We combine recursive strategems in a recursive manner.

It will be useful to be able to count the number of times a set of atoms (in which no element is repeated) appears in a "bag" (in which elements may be repeated). One way to accomplish this would be to remove sets from the bag, counting the number of successful removals. To remove a full set from a bag, each member of the set must be removed from the bag. We must define a function REMBER (ANATOM BAG) which will remove a single atom from a bag.

```
FUNCTION REMBER (ANATOM BAG),
WHEN EMPTY(ANATOM) BAG EXIT,
WHEN EMPTY(BAG) F EXIT,
EQ(FIRST(BAG), ANATOM) REST(BAG) EXIT,
ADJOIN (FIRST(BAG), REMBER(REST(BAG)),
ENDFUN:
```

The ADJOIN was required so the early elements of BAG would not be lost. REMBER must be applied for each element in ASET, which we manage by EXSET(ASET, ABAG).

```
DEFINE FUNCTION EXSET(ASET, ABAG),
WHEN EMPTY(ASET) ABAG EXIT,
EXSET(REST(ASET), REMBER(FIRST(ASET),ABAG)),
ENDFUN;
```

Finally, it will be necessary to count the number of times this excision of a set from a bag is successful.

```
DEFINE FUNCTION COUNTSET(ASET, ABAG),
WHEN EMPTY(ABAG) O EXIT,
WHEN EQ (EXSET(ASET, ABAG), ABAG) O EXIT,
ADD(1, COUNTSET(EXSET(ASET, ABAG)),
ENDFUN:
```

This permits the definition of the class constant matrix $A(i,j:k)$, which occurs in the equation $C_i\, C_j = \Sigma(k)\, A(i,j:k)\, C_k$, if we can produce the bag by the group's combination rule.

153

Representation of Permutations

Our major task in molecular symmetry groups is to represent the combination rule for permutation groups, the permutation operation itself [12]. A function PERMUTE (A P B) will be different from all the functions described so far, since it must return a list B which is the result of shuffling A according to permutation P. The list P may be integers in any order. The permutation (p1 p2 p3 ... pk) could be read: pick the ai (i-th) element of A and put it in position pi in the list under construction B. The function defines and returns the list B.

This is a bulky expression of a permutation, and would lead the program to spend a lot of time not exchanging list elements. It would be preferable to represent a permutation by reference only to the altered objects, such as

P=((1 3 2)(5 6)(17 18 19)).

This should be read "1 takes the place of 3, 3 takes the place of 2, 2 takes the place of 1; 5 takes the place of 6, ..." P is a list of lists. The identity is (), the empty list. Each sublist is a separate permutation problem. Now we need a function to accomplish a transformation from the full expression of a permutation to a (generally more compact) cycle list.

B=(5 3 1 2 4 6 8 7 ...)

(1 5 4 2 3)(6)(7 8) ...

The procedure is to construct a list beginning as (1); then starting with the first element of list B, i.e., B(1)=5, identify the corresponding member of B, i.e., B(5)=4, and continue until the first member of the cycle is found.

We must find a particular element of a list;

```
DEFINE FUNCTION FIND(I,B)
WHEN EMPTY(B) F EXIT,
WHEN EQ(I,1) FIRST(B),
FIND(DIFFERENCE(I,1),REST(B))
ENDFUN;
```

Then a list of cycles may be recovered. The strategy is to trace a cycle
until we encounter an element a second time; then we process the sublist
remaining after the members of the first cycle are removed.

```
DEFINE FUNCTION CYCLE(B),
    WHEN EMPTY(B) F EXIT,
    CYONE:FIRST(B),
    LATEST:FIRST(B),
  LOOP
    NEXT:FIND(LATEST,B),
    WHEN MEMBER(NEXT,CYONE), CYONE EXIT,
    CYONE:ADJOIN(CYONE,NEXT)
    LATEST:NEXT
  ENDLOOP
  LIST(CYONE,CYCLE(EXSET(CYONE,B)))
ENDFUN;
```

If P has been put into cycle form, we can effect the operation easily.

```
DEFINE FUNCTION TOPPERMUTE(A,P),
WHEN EMPTY(P) A EXIT,
TOPPERMUTE(PERMUTE(A,FIRST(P)), REST(P)),
ENDFUN;
```

Now we need a function to accomplish a single (sub)permutation represented
by a list of atoms, a LAT. Any permutation such as (1 3 2 5) can be expressed as a sequence of pair exchanges; in this case, (1 3 2 5) =
(13) (32) (25); the pattern is that each element except the first and
last appears as the last and first element of two adjacent pair exchanges.
To effect the transform from a permutation of form (a1 a2 a3 ... ak) to
a list of pairs (a1 a2)(a2 a3) ... (ak-1 ak), we need a function PAIRS (PER).

```
DEFINE FUNCTION PAIRS(PER),
WHEN EMPTY(PER) PER EXIT,
WHEN EQ(1, COUNT(PER)) F EXIT,
WHEN EQ(2, COUNT(PER)) PER EXIT,
LIST(
     LIST( FIRST(PER), FIRST( REST(PER)),
     PAIRS( REST(PER))
     ),
ENDFUN;
```

Now we have expressed the permutation as a sequence of pair exchange

operations. Let the pair exchange FIRST(PAIRS(PER)) be called (I J).

I = FIRST (I J) and J = (FIRST(REST((I J))). To complete the work we need

to effect the pair exchange, by PAIREX(I, J, LAT). This task is simplified

by the fact that for i<j any pair exchange (i j) can be effected by the

sequence

$$(i+2 \ i+1) \ (i \ i+1) \ (i+2 \ i+1) = (i \ i+2)$$

$$\dots\dots$$

$$(j \ j-1) \ (i \ j-1) \ (j-1 \ j) = (i \ j).$$

This is recursive in spirit, and easily implemented in LISP with the function

SWAP which accomplishes (i i+1). SWAP has to find the first object of the

two to exchange before it acts.

```
DEFINE FUNCTION SWAP(I,LAT),
WHEN GREATER (I, COUNT(LAT)) F EXIT,
WHEN EQ(I, 1) FLIP(LAT) EXIT,
ADJOIN(FIRST(LAT), SWAP(DIFFERENCE (I,1),REST(LAT))
ENDFUN;
```

Now the actual exchange occurs, of the first two elements of the paired list.

```
DEFINE FUNCTION FLIP(LAT),
WHEN GREATER(2, COUNT(LAT)) F EXIT,
ADJOIN( FIRST(REST(LAT), ADJOIN( FIRST(LAT), REST(REST(LAT))))),
ENDFUN;
```

Now we are ready:

```
DEFINE FUNCTION PAIREX(I,J,LAT),
WHEN EMPTY(LAT) F EXIT,
SWAP(J-1, PAIREX(I,J-1, SWAP(J-1,LAT))
ENDFUN;
```

The permutation is now completed by:

```
DEFINE FUNCTION PERMUTE (PER, LAT),
WHEN EMPTY(PER), F EXIT,
I:FIRST(FIRST(PER),
J:FIRST(REST(FIRST(PER)))),
PERMUTE(REST(PER), PAIREX(I,J,LAT))
ENDFUN;
```

This function must be supplied the standard form of the permutation PAIRS(PER).

Our next task is to construct the multiplication table of the group. For small groups it may be possible to store a look-up table. However it is more generally the case that we store only the sub-tables expressing the combination rules of the factor groups which define the full group through semidirect products. To provide a small concrete example, one might store the multiplication tables for C_3 and C_s, from which all the group theoretic results for C_3 could be constructed as required.

Classes, Cosets, and Double Cosets

Given a multiplication table expressed either as a single array or as a sequence of table inspections, it is a straightforward process to identify groups of operations within the same conjugacy class. This relation links any two operations (R, S) satisfying $g^{-1}Rg = S$ (all g) in group B.

A group G, which has a subgroup H, may be written as the union of left cosets $\Sigma(g)$ gH, in which g is an element of G not in H. There are $O(G)/O(H)$ such cosets, where $O(G)$ is the number of operators in group G. If the group G contains subgroups A and B, G may alternatively by represented by a union of double cosets AgB composed of all products of the form a x g x b, a in A and b in B. There are $O(G)/O(A)xO(B)$ such double

cosets. Double cosets have a variety of applications, identifying equivalent integrals in molecular orbital calculations [13] and enumerating
permutational isomers and isomerization pathways [14]. It is a straightforward matter to generate such cosets and double cosets automatically,
once the combination table is available.

Characters and Representations

The key to the definition of characters and representations of point
groups is the definition of the class multiplication matrices. That is,
the class product $Cp \times Cq = A(pq;k) Ck$. The characters X_i are solutions
to $R_i \times R_j \times X_i \times X_j = \Sigma(k) A(ij;k) R_k X_k$, where R_k is the number of operators
in class k. We can recover characters from this equation by a method of
Blokker and Flodmark [15]. Using the orthogonality theorem in the form
$\Sigma(g) X_i \times X_j = h \delta(i,j)$, h being the dimension of the irreducible representation, we find $\Sigma(k) [A(ij;k) - L \delta(jk)] R_k \times X_k = 0$. The solutions L
to the equations for characters are the (generally complex) roots of polynomials in L. Here symbolic mathematics is particularly helpful.

Factorization of Complex Polynomials

Here is our first major application of computer-aided symbolic
algebra [16]. We need to code a routine which will extract factors from
a polynomial exactly and directly if it can. But if it can't, we would like
to try the already established factors in hope of simplifying the problem
to one which can be solved exactly. Our symbol manipulator permits the
exact solution of any linear, quadratic, cubic, and quartic equations. (As
Galois proved [17], a solution in closed form for the general equation of
fifth order and greater is not feasible in general.) In the event that a
class-constant equation is of a higher degree, our strategem is to solve

the smaller systems first, which will yield several of the characters.
These characters may be roots of the high-degree polynomials, since there
is considerable redundancy in the class equations. With these trial roots
in hand, one might be able to reduce a high order polynomial by synthetic
division. We use Cope's public domain routine for synthetic division in
Mu [18], which we have extended slightly to produce simpler non-numerical
coefficients of the expression resulting from the synthetic division. (It
is possible, but rarely occurs, that this step fails to produce a full set
of characters. If this occurs, an alternative factorization of the group
may be needed.) This direct method will not succeed when there are more
than four classes and the roots are not simple. For large groups with
high-order class constant matrices, we need a more powerful method for
recovering characters.

An Alternative Method for Characters due to Dixon, Which Profits
from High Precision Symbolic Arithmetic

Dixon has made the search for roots of the class constant matrix much
more efficient by establishing a restriction on the potential values of the
characters [19]. We define n for each operator g in the group, where
g^n = the identity. If e is the least common multiple of the orders of
operators in the group, the characters are expressed by polynomials in the
(complex) e-th roots of unity. $j_{X(a)} = \Sigma[0 < s < e-1] \, m_j(s) \, z^s$. The numbers
$m_j(s)$ are integers between 0 and 1_j, where 1_j is the dimension of the j-th
irreducible representation. Without any effort at explanation, I quote
the result of Blokker and Flodmark's discussion [20]. We must solve for
the roots of the matrix of class constants, A(ij:k). However this time we
express all arithmetic modulo p, where p is a prime number such that

$p > 2\sqrt{g}$ and $(p-1)/e$ = an integer. Here g is the order of the group and e is
the least common multiple defined above. Instead of searching the complex
plane for roots of the polynomial obtained from the secular determinant,
it is sufficient to limit investigation to the set of p integers $[0, 1, \ldots$
$p-1]$. Now we need not solve high-order equations, but merely substitute
integer estimates of eigenvalues x (k) in an array and evaluate the polynomial arising from the secular determinant. The MuMATH system handles MOD
arithmetic with ease. Alternatively we could use the powerful techniques
for polynomial factorization in modulo arithmetic described by Knuth [21],
though one must beware of repeated roots in that case. (The polynomial is
in general not "square-free".) Given the proper normalized eigenvalue
$X(ij) = (1_j/h_i)$ x (ij), we find the $m_j(s)$ by $m_j(s) = (1/e) \Sigma [0 < n < e-1]$
$X (j:a^n) r^{-sn}$. Here r is an integer such that $r^e = 1$ (modulo p) and $r^k <>$
1 for k = 0 to e. Since this description may seem terse, an example application is worked through in the appendix.

Dixon's method permits construction of characters for extremely large
groups. Once the characters are known, the representations may be recovered
from the eigenvectors of the class constant matrix. Details are given by
Blokker and Flodmark [20]. However the method proposed by Blokker and Flodmark
for the irreducible representation matrices requires the construction of the
regular representation matrix. The regular representation matrices are
square, of order $O(G)$ x $O(G)$, and are $O(G)$ in number. (There are only
$O(G)$ x $O(G)$ nonzero elements, however.) This becomes unmanageable for large
groups; we will take a different tack for the representation matrices.
Fortunately we can induce representations of the large groups, from representations of their factors.

Problems of Scale: Inducing Representations

If we have a group G which contains a subgroup H, and a representation of G called R(G:all g), one may recover a representation of H by restricting R to those g within H.   The new representation has been subduced from G to H. This problem is not too hard, though generally the subduced representation is not irreducible even if R is.   It is tougher to go the other way.

Let G be a group, which contains subgroup H.   H has a representation expressed in a basis (Bj).   G may be generated from H by certain of its operations s(k), according to $G = \Sigma(k)\ s(k)H$.   There are $O(G)/O(H)$ of these operations.   Then the set s(k)B(j) constitutes the basis for representation of G.   The representation itself can be obtained by a description of the action of some g (in G) on the basis member s(k)B(j)

$$g\ s(k)B(j) = \Sigma(pq)\ s(p)B(q)\ R(G:\ pq,kj)$$

If G is constructed as   (k) s(k)H, the sum of left cosets of H in G, the cosets compose a basis for a representation of G which is called the "ground" representation *G.   Since an element g in G can only exchange members of the set of left cosets, *G is a permutation matrix, such that *G(kl) = 1 if g [s(k)H] = [s(l)H], and is zero otherwise.   In other words g s(l) = s(k) h for some h in H, called "the subelement of g by s(l)", h(l:g).   Thus h(l:g) = s(k)* g s(l).

Now we can construct R(G: pq,kj) by processing the ground representation *G,(pk)   replacing each nonzero element (qj) by the matrix R(H:qj) representing h(l:g).   The character of the induced representation is the trace of this supermatrix, or alternatively:

$$\Sigma*G(kk)\ X(s(k)*\ g\ s(k)).$$

The representation obtained in this way is generally reducible. For example, if we induce a representation of C_{3v} from C_3, eH and σH are the cosets making up a basis for the 2X2 ground representations; and even if irreducible representations of C_3 are employed, the "supermatrix" representation matrices of C_{3v} must be 2X2 [22]. We have spoken of "supermatrices" but it will be clear that *G is sparse; all the information in *G can be stored as a permutation list.

Altmann describes how the induced representations defined above may be reduced [2]. The irreducible representations of H, IF(H:st), play a role. However the passage from an irreducible representation of H to an irreducible representation of G is indirect. In brief, we must find the sets of mutually conjugate representations, called "orbits". (Two of H's irreducible representations are called "conjugate representations" if they represent conjugate elements. Elements j and h are conjugate if $g^{-1}hg = j$.) The operators g in G which transform a particular representation *Hj into itself constitute a number of cosets gH of G (including H), and compose the "little group" Lj of Hj. The little group is a necessary intermediate stage in the induction.

Our major computational tasks are

(1) Generate the irreducible representations of the invariant subgroup; this can be done by table look-up, or direct application of Blokker and Flodmark's matrix method.

(2) Establish conjugacy relations among these irreducible representations.

(3) Generate the set of operations in G which produce identical irreducible representations (this is the "little group" Lj).

162

(4A) Define the irreducible representations of the
little group, if it is of convenient size; or

(4B) define the "little co-group."

If G is the semidirect product HvK then the little group Lj is the semi-

direct product Hv(Kj) where Kj is a subgroup of K. Kj is called the little

co-group of *Hj. Woodman states "starting with one irreducible representa-

tion of H and one irreducible representation of its little co-group, we

generate an irreducible representation of G" [23]. This is accomplished

by a construction of a supermatrix: beginning with the ground representa-

tion defined by the little co-group, we insert the irreducible representa-

tions of H at the non-zero positions of that ground representation. This

is particularly simple when the factors H and K are cyclic groups, and

important simplifications obtain when H is an invariant cyclic subgroup.

Our work is limited to this frequently encountered case.

In the most favorable (but rare) case, the little group Lj is identical

with H, and the irreducible representations of H correspond to those of G.

In the least favorable case Lj is identical with G, and cannot easily be

represented. But in that case the little co-group is K, which is guaranteed

to be manageable in size.

(5A) Given the i-th representation of the little group
K and that this representation produces the irreducible
representation k of H upon subduction, we place this (ik)
labeled matrix in the st location of the ground representation of G. The ground representation of G is based
on the coset expansion of G by K, and the st location is
nonzero for operations sgt in the little group K.

(5B) When the little group is too large to be manageable, the little co-group can be used as in (5A).

Point Groups are (Semi)direct Products

It is often the case that molecular symmetry groups can be expressed as $G = L[X]M$, a direct product of subgroups L and M. This occurs for two groups L and M which share only the identity element and for which the product (lk x mj) is meaningful, if all elements of L commute with all elements of M. If one relaxes the commutation requirement, the set of (lk mj) forms a group called the semidirect product LvM. $C_{3v} = (C_3)v(C_s)$ for example. Altmann shows that the familiar point groups up to O_h can be expressed as a sequence of (at least) semidirect products of groups no larger than D_2. This permits a means of computation of properties of (say) O_h which is very sparing of machine memory; in fact a modest programmable pocket calculator can do this job. But the real advantage of the semi-direct product picture lies in its application to the large groups which are required to describe easily rearranged molecules.

Molecular and NMR Symmetry Groups are Often Semidirect
Products of Permutation Groups.

Organic molecules have various symmetries, applicable on different time scales. On the slowest spectroscopic time scale, defined by the NMR transition, protons within methyl groups become equivalent. On the average, the system is properly described by permutation symmetries. Woodman has shown that when a molecule can be viewed as a rigid frame supporting groups which may execute internal rotations, its molecular symmetry group may be written as a semi-direct product of the group of the frame and groups for each internal rotor [23]. For example the molecular symmetry group for boron trimethyl is the semidirect product of D_{3h} (frame) and $(C_3)^3$, cor-responding to the cyclic permutations of protons within each of the three freely rotating methyl groups [24].

In the NMR problem, we are concerned with the symmetry of the typical Hamiltonian which has the form

$$(\Sigma:i) \; g\beta H(1-\sigma(i))I_{zi} + (\Sigma:i,j) \; I:J:I + \dots$$

The fact that the first term is a sum guarantees full permutation symmetry Sn for interchange of any two terms referring to spins with like shielding tensors $\sigma(i)$. The second term is invariant to any exchange among those particles which share shielding tensors, are equally coupled to one another, and are equally coupled to every other nucleus in the system. The symmetry group of the NMR Hamiltonian contains factors of the symmetric group (all possible permutations within such a set) for each set of equivalent nuclei. (The result of applying this group to factoring the matrix representation of the spin Hamiltonian is equivalent to considering each set as a "composite particle" with a redefined spin angular momentum [23].)

NMR symmetry groups are clearly expressible as semidirect products of the permutation groups for each set of equivalent nuclei [23]. NMR symmetry groups are often larger than groups relevant to other types of motion. More interesting chemical exchanges also may be represented as factors helping to compose the full group. For example, the effective symmetry of a system such as BeB_2H_8 where a set of eight protons is rapidly exchanged between a D_{2d} and two C_{3v} structures can be shown to be $(S_4)v(S_4)$ [1]. The feasible chemical permutations add new generators to define the total group.

Symmetry Adaptation of Bases and Coupling Coefficients for
Semidirect Products

It will often be the case that we have use for symmetry-adapted functions for a system whose full symmetry can be expressed as a semi-direct product

of several factors. The irreducible representations permit the easy
generation of symmetry-adapted combinations.

$\Psi(vw:) = \Sigma(g) \; V(vw:g) \; gR$

Here Ψ is a function guaranteed to have the symmetry labels vw, and g is
an operator in the group. V is an irreducible representation matrix. This
is the familiar symmetry projection process. For irreducible representations
constructed for semi-direct products, the symmetry projection can be accomplished in steps. In acetaldehyde, where the full group can be expressed
as the semi-direct product of $(S_3)v(C_s)$, one can symmetry adapt the spin
functions for the methyl protons, and then apply the operations and irreducible
representations of the little co-group to express the full symmetry.

It will often be necessary to construct product functions which are
symmetry-adapted.

$\Psi(wv) = \Sigma(pq:rs) \; C(wv:pq:rs) \; M(pq:a) \; N(rs:b)$.

Here M and N are functions symmetry adapted to species pq and rs respectively,
and the resulting combination ψ is adapted to species wv. The elements of
the matrix C are called vector-coupling coefficients. They are most familiar
as Clebsh-Gordon, or Wigner coefficients in angular momentum problems.
These vector coupling coefficients contain the full geometric structure of
any problem, and appear in the Wigner-Eckart expression for integrals of
operators expressed as tensors of the molecular symmetry group. We may
recover explicit values of these by operating on the expression by g.

$$g \; \psi(wv:a) = \Sigma(b) \; R(wv:ab:g) \; \psi(wv:b)$$
$$= \Sigma(pq:rs) \; \Sigma(de) \; C(wv:pq:rs) \; R(pq:cd:g) \; M(pq:d) \; R(rs:ef:g) \; N(rs:f).$$

We can recover the values of C by means of the orthogonality relations for
irreducible representations.

$$\Sigma(g) \ R^*(wv:ab:g)R(wv:ab:g) = (O(G)/1(wv))$$
$$= C(wv:pq:rs) \quad (g) \quad R^*(wv:ab:g)R(pq:cd:g)R(rs:ef:g)$$

The coupling coefficients are easily defined for direct products, where $G = G \ [X] \ H$ and operations h in H commute with operations f in F. Then each member of the product set is composed of two factors, each symmetry-adapted in either F or G. All vector coupling coefficients are then 0 or 1. But more generally, when commutativity can no longer be assumed, so $G = FvH$, the results will not be so simple.

Wigner-Eckart Theorem for Semidirect Products

The Wigner-Eckart theorem applies to integrals of operators which may be classed as irreducible tensors in the group. In order to display the Wigner-Eckart theorem for composed groups, we must first show that the operators such as the nuclear spin coupling terms can be symmetrized according to the full group. Our goal is to develop tensor expressions for Hamiltonian terms, so that general integrals can be expressed by numerical values dictated by group systems. That is, for acetaldehyde, the Hamiltonian referring to Zeeman terms, coupling among nuclei, external perturbations due to fields, and motion permitting relaxation, should appear as

$$Ho(AA) + V(AEx) + V(AEy) + V(BEx) + V(BEy)$$

where initially the perburbations can all be classed as adapted to the symmetries of the frame and rotor. It is almost always the case that perturbations are totally symmetric with respect to permutations, and can mix states of the system which match in permutation symmetry.

And In Conclusion:

The full description of symmetries in systems which are easily rearranged requires treatment of rather large groups. The algebraic techniques of

induction of representations reduce the bulk of the problem, but require
computational aid if their application is to become widespread. Computer
algebra systems, even at the smallest scale, can ease the application of
these algebraic methods.

References:

[1] C. Trindle, J. Comput. Chem. 5, 162 (1984)
[2] S. L. Altmann, Induced Representations in Crystals and Molecules,
Academic Press (New York, 1977); see also S. L. Altmann, Rev. Mod. Phys.
35, 641 (1963)
[3] GPTHEORY: FORTRAN Computer Program for Determining Molecular Symmetry
Properties, T. D. Bouman and G. L. Goodman, Argonne National Laboratory
Tech. Rept. ANL-7803, 1971, and S. Flodmark and E. Blokker, Int. J. Quantum
Chem. 1S, 703 (1967) are examples; a survey is given in [1].
[4] Augusta Ada, Lady King, later Countess of Lovelace, provided these
remarks in a commentary on Babbage's report; reproduced in V. Bowden's
Faster Than Thought; the occasion described in Charles Babbage: Pioneer
of the Computer, A. Hyman, Princeton U. Press (Princeton, NJ, 1983)
[5] A clear discussion of the Turing machine is given by A. Hodges,
Alan Turing: The Enigma, Simon and Schuster (New York, 1983)
[6] P. McCorduck, Machines Who Think, Freeman (San Francisco, 1978), recounts
the history of the AI movement.
[7] A. Turing, "Computing Machinery and Intelligence," reprinted in Minds
and Machines, A. R. Anderson ed., Prentice Hall (Englewood Cliffs, NJ, 1963)
[8] J. Slagel, J. ACM, 10, 507 (1963)
[9] E. Feigenbaum and A. Barr, Handbook of Artificial Intelligence, Vol. I,
II, W. Kaufmann, Inc (Los Altos, CA 1982)
[10] J. McCarthy, et. al., LISP 1.5 Programmer's Manual, MIT Press (Cambridge,
MA, 1963); a recent incarnation is described in G. L. Steele, Common LISP:
The Language, Digital Press (Burlington, MA, 1984)
[11] MuMATH is available from the Soft Warehouse P.O. Box 11174, Honolulu,
HI 96828; described by D. Yun and R. Stoutemyer, in "Symbolic Mathematical
Computation," Encyclopedia of Computer Science and Technology V15;
M. Dekker (New York, 1980)
[12] For definitions and properties of permutations, see R. Pauncz, Spin
Eigenfunctions: Construction and Use, ch. 6, Plenum (New York, 1979); for
algorithms for multiplying permutations see D. Knuth, The Art of Computer
Programming, Vol. I ch. 1, p 160ff Addison Wesley (Reading, MA, 1968)
[13] E. Davidson, J. Chem. Phys. 62, 400 (1975)
[14] E. Ruch, W. Hasselbarth, and B. Richter, Theoret. chim. Acta 19, 288
(1970); W. Hasselbarth and E. Ruch, Theoret. chim. Acta, 29, 259 (1973)
and W. Klemperer, J. Chem. Phys. 56, 5478 (1972)

[15] S. Flodmark and E. Blokker, Int. J. Quantum Chem. 1S, 703 (1967)
[16] A historical overview of computer algebra is given in reference [9];
a clear discussion is found in R. Pavelle, M. Rothstein, and J. Fitch,
Scientific American, 136 (Dec. 1981); for a current review, see Computer
Algebra: Symbolic and Algebraic Computation (Second Ed.), B. Buchberger,
G. E. Collins, and R. Loos, with R. Albrecht, Springer-Verlag (New York,
1983); calledCASAC in the following references. A substantial effort is
described in J. Neubueser's "Computing With Groups and Their Character
Tables," CASAC p 45. J. Calmet and J. A. van Hulzen, in "Computer Algebra
Applications," CASAC p 245 remark that "[In biology and chemistry] ...
computer algebra has almost no impact at all." This observation may soon
become dated.
[17] Galois' story is recounted by D. M. Bishop, Group Theory and Chemistry,
Oxford U. Press (London, 1973)
[18] W. Cope, "Synthetic Division of Polynomials," in The Soft Warehouse
Newsletter #7, P.O. Box 11174, Honolulu, HI 96829
[19] J. Dixon, Numer. Math. 10, 446 (1967), and Math. Comput. 24, 707 (1970)
[20] E. Blokker and S. Flodmark, Int. J. Quantum Chem. 4, 463 (1971)
[21] D. Knuth, The Art of Computer Programming, Vol. II ch. 4 Addison Wesley
(Reading, MA, 1981); see also E. Kaltofen, "Factorization of Polynomials,"
in CASAC [16], p 95
[22] Altmann, [1] p 143
[23] C. M. Woodman, Mol. Phys. 19, 753 (1970)
[24] H. C. Longuet-Higgins, Mol. Phys. 6, 445 (1963)

Appendix

It may be worthwhile to work through an example of Dixon's method. Take
the C_3 group; the classes are E, C_3, and C_3^2. The class multiplication matrices
are constructed from E*K1 = K1*E =K1, K2*K2 = K3, K2*K3 = E, K3*K3 = K1.
The common multiple of the orders (1,3,3) is 3. Therefore the characters
will be expressed in the cube roots of unity, namely (1, q, q^2), where
q = exp(2 Π/3). The prime number p>2(1.732), with (p-1)/3 = an integer,
so p = 7. The integer r such that r^3 MOD p = 1 must be 2. (Note 2^1 MOD
p = 2 and 2^2 MOD p = 4.) The polynomials for classes E, C_3, and C_3^2 are
$(L-1)^3$, (L^3-1), and (L^3-1). In MOD 7 arithmetic, the eigenvectors are
(1,1,1), (1,2,4), and (1,4,2). These are the x's and since all classes are
of dimension 1x1, they are also the X's. We can verify that these obey the
secular equation.

Consider the C_{3v} group; the classes are K1 = [E], K2 = [C_3, C_3^2],
K3 = [$\sigma1$, $\sigma2$, $\sigma3$] and the class equations are ExE = E, ExKj = Kj, K2xK2 =
2E + K2, K2xK3 = 2K3, K3xK3 = 3E + 3K2. The roots MOD 7 are (1,1,1),
(2,2,6), and (0,3,4).

8

POLYMER MODELING APPLICATIONS OF SYMBOLIC COMPUTATION

JOHN T. BENDLER

Polymer Physics and Engineering Branch, General Electric Corporate Research and Development, Schenectady, NY 12301

and

MICHAEL F. SHLESINGER

Physics Division, Office of Naval Research, 800 North Quincy Street, Arlington, Virginia 22217

ABSTRACT

Statistical models are employed to describe (a) the geometry and (b) kinetics of local conformational backbone motions in glassy polymers. First, the chemical shift anisotropy (CSA) tensor line shape of an aromatic ^{13}C is modeled using a double-well potential with temperature-dependent flips and oscillations modulating the experimental principal components. MACSYMA is used to perform matrix multiplications and subsequent Boltzmann averages. Second, bond-defect diffusion models lead to Levy-stable laws for conformer orientational survival probability densities. MACSYMA assists in the analysis and resummation of series expansions of stable laws needed for data processing.

INTRODUCTION

The physical laws which govern microstructure and dynamics in amorphous disordered solids and liquids generally lead to statistical distributions of subsystem parameters. Phenomenological "statistical mechanical" kinetic models that postulate probability concepts from the start are useful if they result in descriptive methods applicable to different classes of materials, and par-

169

ticularly if they focus attention on underlying mechanisms which generate the observed distributions. Stochastic modeling has a long and successful history in polymer physics, starting from the random-walk model of a polymer chain introduced by Kuhn (1). The size and shape distributions of real polymer molecules in solution at the theta point are accurately accounted for in terms of the Gaussian statistics of a three-dimensional random-walk. Self-avoiding and interacting walks have become the de-facto models for poor and good solvents as well. The discovery from low-angle neutron scattering of isotopically labelled chains that molecular dimensions in melts and solids also follow Gaussian statistics (2) adds to the theoretical importance of such simple models.

Local geometries and motions in polymer chains are dependent on details of chemical structure and packing, so that useful generalizations have been slower to develop. An important empirical finding of recent years is that macroscopic relaxation (eg., mechanical and dielectric) in glassy materials appears to obey the Kohlrausch-Williams/Watts (KWW) function;

$$\varphi(t) = e^{- (\frac{t}{\tau})^{\alpha}} \tag{1}$$

A variety of theoretical ideas have been advanced to account for this special form, and such investigations remain an area of active interest. Our work has emphasized a stochastic interpretation of eq 1, and in particular, Shlesinger and Montroll (3) used the ideas of Glarum (4) to derive the KWW function from a defect diffusion picture, with defect motion occurring intermittently in bursts and pauses, the event-time probability density behaving at long-time as $\sim t^{-1-\alpha}$. The surprising appearance of power-law decay implies non-Gaussian

statistics, and therefore the stable momentless probability densities which
are generalizations of the Gaussian and which play a fundamental role in the
theory of random "fractals".(5)

The stable densities are of significance in the theory of probability
where they appear as the limit laws for sums of random variables whose individual distributions have no moments. (5,6) The limit theorems are generalizations of the Central Limit Theorem, and the stable densities generalize the
normal law. Only three examples are known in closed form, and thus numerical
efforts concentrate on integral representations and series expansions. (7,8)
In the second part of this article we present some results of the symmetric
stable density of a complex argument which has practical application for
parallel decay mechanisms. MACSYMA's ability to combine numerical evaluation
with flexible methods of series representation makes it well-suited to this
application.

Solid-state NMR relaxation and line shape analysis is the primary source
of molecular information about structure and motion in the glassy solid. The
work of Inglefield, Jones and co-workers (9-12) studying proton spectra in
chemically-modified polycarbonates, as well as polycarbonates isotopically
enriched in ^{13}C and deuterium has established the prominent role of phenylene
π flips which take place in the solid state of the polymer more than 200
degrees C below the glass transition. The fact that the rings rotate about
their 1-4 axis (which lies in the backbone) with very little re-orientation of
the inertial axes was the unexpected and unambiguous conclusion from the earliest of these studies. (9) Details of the temperature dependence of the rate
of flipping and oscillation within a single well are now available from
temperature-dependent studies of a ^{13}C enriched high polymer with the label

>95% ortho to the carbonate unit. (12) In the first section below we describe some aspects of a MACSYMA-aided analysis associated with modeling the temperature-dependent chemical shift anisotropy (CSA) tensor for the ring motion.

RING FLIPS AND OSCILLATION IN GLASSY POLYCARBONATE

Unlike the dipole proton-proton coupling in the phenylene unit, which is parallel to the 1-4 axis and hence to the polymer backbone, the ^{13}C CSA tensor σ has all three of its principal components off axis. The components lie roughly along the ^{13}C H bond, perpendicular to the bond and in the plane of the ring, and perpendicular to the ring. In this principal axis molecule frame the tensor is diagonal;

$$\sigma \;=\; \begin{matrix} \sigma_{11} & 0 & 0 \\ 0 & \sigma_{22} & 0 \\ 0 & 0 & \sigma_{33} \end{matrix} \qquad (2)$$

In the glassy solid near absolute zero one would observe a powder average of these tensor components arising from a random static distribution of molecular orientations. As the temperature rises, molecular motion averages the interactions to a degree that depends on both the amplitude and speed of the motion. For simplicity, we consider the high-temperature limit when the speed of the motion is "fast" compared to the total signal width. In this case, we need consider only the amplitude of the motion. (Spin relaxation studies indicate that this is true for polycarbonate ring flips from about 0°C to the glass transition at 150 $^{\circ}$C . (12)) Further, since it is known that the predominant motion is along the 1-4 axis, we only consider flips and oscillations around this axis. To examine the influence of such motions, we first transform to the

1-4 frame by introducing a rotation operator M_θ which rotates the diagonal CSA tensor σ to a reference frame rotated by an angle θ around the z axis perpendicular to the ring plane;

$$M = \begin{array}{ccc} \cos\theta & -\sin\theta & 0 \\ \sin\theta & \cos\theta & 0 \\ 0 & 0 & 1 \end{array} \qquad (3)$$

(In the case of benzene's CSA, which is a useful model for polycarbonate, $\theta = \pi/3$.) The tensor in the 1-4 frame is thus

$$\sigma_{14} = M.\sigma.M^{-1} \qquad (4)$$

MACSYMA's ability to do matrix multiplication (by using a period between matrices) and to compute matrix inverses by exponentiating the matrix to the -1 power allows equation 4 to be performed with a simple one-line command. The output is voluminous, but is simplified subsequently by user-supplied trigonometric substitutions.

The CSA tensor of equation 4 may now be subjected to an oscillation of a specified angle α about the 1-4 axis by application of the rotation operator

$$R_\alpha = \begin{array}{ccc} 1 & 0 & 0 \\ 0 & \cos\alpha & -\sin\alpha \\ 0 & \sin\alpha & \cos\alpha \end{array} \qquad (5)$$

An oscillation of a fixed amplitude is applied to σ_{14} by the similarity transformation;

$$\sigma_{14}(\alpha) = R_\alpha . \sigma_{14}.R_\alpha^{-1} \qquad (6)$$

The ring oscillations take place within a potential well with period π. The potential energy may be modeled, as a first approximation, by

$$V(\alpha) = \frac{1}{2}V_0[1 - \cos 2\alpha] \sim V_0\alpha^2 \tag{7}$$

The Boltzmann factor for finding thermal averages of the oscillation-averaged tensor $\sigma_{14}(\alpha)$ is

$$\exp \frac{-V(\alpha)}{kT} \sim e^{-\rho\alpha^2} \tag{8}$$

where kT is Boltzmann's constant times the absolute temperature and $\rho = \frac{V_0}{kT}$. The quadratic approximation of equations 7 and 8 is satisfactory for $\rho > 3$ or so, which is the case for polycarbonate. If the integrals over α are taken to ∞, the final expressions reduce to simple algebraic form. The error functions are known to MACSYMA and may be used, but the final results are much more complicated. This amount of computation is probably not justified since the true shape of the potential is only approximately given by the simple trigonometric form of equation 7. The thermal average over the oscillation amplitudes α is found from

$$\langle\sigma_{14}(\alpha)\rangle_\alpha = Q^{-1} \int\limits_{-\infty}^{\infty} e^{-\rho}\sigma_{14}(\alpha) \, d\alpha \tag{9}$$

where Q^{-1} is the partition function of the angle integration. The integrals over the tensor in equation 9 may be done directly in MACSYMA, though on a VAX it is a long calculation, and a more satisfactory procedure is to integrate each matrix element separately.

In addition to oscillating in a single well, the rings occasionally flip by about 180 degrees, then spending time in the second well oscillating about its potential minimum. The geometric effect of these motions may be reproduced by concatenating a flip transformation of angle $\beta \sim \pi$ with an oscillation of angle γ;

$$R_{\gamma;\beta} = R_\gamma \cdot R_\beta \qquad (10)$$

so that the CSA tensor becomes;

$$\sigma_{14}(\gamma;\beta) = R_\gamma \cdot R_\beta \cdot \sigma_{14} \cdot R_\beta^{-1} \cdot R_\gamma^{-1} \qquad (11)$$

To complete the calculation of the π flips to the line shape, thermal averages of $\sigma_{14}(\gamma;\beta)$ are also needed. In the simplest case of a flip angle β of exactly π and identical potential curvature and depth in both wells, the thermally averaged CSA tensor $\langle\langle\langle\sigma_{14}\rangle\rangle\rangle$ is diagonal and has components;

$$\langle\langle\langle\sigma_{14}\rangle\rangle\rangle_{11} = .25(\sigma_{11} + 3\sigma_{22}) \qquad (12a)$$

$$\langle\langle\langle\sigma_{14}\rangle\rangle\rangle_{22} = \sigma_{33}\langle\sin^2 \alpha\rangle + .25(\sigma_{22} + 3\sigma_{11})\langle\cos^2 \alpha\rangle \qquad (12b)$$

$$\langle\langle\langle\sigma_{14}\rangle\rangle\rangle_{33} = \sigma_{33}\langle\cos^2 \alpha\rangle + .25(\sigma_{22} + 3\sigma_{11})\langle\sin^2 \alpha\rangle \qquad (12c)$$

where the thermal averages of the trigonometric functions using equation 8 are

$$\langle\cos^2 \alpha\rangle = 1 - \langle\sin^2 \alpha\rangle = \tfrac{1}{2}[1 - e^{-\rho^{-1}}] \qquad (13)$$

A more realistic model incorporates a distribution of flip angles β around 180 degrees, with a temperature-dependent dispersion. Such a simulation using a

square well potential has been carried out by Jones, Inglefield and co-
workers, and we have found analytical expressions for harmonic oscillator
potentials. These results will be reported elsewhere along with comparisons to
experiment.

A MODEL FOR PARALLEL RELAXATION MECHANISMS

Viscoelastic responses in solid polymers are characterized by hereditary
"after-effects" and non-linearity, and one does not expect that the theory of
Markov processes will be useful in describing or understanding them. Nevertheless, a certain class of relaxation properties of polymers and glasses can be
modeled using a continuous- time random-walk (CTRW) formalism. (13,14) A basic
equation for the analysis of stationary stochastic processes is Chapman-
Kolmogorovs';

$$P(y_1, y_2; t) = \int P(y_2, y; t_1) P(y, y_1; t-t_1) dy \qquad (14)$$

where $P(y_1, y_2; t)$ is the probability density that the variable y suffers a
transition from y_1 to y_2 in time t. If the process y has translational
invariance and the unbounded range ($-\infty < y < \infty$), equation (14) becomes

$$P(y_1-y_2; t) = \int_{-\infty}^{\infty} P(y_2-y; t_1) P(y-y_1; t-t_1) dy \qquad (15)$$

and introducing the Fourier transform (FT)

$$\tilde{p}(k, t) = \int_{-\infty}^{\infty} P(y; t) e^{iky} dy \qquad (16)$$

equation 15 is reduced to algebraic form

$$\tilde{p}(k,t) = \tilde{p}(k, t - t_1) \; \tilde{p}(k, t_1) \tag{17}$$

The functions e^{-Dtk^2} and e^{-akt} satisfy equation 17 and are the characteristic functions (or FTs) of the Gauss and Cauchy densities respectively. The more general case was investigated by Paul Levy (14);

$$\tilde{p}(k,t) = e^{-btk^{\alpha}} \quad 0 < \alpha \leq 2 \tag{18}$$

If b is real and positive, then $\tilde{p}(k,t)$ in equation 18 is related by Fourier inversion to the symmetric stable density of characteristic exponent α . (Equation 18 is formally identical to equation 1, so that the KWW relaxation function is the characteristic function of the symmetric stable densities.) A change of variable allows the stable symmetric density to be written

$$Q_{\alpha}(z) = \frac{1}{\pi} \int_0^{\infty} e^{-u^{\alpha}} \cos zu \; du \tag{19}$$

Closed-form expressions for $Q_{\alpha}(z)$ exist only for $\alpha = 2$, 1 and $\frac{1}{2}$. Empirically, it is found to successfully describe dielectric, mechanical and magnetic relaxation in polymers with $0.3 < \alpha < 0.8$ near the glass transition, with much smaller exponents in the glassy solid itself.(7,8) Electric and mechanical dispersion may be fitted using the sine transform, and transient stress, volume and magnetic decay are described by the characteristic function $e^{-(\frac{t}{\tau})^{\alpha}}$. (15) Numerical evaluations can be difficult owing to slow convergence and because the physically important values of z are in the midrange near 1. Series expansions for $Q_{\alpha}(z)$ for large and small z are known. For small z ;

$$Q_\alpha(z) = \frac{1}{\pi\alpha} \sum_{n=0}^{\infty} \frac{(-)^n z^{2n}}{2n!} \Gamma(\frac{2n+1}{\alpha}) \qquad (20)$$

This asymptotic series for small z is useful for high temperature solids where α is small but the relaxation rate τ^{-1} is large. A convergent large z series for $Q_\alpha(z)$ was found by Wintner;[7]

$$Q_\alpha(z) = \frac{\alpha}{\pi} \sum_{n=0}^{\infty} \frac{(-)^n}{(n! z^{\alpha(n+1)+1})} \Gamma(\alpha[n+1]) \sin[\frac{\pi\alpha(n+1)}{2}] \qquad (21)$$

Equation 21 is convergent for $0 < \alpha < 1$ which is the range of interest for glassy solids. Unfortunately, several hundred terms of the series may be needed for α as large as 0.2, necessitating a search for better methods. Alternative expansions have previously been found, partly with the aid of MACSYMA. Here we consider a slightly more general case.

Most theories (eg., defect diffusion (3)) which have been proposed to explain the physical basis of the KWW relaxation function of equation 1, also allow the possibility of simultaneous parallel decay mechanisms approximately modeled here by an additional (exponential) decay factor;

$$\varphi(t) = e^{-(\frac{t}{\tau})^\alpha} e^{-\frac{t}{\tau'}} \qquad (22)$$

In fact, in a polymer glass above Tg we might expect to find a temperature interval in which backbone conformer fluctuations would decay with roughly equal contributions from the KWW and exponential processes. It is of practical interest then to examine the spectral density $J_\alpha(\omega)$ corresponding to the relaxation function of equation 22;

$$J_\alpha(\omega) = \int_{-\infty}^{\infty} e^{-\left(\frac{t}{\tau}\right)^\alpha} e^{-\frac{t}{\tau'}} e^{-i\omega t} \, dt \qquad (23)$$

By changing variables and keeping track of absolute value signs on the decay functions, equation 23 may be written;

$$J_\alpha(\omega) = \left(\frac{\tau}{2\pi}\right) \left[\int_0^\infty e^{-x^\alpha - qx} \, dx + \int_0^\infty e^{-x^\alpha - q^+x} \, dx \right] \qquad (24)$$

where $q = i\omega\tau - \frac{\tau}{\tau'}$. The generalization of equation 20 is found by expanding the exponential of the linear term and integrating the resulting series term by term;

$$J_\alpha(\omega) = \left(\frac{\tau}{\pi\alpha}\right) \sum_{n=0}^{\infty} \frac{(-)^n \rho^n}{n!} \Gamma\left(\frac{n+1}{\alpha}\right) \cos(n\phi) \qquad (25)$$

where $\rho = [\omega^2\tau^2 + (\frac{\tau}{\tau'})^2]^{1/2}$ and $\phi = \tan^{-1}(-\omega\tau')$.

The analog of the convergent expansion, equation 21, is obtained by expanding the exponential of the fractional powers in equation 24 and integrating term by term;

$$J_\alpha(\omega) = \left(\frac{\tau}{\pi}\right) \sum_{n=0}^{\infty} \frac{(-)^n}{n! \rho^{\alpha n+1}} \Gamma(\alpha n + 1) \cos[\phi(\alpha n + 1)] \qquad (26)$$

The series of equations 25 and 26 are programmed easily in MACSYMA and since 20 to 30 terms can be taken directly, checks on convergence behavior are straightforward. (They reduce to equations 20 and 21 respectively for $\tau' = \infty$ or $\phi = \frac{1}{2}\pi$.) The gamma functions are the limiting factor in the number of terms able to be taken, and judicious use of Sterling's approximation can be used to side-step this limitation.

CONCLUSIONS

MACSYMA seems to be especially useful for <u>developing</u> theoretical models
since one has the flexibility to go between the analytical and numerical work
quickly. This is evident from the matrix analysis/Boltzmann averaging involved
the NMR shift-tensor model. The final expressions for the principal components are nice to have, of course, but could be derived by hand in a day or
two. The numerical comparison to experiment (not discussed in this paper)
could have been done numerically with some non-linear fitting methods. With
MACSYMA, the exact expressions are evaluated against experiment with Newton-
Raphsons method. The point is that it is not known ahead of time what combination of flipping and oscillation suffices to account for the temperature
dependence. In fact dozens of variants were evaluated and rejected before the
results reported here were found. Alternative models with unequal energies and
oscillation frequencies in each well were examined. This would have been more
tedious to do with numerical simulations in each case, and less conclusive
since failures could be attributed to numerical problems! Numerical difficulties do occur with MACSYMA, especially if a large amount of floating-point
arithmetic is involved, but problems are easier to spot, and simple cases can
be tried to verify the formulae.

The ability to deal systematically with Taylor and Laurent expansions is
crucial to working with the stable "fractal" functions, since closed-forms
almost never exist. Once again, the ability to try many different forms of a
series is important since one does not know ahead of time which will be best

suited to the application.

ACKNOWLEDGMENTS

One of us (JTB) wishes to thank Professors Paul Inglefield and Alan Jones for numerous discussions and explanations of the physics behind CSA tensor line shapes. Also we thank Mr. John Connolly for bringing the problem of parallel mechanisms to our attention. Finally we are grateful to Dr. M.A. Hussain for introducing us to MACSYMA and helping us on many occasions. JTB's participation in this research is part of a collaborative NSF program between General Electric Company and Clark University (DMR-8108679).

REFERENCES

1. W. Kuhn, Kolloid Z., 76, 258 (1936).

2. P.G. de Gennes, "Scaling Concepts in Polymer Physics", (Cornell University Press, Ithaca, 1979), Chapter II.

3. M.F. Shlesinger and E.W. Montroll, Proc. Natl. Acad. Sci., 81, 1280 (1984).

4. S.H. Glarum, J. Chem. Phys., 33, 1371 (1960).

5. B.B. Mandelbrot, "The Fractal Geometry of Nature", (Freeman, New York, 1982).

6. W. Feller, "An Introduction to Probability Theory and Its Applications", (John Wiley, New York, 1966), Volume II.

7. E.W. Montroll and J.T. Bendler, J. Statistical Physics, 34, 129 (1984).

8. M. Dishon, G.H. Weiss, and J.T. Bendler, J. Res. Natl. Bur. Stand., (In Press).

9. P.T. Inglefield, A.A. Jones, R.P. Lubianez, and J.F. O'Gara, Macromolecules, 14, 288 (1981).

10. A.A. Jones, J.F. O'Gara, P.T. Inglefield, J.T. Bendler, A.F. Yee, and K.L. Ngai, Macromolecules, 16, 658 (1983).

11. P.T. Inglefield, R.M. Amici, J.F. O'Gara, C.C. Hung, and A.A. Jones, Macromolecules, 16, 1552 (1983).

12. J.F. O'Gara, A.A. Jones, C.C. Hung, and P.T. Inglefield, Macromolecules, (In Press).

13. E.W. Montroll and G.H. Weiss, J. Math. Phys., 6, 167 (1965).

14. A summary of results on Levy densities (prior to the discovery of their connection with glass-relaxation and polymers) with references to early literature is found in E.W. Montroll and B.J. West, Studies in Statistical Mechanics, Vol. VII (Eds. E.W. Montroll and J.L. Lebowitz) (North Holland, New York, 1979) chapter 2.

15. J.T. Bendler, J. Statistical Physics, 36, 625 (1984).

9

STABILITY ANALYSIS AND OPTIMAL CONTROL OF A PHOTOCHEMICAL HEAT ENGINE

Stanley J. Watowich, Jeffrey L. Krause, R. Stephen Berry

Department of Chemistry and the James Franck Institute,
The University of Chicago,
Chicago, Illinois 60637

ABSTRACT

We examine a class of heat engines in which selectively absorbed radiant energy drives an exothermic reaction. The chemical reactor, a cylinder fitted with a piston, incorporates the dissipative losses of friction and heat conduction. Analysis by computer algebra yielded an algorithm for performing a general linear stability analysis of the system. Bifurcation sets mapping regions of single and multiple steady states are generated. In regions sustaining multiple steady states, driving the engine in a cycle about an unstable steady state generates net power output. Optimal control analyses determine piston trajectories yielding maximum power. A linear stability analysis of the optimally controlled system divides the parameter space into regions where the behavior of a steady state moves from an unstable focus to an unstable node. Using parameter sets which map to the unstable focus, the explicit optimal piston trajectory is determined numerically.

INTRODUCTION

This paper describes the incorporation of computer algebra in recent studies of finite-time thermodynamic processes. As a case specific example, we discuss how the determination of the optimal performance of light-driven, dissipative heat engines is aided by the computer algebra program, MACSYMA [1]. We analyze these engines in five steps: Step 1 reduces the real-time system to a generic model, whose salient time-dependent features prove applicable to a broad class of systems. Step 2 determines the stability of the steady state point about which the uncontrolled, free-piston engine operates. Step 3 employs the methods of optimal control theory to construct the system of differential equations defining the piston trajectory that maximizes work output along each stroke of the heat engine's cycle. Step 4 determines the stability of the steady state point about which the optimally controlled engine operates. Step 5 integrates the system of equations to provide a numerical specification of the optimal trajectory. Computations in steps 2 - 4 are performed almost exclusively by computer algebra routines.

We use the tools available in the symbolic algebra package MACSYMA to (i) perform the linear stability analysis, (ii) generate the parameter sets mapping the stability boundaries of the system, (iii) construct the equations for the optimally controlled system and (iv) output the optimal control equations to numerical routines. MACSYMA commands and the intermediate outputs generated by MACSYMA are not recorded in the text of this paper. Instead, the reader is referred to Appendix 1, where we list the computer algebra algorithm supplying the results incorporated in Secs. 4, 6 and 7.

FINITE-TIME THERMODYNAMIC STUDIES

Thermodynamic studies in chemistry, physics and related engineering disciplines have long been concerned with determining the performance of real processes. A central criterion for judging performance has been the comparison of real-time systems to ideal, reversible processes. The reversible case provides an easily calculable measure of the upper bound on the work a process can supply or the lower bound on the work needed to drive a process. The reversible and irreversible work which a system produces can be quite different. For example, a coal fired steam plant in West Thurrock (U.K.) operates at an observed efficiency of 36% [2]. The efficiency for the same plant, based on an ideal Carnot engine operating between the same two heat reservoirs, would be 64.1%. The difference between the observed and ideal performance is considerable, suggesting that substantial improvements in efficiency might be possible. However, to obtain the efficiency implied by the reversible process, the plant would necessarily run infinitely slowly. Real processes proceed at finite rates, so one might imagine that there is a bound, more realistic than the efficiency of the reversible Carnot engine, for the efficiency of the heat engine constrained to operate at the actual rate of the real engine.

During the past decade, research in finite-time thermodynamics has sought to incorporate time and rate constraints into the analysis of process performance. One goal of this work has been to provide more realistic bounds on the performance of real processes than those obtained from reversible limits [3]. In many processes the detailed time path that yields optimal performance is desired. Optimal control theory [4] is used to determine the trajectory which yields the optimal performance. The optimal path contains all the information about the temporal evolution of the modelled process, including upper bounds on the work that the system can provide. Unfortunately, this information typically may only be obtained numerically, after solving a set of coupled, nonlinear ordinary differential equations with mixed boundary conditions.

Some of the goals of a finite-time thermodynamic study of a process are similar to those of an engineering study. However, the approaches of the practicing engineer and theoretical thermodynamicist differ in two ways. The engineer typically attempts to construct a model which is as exact a representation as possible for the system under study. The result is a simulation that is both detailed and specific, which can then be used to optimize the system's parameters to maximize the performance of the real-time process. By contrast, the finite-time approach is concerned with generic models that represent broad classes of real-time processes and describe the dominant features of real systems. The generic models are simplified abstractions that do not contain all the individual details of a specific process. Rather they incorporate only those temporal features or irreversibilities which are central to the class of systems under study. Including in the model only the main sources of dissipation leads to performance bounds which are more realistic and more useful than bounds based on reversible processes. Further, bounds for simplified generic systems are sometimes far more economical to calculate than bounds based on detailed, realistic models. Effective directions for further, more detailed, studies can be determined by comparing the relative importance of the loss terms in the finite-time model. The other difference is that in carrying out a finite-time analysis of a thermodynamic process path, one typically optimizes a control variable all along the path. While some engineering optimizations do involve optimal control analyses, it is more typical for the engineering optimization to assume a path and optimize a set of design parameters.

Recent work in our group [5] has concentrated on a class of light-driven heat engines incorporating the dissipative losses of heat leak and friction. The heat engine considered here is simultaneously coupled to a high and low temperature external reservoir. Since the system is not in equilibrium with both reservoirs, it operates irreversibly; such systems, which would yield no work at all if they where operated reversibly, are called "dissipative heat engines". One realization of such an engine is the thermoacoustic engine of Wheatley et al [6].

We are interested in calculating the maximum work obtainable from such a light-driven heat engine and in elucidating the time-path providing maximum work. Sec. 3 of this paper describes a model for the dissipative heat engine. Sec. 4 contains a stability analysis on this model. Sec. 5 reviews optimal control theory as expressed by the Maximum Principle of Pontryagin. Sec. 6 applies the results of the Maximum Principle to construct the differential equations governing the piston trajectory that provides maximum work. Sec. 7 utilizes stability theory to investigate the behavior of solutions to these differential equations in the various regions of the parameter space. Finally, Sec. 8 shows numerical solutions for the piston trajectory that enables the light-driven engine to deliver maximum work.

187

DESCRIPTION OF THE MODEL

Consider the system shown in Fig. 1. A cylinder of radius r is fitted at one end with a piston of mass m and capped at the opposite end with a transparent window. The cylinder is in contact with an environment of constant temperature T_0 and constant pressure P_0. As usual, the environment serves as an infinite low temperature heat reservoir. Heat flow through the cylinder walls obeys the Fourier cooling expression

$$\frac{dQ_F}{dt} = -\kappa (T - T_0) \tag{1}$$

where T is the temperature of the fluid contained in the cylinder and κ is the thermal conductance of the cylinder walls. The chemical equilibrium

$$A \longleftrightarrow B$$

exists within the cylinder. Additionally, species A enters into the following reaction: $A + light \rightarrow A^* \rightarrow A + heat$. Species A selectively absorbs incident light

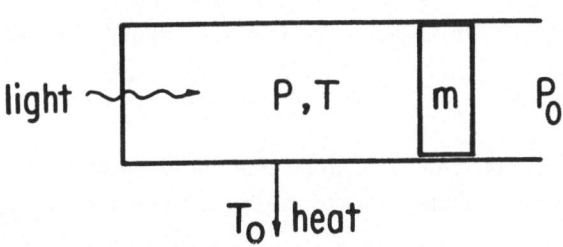

Fig. 1. Light-driven heat engine.

to form some excited species A^*, which then rapidly relaxes back to the ground state of A. We assume that fluorescence is negligible. The absorbed radiant energy is released as heat, which raises the temperature of the system. This mechanism allows the light source to be treated as a high temperature heat reservoir in continuous, reversible contact with the contained fluid. The fluid is spatially homogeneous, in chemical equilibrium and described by the state equation for an ideal diatomic gas. To this point our model resembles the dissipative system studied by Ross and coworkers [7] and the engine used by Mozurkewich and Berry [5]. We extend this model to include the radiant energy contribution to net heat flow as given by the step-heating function

$$\frac{dQ_R}{dt} = \beta \tan^{-1}(\frac{T-T_1}{T_2})$$
(2)

where β is proportional to the radiant heat flux or light intensity. The equilibrium of this system lies far to the right when $T < T_1$ and far to the left when $T > T_1$. Setting $T_1 = T_0$ and $T_2 = 0.1$, the net heat flow, h, given by the sum of Eq. 1 and Eq. 2, reproduces the form of the net heating function for reactions where the higher enthalpy species selectively absorbs the incident light. Examples of such reactions include NO_2 dimerization and SO_3F dimerization.

We assume frictional forces to be proportional to piston velocity [8], $\Upsilon/\pi r^2$, where Υ is the time rate of change of the displacement of the cylinder, V. In time t, the frictional work, W_f, is expressed as

$$W_f = \int_0^t \alpha (\frac{\Upsilon}{\pi r^2})^2 dt$$
(3)

where α is the friction coefficient along the piston. Equation 3 corresponds to the work done by a well-lubricated sliding surface.

The dynamical equations governing this light-driven heat engine are

$$\frac{dT}{dt} = - \frac{1}{NC_V} (\frac{NRT\Upsilon}{V} - h)$$
(4a)

$$\frac{dV}{dt} = \Upsilon$$
(4b)

and

$$\frac{d\Upsilon}{dt} = \frac{1}{m} ((\pi r^2)^2 (\frac{NRT}{V} - P_0) - \alpha \Upsilon)$$
(4c)

The mole number, N, and the heat capacity, C_V, of the fluid are assumed constant.

Using the dimensionless variables $\xi = \dfrac{V}{\pi r^2 x_0}$, $\Theta = \dfrac{T}{T_0}$, $v = \dfrac{\Upsilon N C_V}{\pi r^2 \kappa x_0}$

and $\tau = \dfrac{t \kappa}{N C_V}$, we rewrite Eqs. 4 as

$$\dot{\Theta} = \frac{-\beta_1 \Theta v}{\xi} + \beta_2 \tan^{-1}\left(\frac{\Theta - 1}{\Theta_2} \right) - \Theta + 1 \tag{5a}$$

$$\dot{\xi} = v \tag{5b}$$

and

$$\dot{v} = \frac{\beta_4 \Theta}{\xi} - \beta_5 - \beta_6 v \tag{5c}$$

The dimensionless parameters are $\beta_1 = \dfrac{R}{C_V}$, $\beta_2 = \dfrac{\beta}{\kappa T_0}$,

$\beta_4 = \dfrac{(NC_V)^2 NRT_0}{\kappa^2 x_0^2 m}$, $\beta_5 = \dfrac{\pi r^2 (NC_V)^2 P_0}{\kappa^2 x_0 m}$ and $\beta_6 = \dfrac{\alpha N C_V}{\kappa m}$. The dot nota-

tion signifies the time derivative $\dfrac{d(\ )}{d\tau}$.

STABILITY CONDITIONS FOR THE UNCONTROLLED ENGINE

Ross, et. al. [7] showed that several steady states can exist for a dissipative system similar to the one described above. Mozurkewich and Berry [5] applied this analysis to a frictionless, light-driven engine. They showed that, given the temperature, T, entropy S, and heating function, h, of a system, the necessary condition for a cyclic process to be capable of producing work,

$$\left(\frac{\partial h}{\partial T} \right)_S > 0 \tag{6}$$

is simultaneously satisfied when the middle steady state point of the dissipative system exhibits unstable oscillations. In the engine studied by

Mozurkewich and Berry, radiant energy drives the piston through ever increasing cycles about the middle steady state point. In the course of each cycle, heat flows into the engine at high temperatures and out at low temperatures.

Since the heating function for our system depends on temperature alone, Eq. 6 becomes

$$\left(\frac{\partial h}{\partial T}\right)_S = \left(\frac{\partial h}{\partial T}\right)_V - \left(\frac{C_V \, \kappa_T}{T \, \alpha}\right) \left(\frac{\partial h}{\partial V}\right)_T$$

$$= \left(\frac{\partial h}{\partial T}\right)_V > 0$$

Expressing the above equation in dimensionless variables gives

$$\left(\frac{\partial h}{\partial \Theta}\right)_\xi > 0 \tag{7}$$

as the necessary condition for the engine to be capable of providing work. This condition is met at the steady state point $\Theta = 1.0$. Being a steady state point, heat flow, h, vanishes when $\Theta = 1.0$. Operating as an engine, the system will cycle about this steady state point alternating between temperature regions where heat flows into the system at $\Theta > 1.0$ and out of the system at $\Theta < 1.0$.

Linear stability analysis provides a first-order approximation to the behavior of the step-heated engine about the point $\Theta = 1.0$. A program based on MACSYMA routines (see Appendix 1) performs a general linear stability analysis. The program linearizes a set of n first-order differential equations, determines the characteristic polynomial of the linearized equations about a steady state point and then constructs the Lienard-Chipart inequalities necessary and sufficient for stability. The results presented below are transcribed from MACSYMA output.

The behavior of the engine is indicated by the response of the system to small perturbations about the steady state. We describe this response by the characterization given to the steady state point. For example, we characterize a steady state as unstable if, in the presence of small perturbations, the temporal evolution of the equations defining the system diverge from the steady state position.

The system is at a steady state when the first time-derivatives of the equations describing the dynamical system vanish. The steady state points for our system are

$$\beta_2 \tan^{-1}(\frac{\Theta - 1}{\Theta_2}) - \Theta + 1 = 0 \qquad (8a)$$

$$v = 0 \qquad (8b)$$

and

$$\xi = \frac{\beta_4 \Theta}{\beta_5} \qquad (8c)$$

Three positive solutions are found for Eq. 8a when $\Theta_2 = 0.1$ and $0.1 < \beta_2 < 0.6798$. Linearizing Eq. 5 about a steady state point gives

$$\delta\dot{\Theta} = \left[\frac{\beta_2}{((\frac{\Theta-1}{\Theta_2})^2 + 1) \Theta_2} - \frac{\beta_1 v}{\xi} - 1 \right]_{s.s.} \delta\Theta + \left[\frac{\beta_1 \Theta v}{\xi^2} \right]_{s.s.} \delta\xi +$$

$$\left[\frac{- \beta_1 \Theta}{\xi} \right]_{s.s.} \delta v \qquad (9a)$$

$$\delta\dot{\xi} = \delta v \qquad (9b)$$

and

$$\delta\dot{v} = \left[\frac{\beta_4}{\xi} \right]_{s.s.} \delta\Theta + \left[\frac{- \beta_4 \Theta}{\xi^2} \right]_{s.s.} \delta\xi + [- \beta_6] \delta v \qquad (9c)$$

The δ notation means a small perturbation about the steady state. The subscript $s.s.$ implies that the coefficients are evaluated at the steady state points. At the middle steady state point, $\Theta = 1.0$ and $\xi = \frac{\beta_4 \Theta}{\beta_5}$, and at $\beta_1 = 0.4$, the characteristic polynomial of the above linear system of equations is

$$\Lambda^3 + (\beta_6 - 10 \beta_2 + 1) \Lambda^2 + (- 10 \beta_2 \beta_6 + \beta_6 + 1.4 \epsilon) \Lambda +$$

$$(\epsilon - 10 \beta_2 \epsilon) = 0 \qquad (10)$$

where $\epsilon = \dfrac{\beta_5^2}{\beta_4}$.

Necessary and sufficient conditions for all roots of the characteristic polynomial to be stable are provided by the Lienard-Chipart criteria [9] . For the characteristic polynomial

$$F(\Lambda) = a_0 \Lambda^n + a_1 \Lambda^{n-1} + \cdots + a_{n-1}\Lambda + a_n = 0$$

these criteria are given by any of the following relations

$$a_n > 0 , a_{n-2} > 0 , \cdots ; \Delta_1 > 0 , \Delta_3 > 0 , \cdots \qquad (11a)$$

$$a_n > 0 , a_{n-2} > 0 , \cdots ; \Delta_2 > 0 , \Delta_4 > 0 , \cdots \qquad (11b)$$

$$a_n > 0 , a_{n-1} > 0 , a_{n-3} > 0 \cdots ; \Delta_1 > 0 , \Delta_3 > 0 , \cdots \qquad (11c)$$

and

$$a_n > 0 , a_{n-1} > 0 , a_{n-3} > 0 \cdots ; \Delta_2 > 0 , \Delta_4 > 0 , \cdots \qquad (11d)$$

The expressions, $\Delta_i > 0$, are the Hurwitz inequalities, where $\Delta_i = \det(B_i)$ for $1 \leq i \leq n$. The elements of the matrix B_i are defined as $b_{jk} = a_{2k-j}$ for $1 \leq j, k \leq n$ and $b_{jk} = 0$ for $2k-j > n$ or $2k-j < 0$.

The Lienard-Chipart criteria given by Eq. 11b yield

$$a_3 = \epsilon - 10 \beta_2 \epsilon > 0 \qquad (12a)$$

$$a_1 = \beta_6 - 10 \beta_2 + 1 > 0 \qquad (12b)$$

and

$$\Delta_2 = 100 \beta_2^2 \beta_6 - 10 \beta_2 \beta_6^2 - 20 \beta_2 \beta_6 - 4 \beta_2 \epsilon +$$

$$\beta_6^2 + 1.4 \beta_6 \epsilon + \beta_6 + 0.4 \epsilon > 0 \qquad (12c)$$

as the necessary conditions for the steady states of the uncontrolled light-driven heat engine to be stable.

Of particular interest are the stability boundaries where the qualitative behavior of the steady state points of the system change. When $a_n = 0$ and all remaining stability criteria are satisfied, a bifurcation from a stable steady state to an unstable node occurs. When $\Delta_{n-1} = 0$ and all remaining criteria are satisfied, the bifurcation is from a stable steady state to an unstable focus. The form of the stability boundaries are conveniently envisioned by viewing cuts of constant β_2 through the (β_2, β_6, ϵ) parameter space. Fig. 2 shows a representative (β_6, ϵ) plot of the stability boundary.

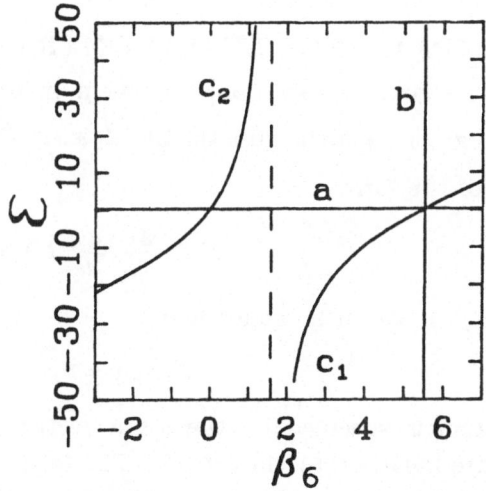

Fig. 2. Stability boundaries in parameter space for the middle steady state of the uncontrolled engine. Steady state values: $\Theta_{s.s.} = 1.0$, $\xi_{s.s.} = \dfrac{\beta_4 \, \Theta_{s.s.}}{\beta_5}$. Parameter values: $\beta_1 = 0.4$, $\beta_2 = 0.65$. Curves a, b and c are the stability boundaries corresponding to the Lienard-Chipart criteria $a_3 = 0$, $a_1 = 0$ and $\Delta_2 = 0$, respectively.

The boundaries, $a_3 = 0$, $a_1 = 0$ and $\Delta_2 = 0$, are represented by curves a, b and c, respectively. For the middle steady state point of the uncontrolled engine to be stable, the (β_6, ϵ) parameter set must simultaneously lie below below line a, to the right of line b and between curves c_1 and c_2. For any given $\beta_2 > 0$, no parameter set (β_0, ϵ) can be chosen such that the middle steady state of the light-driven engine described by Eq. 5 is stable. The middle steady state point is either an unstable node or an unstable focus. Light-driven engines characterized by parameter sets which map into regions of an unstable focus exhibit growing oscillations. In these systems radiant energy drives the engine in a cycle that is capable of providing work.

MAXIMUM PRINCIPLE OF PONTRYAGIN

The general variational problem in optimization theory involves optimizing the known functional $I(\vec{x}, \vec{v}, \frac{d\vec{x}}{dt}, \frac{d\vec{v}}{dt}, t)$, subject to constraints of the form

$$\vec{G}(\vec{x}, \vec{v}, \frac{d\vec{x}}{dt}, \frac{d\vec{v}}{dt}, t) = 0 \tag{13}$$

given the known initial conditions

$$\vec{x}(t_0) = \vec{x}_0 \tag{14}.$$

The system is specified by the state vector $\vec{x}(t)$. The dynamics of the system are incorporated in Eq. 13. The form of the control vector $\vec{v}(t)$ is unknown. The known functional that we optimize is called the objective function.

The maximum principle developed by Pontryagin is well suited to solving variational problems which have only differential constraints. To apply Pontryagin's principle, we determine a control vector $\vec{v}(t)$ that maximizes the objective function

$$I(\vec{x}, \vec{v}) = \int_{t_0}^{t_f} \Phi(\vec{x}, \vec{v}) \, dt + F(\vec{x}(t_f), \vec{v}(t_f))$$

subject to the constraints dictated by the dynamic state equations of the system

$$\frac{d\vec{x}}{dt} = \vec{g}(\vec{x}, \vec{u})$$

The control vector is defined so that it is constrained to the space

$$\vec{U}(\vec{u}) \leq 0 \tag{15}$$

The conditions on the state variables are given by Eq. 14. A new function, called the Hamiltonian after its direct analogy with the Hamiltonian of classical mechanics, is defined by a Legendre transformation

$$H = \Psi + \vec{\lambda} \cdot \vec{g}$$

The control vector that maximizes H simultaneously maximizes the objective function I. The components of the multiplier vector $\vec{\lambda}$ are called adjoint variables. The optimally controlled vector, $\vec{u}(t)$, that maximizes the Hamiltonian satisfies the following necessary conditions:

$$\vec{\lambda}(t_f) = 0 \tag{16}$$

$$\frac{d\vec{x}}{dt} = \frac{\partial H}{\partial \vec{\lambda}} = \vec{g}(\vec{x}, \vec{v}) \tag{17}$$

$$\frac{d\vec{\lambda}}{dt} = -\frac{\partial H}{\partial \vec{x}} \tag{18}$$

and

$$\frac{\partial H}{\partial \vec{v}} = 0 \tag{19}$$

The natural boundary conditions, Eq. 16, are satisfied when $\vec{x}(t_f)$ is not specified. Equations 17 and 18 are the canonical equations of motion that describe the optimal trajectory. The differential equation, Eq. 19, is solved for $\vec{u}(t)$ to obtain an expression for the control vector providing the ($\vec{x}$, $\vec{v}$) trajectory that maximizes H. This is called the optimal path or trajectory. If $\vec{u}(t)$ violates the constraint expressed by Eq. 15, then the necessary condition given by Eq. 19 is replaced by the inequality

$$H(\vec{x}, \vec{u}^*, \vec{\lambda}) \geq H(\vec{x}, \vec{u}, \vec{\lambda}) \tag{20}$$

The Maximum Principle requires this expression to be true for every point along the optimal path. $\vec{u}^*$ is the control vector that maximizes the Hamiltonian and $\vec{u}$ is any control vector permitted by Eq. 15.

OPTIMIZATION OF THE STEP-HEATED ENGINE

We view the step-heated, dissipative engine as a two-stroke cycle composed of a power stroke and a compression stroke. Our objective is to determine the time path of the piston which maximizes work along the cycle. We use the piston velocity, v, as the control variable, and calculate the optimal time path that the piston follows during its cycle. Total cycle time is fixed at a constant value. The initial and final piston positions and fluid temperatures are fixed.

The work output for a fixed cycle time interval, $t_f - t_0$, can be expressed as

$$W = \int_{t_0}^{t_f} \left(\frac{N R T \Upsilon}{V} - \alpha \left(\frac{\Upsilon}{\pi r^2} \right)^2 \right) dt \tag{21}$$

Rewriting Eq. 21 in the dimensionless variables introduced earlier gives

$$\Gamma = \int_{\tau_0}^{\tau_f} \left(\frac{\Theta v}{\xi} - \beta_3 v^2 \right) d\tau \tag{22}$$

where $\Gamma = \dfrac{W}{NRT_0}$ and $\beta_3 = \dfrac{\beta_6}{\beta_4}$. The state variables, Θ and ξ, are subject to the differential constraints given by Eqs. 5a and 5b, respectively. The boundary conditions specifying the cycle are

$$\Theta(\tau_0) = \Theta_0 \tag{23a}$$

$$\xi(\tau_0) = \xi_0 \tag{23b}$$

$$\Theta(\tau_f) = \Theta_f = \Theta_0 \tag{23c}$$

and

$$\xi(\tau_f) = \xi_f = \xi_0 \tag{23d}$$

The Hamiltonian for this system is

$$H = \frac{\Theta v}{\xi} - \beta_3 \xi^2 + \lambda_1 \left(\frac{- \beta_1 \Theta v}{\xi} + \right.$$

$$\left. \beta_2 \tan^{-1}\left(\frac{\Theta - 1}{\Theta_2} \right) - \Theta + 1 \right) + \lambda_2 v \tag{24}$$

The canonical equations for λ_1 and λ_2, conjugate, respectively, to the state variables Θ and ξ, are

$$\dot{\lambda}_1 = \lambda_1 \left(\frac{\beta_1 v}{\xi} - \frac{\beta_2}{\left(\left(\frac{\Theta - 1}{\Theta_2} \right)^2 + 1 \right) \Theta_2} + 1 \right) - \frac{v}{\xi} \tag{25a}$$

and

$$\dot{\lambda}_2 = \frac{\Theta v}{\xi^2} \left(1 - \beta_1 \lambda_1 \right) \tag{25b}$$

Maximizing the Hamiltonian with respect to the control variable v yields the optimal piston trajectory

$$v = \frac{-1}{2 \beta_3 \xi} \left(\beta_1 \lambda_1 \Theta - \lambda_2 \xi - \Theta \right) \tag{26}$$

The time path maximizing work output along each stroke is given by Eqs. 5a, 5b, 25a and 25b subject to the velocity constraint imposed by Eq. 26. We refer to this set of equations as the optimally controlled dynamic system. Analytical solutions are not available for this system of coupled, non-linear ordinary differential equations constrained by mixed boundary conditions.

To solve the system we integrate the equations numerically with an initial-value differential equation solver. A shooting method is used to satisfy the mixed boundary conditions. The values for $\lambda_1(\tau_0)$ and $\lambda_2(\tau_0)$ are initially guessed. The resulting values of $\Theta(\tau_f)$ and $\xi(\tau_f)$ are compared with the desired boundary conditions given by Eqs. 23c and 23d. The guessed initial values, $\lambda_1(\tau_0)$ and $\lambda_2(\tau_0)$, are then modified by a multivariate function optimizing routine, VMCON [10], to minimize the square of the deviation between the desired and resultant final values.

The shooting method does not converge for the nonlinear system described above unless the initial conditions are close to the actual solution. The shooting algorithm performs a finite-step line search about the initial points to determine new initial point values that minimize the square of the deviation between resultant and desired final values. If all values within a finite-step size of the initial points approach a stable steady state as $\tau \rightarrow \tau_f$, then the shooting method fails to find a solution. The shooting method will also fail if, from the guessed initial points, the system exhibits high frequency oscillations as it evolves.

STABILITY ANALYSIS OF THE OPTIMAL SYSTEM

In order for the shooting method to find a solution, we seek trajectories that neither oscillate rapidly nor evolve exponentially towards a stable steady state. Since the trajectory for the optimized cycle operates about the steady state point $\Theta = 1.0$, a linear stability analysis enables us to determine the behavior of the optimal path. The linear stability analysis algorithm employed in Sec. 4 is used to perform the stability analysis presented in this section. The inequalities resulting from the Lienard-Chipart stability criteria allow stability boundaries to be mapped into parameter space. These boundaries delineate regions characterized as an unstable node where the numerical programs will fail to find a solution. Additionally, in regions characterized by an unstable focus the eigenvalues generated by the stability analysis provide a first order approximation to the natural cycle period of the oscillating trajectories. Fixing the cycle time at the system's natural period allows the shooting method to converge rapidly.

The steady state points for the optimal trajectory are

$$\beta_2 \tan^{-1}(\frac{\Theta - 1}{\Theta_2}) - \Theta + 1 = 0 \tag{27a}$$

$$\lambda_1 = 0 \tag{27b}$$

and

$$\lambda_2 = -\frac{\Theta}{\xi} \tag{27c}$$

The steady state temperatures of this system are identical to those found for the underlying system. The steady state conditions do not define a unique equilibrium piston position. Setting $\Theta_2 = 0.1$ we linearize Eqs. 5a, 5b, 25a and 25b, with v defined by Eq. 26, about the steady state point $\Theta = 1.0$, $v = \lambda_1 = 0$ and $\lambda_2 = -\frac{\Theta}{\xi}$. The resulting linear system of equations is given by

$$\delta\dot{\Theta} = \left[\frac{-\beta_1}{2\xi^2\,\beta_3} + 10\,\beta_2 - 1\right]_{s.s.} \delta\Theta + \left[\frac{\beta_1}{2\xi^3\,\beta_3}\right]_{s.s.} \delta\xi +$$

$$\left[\frac{\beta_1^2}{2\xi^2\,\beta_3}\right]_{s.s.} \delta\lambda_1 + \left[\frac{-\beta_1}{2\xi\,\beta_3}\right]_{s.s.} \delta\lambda_2 \tag{28a}$$

$$\delta\dot{\xi} = \left[\frac{1}{2\xi\,\beta_3}\right]_{s.s.} \delta\Theta + \left[\frac{-1}{2\xi^2\,\beta_3}\right]_{s.s.} \delta\xi +$$

$$\left[\frac{-\beta_1}{2\,\xi\,\beta_3}\right]_{s.s.} \delta\lambda_1 + \left[\frac{1}{2\,\beta_3}\right]_{s.s.} \delta\lambda_2 \tag{28b}$$

$$\delta\dot{\lambda}_1 = \left[\frac{-1}{2\,\xi^2\,\beta_3}\right]_{s.s.} \delta\Theta + \left[\frac{1}{2\,\xi^3\beta_3}\right]_{s.s.} \delta\xi +$$

$$\left[\frac{\beta_1}{2\xi^2\,\beta_3} - 10\,\beta_2 + 1\right]_{s.s.} \delta\lambda_1 + \left[\frac{-1}{2\xi\,\beta_3}\right]_{s.s.} \delta\lambda_2 \tag{28c}$$

and

$$\delta\dot{\lambda}_2 = \left[\frac{1}{2\xi^3\,\beta_3}\right]_{s.s.} \delta\Theta + \left[\frac{-1}{2\xi^4\,\beta_3}\right]_{s.s.} \delta\xi +$$

$$\left[\frac{-\beta_1}{2\xi^3\,\beta_3}\right]_{s.s.} \delta\lambda_1 + \left[\frac{1}{2\xi^2\,\beta_3}\right]_{s.s.} \delta\lambda_2 \tag{28d}$$

This system of equations is linear for small perturbations from the steady state point, $s.s.$. For the coefficients in Eqs. 30 subscripted by $s.s.$, ξ refers to the piston position given at the system's steady state. The characteristic polynomial of this system of equations is

$$\Lambda^4 + (10 \, \beta_1 \, \beta_2 \, z - \beta_1 \, z - 100 \, \beta_2^2 + 20 \, \beta_2 - 1) \, \Lambda^2 = 0 \qquad (29)$$

where $z = \dfrac{1}{\xi^2 \, \beta_3}$. When set equal to zero, the term

$$10 \, \beta_1 \, \beta_2 \, z - \beta_1 \, z - 100 \, \beta_2^2 + 20 \, \beta_2 - 1 \qquad (30)$$

describes a stability boundary separating the parameter space (β_1, β_2, z) into regions where the behavior of the steady state point is characterized as either an unstable node, when Eq. 30 is less than zero, or as an unstable focus, when Eq. 30 is greater than zero.

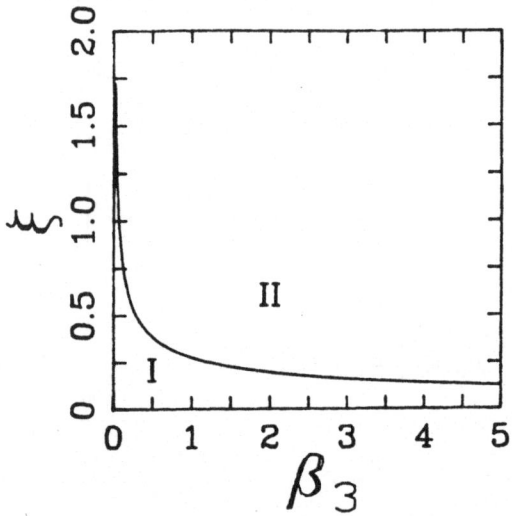

Fig. 3. Stability boundaries in parameter space for the middle steady state of the optimally controlled engine. Steady state values: $\Theta = 1.0$, $\lambda_1 = 0$, $\lambda_2 = \dfrac{-\Theta}{\xi}$. Parameter values: $\beta_1 = 0.4$, $\beta_2 = 0.65$.

The parameter β_1, being the ratio of the ideal gas constant to the constant volume heat capcity, is set equal to 0.4 for the ideal diatomic gas working fluid. The β_2 parameter determines the $\ominus$ separation between the unstable steady state point and the stable steady state points. Setting $\beta_2 = 0.65$ provides a large difference between the $\ominus$ values of the steady state points, yet ensures that these $\ominus$'s remain positive. With fixed β_1 and β_2, the stability boundary formed by setting Eq. 30 equal to zero is shown in Fig. 3. For parameter sets mapping into region I the unstable steady state point of the optimally controlled engine is characterized as an unstable focus. For parameter sets mapping into region II the unstable steady state point is characterized as an unstable node.

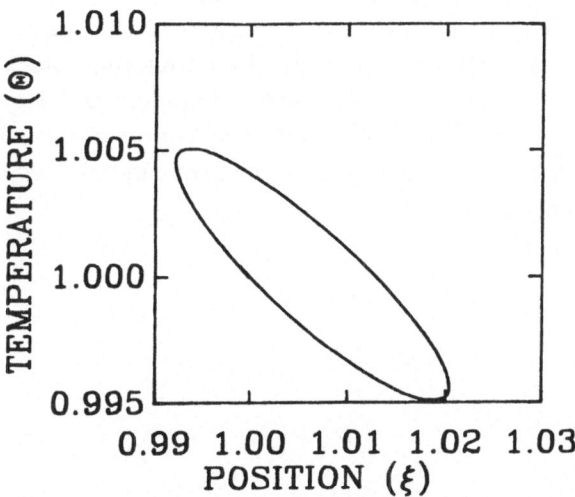

Fig. 4. Optimal piston trajectory for light-driven heat engine. Parameter values: $\beta_1 = 0.4$, $\beta_2 = 0.65$, $\beta_3 = 0.015$. Initial and final cycle constraints: $\ominus(\tau_0) = \ominus(\tau_f) = 1.0$, $\xi(\tau_0) = \xi(\tau_f) = 1.0$.

SUMMARY

Given an arbitrary initial steady state piston position, the stability analysis presented in the previous section guides the choice of parameter values to regions of (β_1, β_2, β_3) space which characterize the steady state as an unstable focus. The trajectory of the optimally controlled engine is oscillatory for small displacements from the unstable steady state. This behavior allows our numerical routine to solve the nonlinear system of differential equations given by Eqs. 5a, 5b, 25a, and 25b with the mixed boundary conditions given by Eqs. 23. The initial and final cycle positions, $\xi(\tau_o)$ and $\xi(\tau_f)$, respectively, are set equal to 1.0 . The initial and final cycle temperatures, $\Theta(\tau_0)$ and $\Theta(\tau_f)$, respectively, are also set equal to 1.0 . The engine parameters are set at $\beta_1 = 0.4$, $\beta_2 = 0.65$ and $\beta_3 = 0.015$. The resulting maximum power trajectory is plotted in Fig. 4. The piston initially compresses the contained working fluid, increasing the internal temperature of the system. This behavior maximizes the engine's potential for providing work output. An in-depth analysis is given by Watowich, Hoffmann and Berry [11] who contrast the trajectories of a light-driven heat engine designed to maximize work output with the trajectories of the heat engine designed to minimize entropy production.

1. EUNICE MACSYMA Release 305, Copyright (c) 1976, 1983 by Massachusetts Institute of Technology, Cambridge, Massachusetts 02139; Enhancements, Copyright (c) 1983 by Symbolics, Inc., 243 Vasser St., Cambridge, Massachusetts 02139.

2. Curzon, F.L. and Ahlborn, B., Am. J. Phys. **43** , 22 (1975).

3 Andresen, B., Salamon, P. and Berry, R.S., J. Chem. Phys. **66** , 1571 (1977); Andresen, B., Berry, R.S., Nitzon, A. and Salamon, P., Phys. Rev. A **15** , 2086 (1977); Gutkowicz-Krusin, D., Procaccia, I. and Ross, J., J. Chem. Phys. **69** , 3898 (1978); Rubin, M.H., Phys. Rev. A **22** , 1741 (1980); Salamon, P., Nitzan, A., Andresen, B. and Berry, R.S., Phys. Rev. A **21** , 2115 (1980).

4. Naslin, P., Essentials of Optimal Control (Boston Technical Publishers, Inc., Cambridge, Massachusetts, 1969); Gottfried, B.S. and Weisman, J., Introduction to Optimization Theory (Prentice-Hall, Inc., Englewood Cliffs, New Jersey, 1973).

5. Mozurkewich, M. and Berry, R.S., J. Appl. Phys. **54** , 3651 (1983).

6. Wheatley, J., Hofler, T., Swift, G.W. and Migliori, A., Phys. Rev. Lett. **50** , 499 (1983).

7. Nitzan, A. and Ross, J., J. Chem. Phys. **59** , 241 (1973); Zimmermann, E.C. and Ross, J., J. Chem. Phys. **80** , 720 (1983).

8. Taylor, C.F., The Internal Combustion in Theory and Practice (MIT, Cambridge, Massachusetts, 1966), Vol. 1, pp. 312-355.

9. Porter, B, Stability Criteria for Linear Dynamical Systems (Oliver and Boyd, Edinburgh, 1967), pp. 80-82.

10. Crane, R.L., Hillstrom, K.E. and Minkoff, M., "*Solution of the General Non Linear Programming Problem with Subroutine VMCON*" (ANL-80-64, Argonne National Laboratory, Argonne, Illinois 1980). The VMCON program was kindly provided by M. Minkoff.

11. Watowich, S., Hoffmann, K.H. and Berry, R.S., manuscript in preparation.

APPENDIX

This appendix contains examples of the routines and
function calls used in the analysis of the light-driven
heat engine. The text is taken directly from MACSYMA
output files. The (c*) lines are inputs to the MACSYMA
program, the (d*) lines are MACSYMA outputs.

The following routine is used to perform the
linear stability analysis of the uncontrolled heat
engine. This proceedure requires the existence of
two files: EQNS containing lines c3 - c7 and STEADY
containing lines C10 - c20. The stability analysis
of the optimally controlled engine likewise utilizes
these files, changing only the system of differential
equations inputted in EQNS.

(c2) BATCH(EQNS) /* input the equations describing the
uncontrolled heat engine */ ;

(c3) H:BETA2*ATAN(((Y[1]-1)/THETA2)-(Y[1]-1);

(d3) $$\text{beta2 atan}\left(\frac{y_1 - 1}{\text{theta2}}\right) - y_1 + 1$$

(c4) 'DIFF(Y[1],T)=-(BETA1*Y[1]*Y[3]/Y[2]-H);

(d4) $$\frac{d}{dt}(y_1) = \text{beta2 atan}\left(\frac{y_1 - 1}{\text{theta2}}\right) - \frac{y_1\, y_3\, \text{beta1}}{y_2} - y_1 + 1$$

(c5) 'DIFF(Y[2],T)=Y[3];

(d5) $$\frac{d}{dt}(y_2) = y_3$$

(c6) 'DIFF(Y[3],T)=BETA4*Y[1]/Y[2]-BETA5-BETA6*Y[3];

(d6)
$$\frac{d}{dt}(y_3) = - y_3 \; beta6 - beta5 + \frac{y_1 \; beta4}{y_2}$$

(c7) EQNLIST:[%TH(3),%TH(2),%TH(1)]$

(d8)   BATCH DONE

(c9) BATCH(STEADY) /* Call the routine to perform the
linear stability analysis of the inputted system of
differential eequations */;

(c10) NEQN:LENGTH(EQNLIST)$

(c11) FOR I:1 THRU NEQN DO ODE[I]:RHS(EQNLIST[I])$

(c12) DEPENDS(ODE,Y)$

(c13) FOR I:1 THRU NEQN DO BLOCK(FOR J:1 THRU NEQN DO
R[J,I]:DIFF(ODE[J],Y[I]))$

(c14) LIN_MATRIX:GENMATRIX(R,NEQN,NEQN);

(d14)

$$
\begin{bmatrix}
\dfrac{beta2}{\left(\dfrac{(y_1 - 1)^2}{theta2^2} + 1\right) theta2} & \dfrac{y_3 \; beta1}{y_2} - 1 & \dfrac{y_1 \, y_3 \; beta1}{y_2^{\,2}} & -\dfrac{y_1 \; beta1}{y_2} \\[4ex]
O & & O & 1 \\[2ex]
\dfrac{beta4}{y_2} & & -\dfrac{y_1 \; beta4}{y_2^{\,2}} & -beta6
\end{bmatrix}
$$

/* evaluate the linearized system of equations at the
middle steady state point */ ;

(c15) MATRIX_SS:EV(LIN_MATRIX,THETA2=0.1,Y[3]=O,

Y[2]=BETA4*Y[1]/BETA5,BETA1=0.4,Y[1]=1,NUMER,EVAL)

$$
\text{(d15)} \quad
\begin{bmatrix}
10.0\ \text{beta2} - 1 & O & - \dfrac{0.4\ \text{beta5}}{\text{beta4}} \\
O & O & 1 \\
\text{beta5} & - \dfrac{\text{beta5}^2}{\text{beta4}} & - \text{beta6}
\end{bmatrix}
$$

(c16) CHAR:EV(EXPAND(CHARPOLY(MATRIX_SS,LAMBDA)),

      NUMER,EVAL)$

(c17) IF COEFF(CHAR,LAMBDA,NEQN)<O THEN CHAR:-1*CHAR
ELSE CHAR:CHAR;

(d17) $\text{lambda}^3 + \text{beta6}\ \text{lambda}^2 - 10.0\ \text{beta2}\ \text{lambda}^2 + \text{lambda}^2$

$- 10.0\ \text{beta2}\ \text{beta6}\ \text{lambda} + \text{beta6}\ \text{lambda} + \dfrac{1.4\ \text{beta5}^2\ \text{lambda}}{\text{beta4}}$

$- \dfrac{10.0\ \text{beta2}\ \text{beta5}^2}{\text{beta4}} + \dfrac{\text{beta5}^2}{\text{beta4}}$

(c18) FOR I:O THRU NEQN DO A[I]:COEFF(CHAR,LAMBDA,NEQN-I)$

(c19) B[J,K]:=IF (2*K-J) < O OR (2*K-J) > NEQN THEN O
ELSE A[2*K-J]$

(c20) HURWITZ[I]:=EXPAND(DETERMINANT(GENMATRIX(B,I,I)))$

(d21)   BATCH DONE

(c22) A[O];

(d22)   1

(c23) A[1];

(d23)   beta6 - 10.0 beta2 + 1

(c24) A[2];

$$(d24) \quad -10.0 \ \text{beta2 beta6} + \text{beta6} + \frac{1.4 \ \text{beta5}^2}{\text{beta4}}$$

(c25) A[3];

$$(d25) \quad \frac{\text{beta5}^2}{\text{beta4}} - \frac{10.0 \ \text{beta2 beta5}^2}{\text{beta4}}$$

(c26) HURWITZ[2];

$$(d26) \quad -10.0 \ \text{beta2 beta6}^2 + \text{beta6}^2 + \frac{1.4 \ \text{beta5}^2 \ \text{beta6}}{\text{beta4}} +$$

$$100.0 \ \text{beta2}^2 \ \text{beta6} - 20.0 \ \text{beta2 beta6} + \text{beta6} -$$

$$\frac{4.0 \ \text{beta2 beta5}^2}{\text{beta4}} + \frac{0.4 \ \text{beta5}^2}{\text{beta4}}$$

The following MACSYMA output generates the system of differential equations for the optimal control analysis. The hamiltonian is contained in lines c7 - c11, which is entered through the batch file HAMILTON.

(c2) BATCH(TESTMAX) /* batch in the routines to form the system of state and adjoint differential equations */ ;

(c3) KEEPFLOAT:TRUE$

(c4) DEPENDS(HAMIL, [Y,U], YPRIME,Y) $

(c5) DEPENDS(H, [Y[1],Y[2]]) $

(c6) BATCH(HAMILTON) /* batch in the hamiltonian for the system under study */ ;

(c7) NDE:2 /* number of state equations in system under study */ $

(c20) BATCH(OUT) /* routine which normally outputs the differential equations directly to fortran numerical programs; for purposes of this example the equations are only displayed in fortran compatible form */ ;

(c21) FORTRAN(H:H) ;

    h = beta2*atan((y(1)-1)/theta2)-y(1)+1

(d21)   done

(c22) FORTRAN(U:A[-2]) /* equation for velocity giving maximum work output */ ;

    u = - (y(1)*y(3)*beta1-y(2)*y(4)-y(1))/(y(2)*beta3)/2.0

(d22)   done

(c23) FOR I:1 THRU 2*NDE DO FORTRAN(YPRIME[I]:YPRIME[I])
/* state equations for maximum work output */

    yprime(1) = -y(1)*beta1*u/y(2)+beta2*atan(((y(1)-1)/theta2)-
    1   y(1)+1

    yprime(2) = u

/* adjoint equations for maximum work output */

    yprime(3) = y(3)*(beta1*u/y(2)-beta2/(((y(1)-1)**2/theta2
    1   **2+1)*theta2)+1)-u/y(2)

    yprime(4) = y(1)*u/y(2)**2-y(1)*y(3)*beta1*u/y(2)**2

(d23)   done

(d24)   BATCH DONE

(c8) DQ:BETA2*ATAN((Y[1]-1)/THETA2)$

(c9) Q:1-Y[1]$

(c10) H:Q+DQ$

(c11) HAMIL:Y[1]*U/Y[2]-BETA3*U**2+Y[3]*(BETA1*Y[1]*

U/Y[2]-H)/(-1)+Y[4]*U

/* hamiltonian to maximize the work output of the light-driven

heat engine */ ;

$$
(d11) \quad -\ beta3\ u^2\ -\ y_3 \left(\frac{y_1\ beta1\ u}{y_2} -\ beta2\ atan\left(\frac{y_1 - 1}{theta2}\right) + y_1 - 1 \right)
$$

$$
+\ y_4\ u + \frac{y_1\ u}{y_2}
$$

(d12)   BATCH DONE

(c13) A[0]:DIFF(HAMIL,U)=0;

$$
(d13) \qquad\qquad -\ 2\ beta3\ u - \frac{y_1\ y_3\ beta1}{y_2} + y_4 + \frac{y_1}{y_2} = 0
$$

(c14) SOLVE(A[0],U);

$$
(d14) \qquad\qquad \left[u = -\ \frac{y_1\ y_3\ beta1 - y_2\ y_4 - y_1}{2\ y_2\ beta3} \right]
$$

(c15) A[-1]:LAST(%)$

(c16) A[-2]:RHS(A[-1])$

(c17) FOR I:1 THRU NDE DO YPRIME[I+NDE]:-DIFF(HAMIL,Y[I])$

(c18) FOR I:NDE+1 THRU 2*NDE DO YPRIME[I-NDE]:DIFF(HAMIL,Y[I])$

(d19)   BATCH DONE

10

FOURIER TRANSFORM ALGORITHMS FOR SPECTRAL ANALYSIS DERIVED WITH MACSYMA

ROBERT H. BERMAN

Massachusetts Institute of Technology, Cambridge, MA 02139 U.S.A.

ABSTRACT

The general problem of computing the power spectrum from N samples of a signal $x(t)$ by Fourier Transform (FT) methods is considered. It is commonly, and erroneously, believed that fast (FFT) methods apply efficiently only when N is a power of 2. Instead, I describe several techniques, using MACSYMA , applicable for computing transforms, correlation functions and power spectra that are faster than radix-2 FFTs. I also mention the possibility of a quasi-analytic B-spline method for computing transforms. For these problems, I will illustrate certain systematic techniques to generate optimized FORTRAN code from MACSYMA .

One technique uses $(1, \exp(2\pi i/3))$ as a basis for complex numbers instead of the more familiar $(1, \exp(\pi i/2))$ to derive radix-3 and radix-6 FFTs *without* multiplications. A second technique rederives the Good algorithm for computing the discrete Fourier transform when N can be factored into relatively prime factors. Finally, I describe the automatic generation of complete FORTRAN programs within MACSYMA that implement the Good algorithm.

Taken together, these automatic program generating examples illustrate several ideas important for further development of general automatic facilities for organizing, documenting, and producing run-time efficient large scale FORTRAN codes.

INTRODUCTION

The purpose of this paper is to describe certain symbol manipulation capabilities of MAC-SYMA for writing complete FORTRAN codes that are used for computing discrete Fourier transforms (DFT) of arbitrary length. In contrast to most of the other papers at this meeting, the emphasis here is on the capabilities for generating optimized FORTRAN code for applications on supercomputers, like the Cray-1, or array processors, rather than on capabilities in MAC-SYMA for algebraic simplification or calculus. However, the MACSYMA techniques that are discussed for promoting automatic program generation facilities are not restricted for use on the Cray architecture. Rather, the Cray is used as a specific target machine to illustrate these ideas that can be of general usefulness in other applications.

There are several important reasons for having a DFT subroutine which works with an arbitrary length signal L, rather than a radix-2 DFT where L is restricted to be a power of 2.

210

It is also important that the resulting FORTRAN code be optimized for execution efficiency. Since code optimization techniques can be very specialized depending on the target machine architecture, it is important to demonstrate that they can be automated, thereby reducing costs to humans in time and error rates. Furthermore, it is important that these techniques be available for systematic application, rather than treated as special case tricks. As an illustrative case, when these techniques are incorporated into a symbolic manipulation environment like MACSYMA they can be fine-tuned to produce efficient code for a variety of machines, since these optimizations are based on mathematics to which compilers typically have no access.

There are at least two distinct viewpoints about what constitutes optimized FORTRAN code. One traditional view of numerical analysts is that optimzing means producing code that is appropriate for stable numerical evaluation and that minimizes the computational complexity (floating point operation count). A classical example of this is using Horner's rule to evaluate finite power series in one variable. There are already primitive capabilities within MACSYMA for Horner's rule and for the extraction of common subexpressions from more complex expressions that generally work in a wide majority of applications [1, 2]. A second view about optimizing code is that opitimizing means reducing the running time or cpu time of the FORTRAN code that implements the algorithm. In a number of important cases, this second viewpoint leads to the surprising result that *more* code rather than less leads to less running time because of certain nonlinear behaviors of the computer hardware [3, 4, and 5]. For example, on the Cray-1, vectorizing certain loops, by writing more FORTRAN statements, leads to factors of 2 to 3 improvement in running actual production code [5, 4]. On a parallel processor, the choice of relaxation algorithms instead of direct methods for solving elliptic equations can be faster even though relaxation algorithms perform more arithmetic operations because of the parallelism in the algorithm and hardware [3, 6, and 7].

There are a number of very important issues associated with the generation of optimized FORTRAN, or some other language, as the output of a symbolic calculation. For example, one class of users are those who develop large physics codes based on finite difference or finite element approximations to partial differential equations. Three important objectives for these users in using a symbolic computing system are: (1) to formulate the physical model with mathematics (*e.g.*, partial differential equations) and simplify the resulting equations; (2) to apply space and time discretization techniques; and (3) to generate FORTRAN, or some other high level language, that is well suited for the hardware of the numerical machine [1, 2, 7, 8, and 6]. It is not within the scope of this paper to discuss certain issues associated with the choice of retaining FORTRAN as a high level programming language for emerging supercomputer machine architectures, but Refs. [7] and [6] discuss several important considerations for a computational physicist to consider.

What is important is to point out that an automatic programming facility embedded in a symbolic computing environment promotes several important advantages [9]. These include program portability, immutability, and flexibility for diverse machine architectures and program development. They also apply to the software control structure of the program, documentation, and program authentication, testing and verification. I will try to illustrate a number of these ideas as they apply to the DFT subroutines that I describe below.

The choice of examining how to optimize the DFT is motivated by the special needs I have in computing models of plasma turbulence on the Cray-1 at National Magnetic Fusion Energy Center [10, 11]. However, the problems that arise and their solutions in MACSYMA are of interest to a general audience. In this research, it is necessary to distinguish coherent fluctuations in the electrostatic potential $\phi(x, t)$, like waves or modes, from incoherent fluctuations, like clumps or holes, by measuring the power spectrum $S(k, \omega)$ where k is the wavenumber and ω is the frequency. A resolution in *phase velocity* ω/k for this distinction is required since plasma normal modes can be easily excited. The potential $\phi(x, t)$ is derived from large-scale particle-mesh simulations lasting thousands of time steps. In addition to diagnostics at each time step, the simulations use Fourier transform methods for solving elliptic equations with a variety of boundary conditions. This calculation can be characterized by the need to compute DFTs with large and small N as efficiently as possible, *i.e.*, as fast as possible for small N, and without excessive memory costs for large N.

Another constraint is the need to compute DFTs of appropriate size. Since most subroutine libraries provide DFTs that work only when N is a power of 2, it is customary to pad the signal with sufficient zeros to make the signal length L compatible with the DFT program, but at the cost of extra memory. In the case of two or three dimensional transforms, this padding can be very expensive for the extra storage it costs. In certain applications, this padding can cost upto a factor of 8 in storage for three dimensional transforms. In other applications, symmetry arguments can reduce this to a factor of $11/2$. Thus, alternatives to radix-2 transforms are desirable for arbitrary N on the grounds of speed and memory costs.

The rest of this paper is organized as follows. First, I will review a number of mathematical properties of the DFT that are relevant to the computation of correlation functions and power spectra. I will also describe briefly the application of B-splines to computing a quasi-analytic transform. Next, I will indicate how very fast DFTs may be be computed when N is small ($N \leq$ 16). Then I will describe the Cooley-Tukey algorithm that forms the basis of most composite N DFT codes. I will then show how it can be improved with the eponymous Good algorithm. [12, 13, 14, 15, 16, 17, 18, and 19]. Next, I describe how to generate complete FORTRAN code from MACSYMA using the Good algorithm. Finally, I conclude with a few remarks about the advantages of having a systematic automatic code generation facility embedded in a symbolic computing environment.

SOME PROPERTIES OF THE DFT RELEVANT TO SPECTRAL THEORY

Suppose that there is a (possibly complex) time series $x(t)$ and that only L of its samples $x_j = x(j\Delta t)$ for $j = 0, 1, \ldots, L-1$ are available. The DFT of the L values in the series $\{x_j\}$ is

$$X_k = \sum_{j=0}^{L-1} \exp\left(\frac{2\pi i k j}{L}\right) x_j. \tag{1}$$

For notational convenience, I will sometimes write $W_N = \exp(2\pi i/N)$.

The inverse discrete transform (IDFT) exists and can be written in a similar form to the DFT:

$$x_k = \frac{1}{L} \sum_{j=0}^{L-1} W_L^{-kj} X_j. \tag{2}$$

Once a subroutine exists for computing the DFT, the IDFT can also be computed since $\text{IDFT}\{x\} = L^{-1}\text{DFT}\{x^*\}^*$ using the symmetry of complex conjugation. Note that the sequence X_k and its inverse x_j defined in this way are *periodic* with period L. This property that the DFT and its inverse represent the periodic extension of $\{x\}$ is true even though the original continuous signal $x(t)$ is not periodic. In particular, $X_{-k} = X_{L-k}$ and $x_{-k} = x_{L-k}$.

Other symmetry properties of the DFT follow naturally. For example, if x_j is real, $X_{-k} = X_{L-k} = X_k^*$.

The product of the DFTs of two sequences is the DFT of the circular convolution of those two sequences. This property follows from direct substitution, since

$$\frac{1}{L} \sum_{k=0}^{L-1} x_k y_{j-k} = \frac{1}{L} \sum_{k=0}^{L-1} \left[\frac{1}{L} \sum_{n=0}^{L-1} W_L^{-nk} X_n\right]\left[\frac{1}{L} \sum_{m=0}^{L-1} W_L^{-m(j-k)} Y_m\right]$$

$$= \frac{1}{L^3} \sum_{n=0}^{L-1} \sum_{m=0}^{L-1} X_n Y_m W_L^{-mj} \left[\sum_{k=0}^{L-1} W_L^{-k(n-m)}\right]. \tag{3}$$

Now the last sum in brackets is equal to L only when $n = m$ and is 0 otherwise. Thus, one obtains

$$\frac{1}{L} \sum_{k=0}^{L-1} x_k x_{j-k} = \frac{1}{L^2} \sum_{n=0}^{L-1} W_L^{-nj} X_n Y_n. \tag{4}$$

In an analogous manner, the DFT of the correlation function or lagged product of two sequences is the product of the DFT of one sequence and the DFT of the other, but with a negative argument:

$$\frac{1}{L} \sum_{k=0}^{L-1} x_{j+k} y_k = \frac{1}{L} \sum_{k=0}^{L-1} x_k y_{k-j} \qquad \text{has the DFT} \qquad X_n Y_{-n}$$

$$\frac{1}{L} \sum_{k=0}^{L-1} x_k y_{k+j} = \frac{1}{L} \sum_{k=0}^{L-1} x_{k-j} y_k \qquad \text{has the DFT} \qquad X_{-n} Y_n. \tag{5}$$

NEED FOR EFFICIENT DFT CALCULATIONS OF CORRELATIONS

For spectral analysis, the circular correlation function C_k^{circular} is estimated by the circular lagged product Eq. (5) above and the power spectrum S_k^{circular} is estimated as the DFT of the correlation function C^{circular}. However, in many applications, there is no physical basis for assuming the signal $x(t)$ or its samples $\{x\}$ have a periodic extension. Thus, it is necessary to compute an aperiodic correlation function and power spectrum.

Through suitable modification, it is possible to use circular convolution to calculate the aperiodic convolution or lagged product of two series when each aperiodic series has zero value everywhere outside some finite interval [20, 19, and 21]. Suppose, x is known for M values and y is known for N values. The result of aperiodically convolving $\{x\}$ and $\{y\}$ can be obtained from the result of circular convolution of suitably augmented sequences. Let the augmented periodic sequences have length $L \geq M + N - 1$ and be called $\{x'\}$ and $\{y'\}$. Let the augmented periodic sequence $\{x'\}$ of period L be defined as

$$x'_j = \begin{cases} x_j, & \text{for } 0 \leq j \leq M - 1 \\ 0, & \text{for } M \leq j \leq L - 1 \end{cases} \tag{6}$$

and the augmented periodic series $\{y'\}$ be defined analogously. The zero values allow the two series to be totally non-overlapped for at least one lagged product even though the lag is a circular one. While the result is a periodic series, each period is the desired aperiodic calculation.

Therefore, the typical calculation to estimate the (aperiodic) power spectrum of a signal $x(t)$ by direct methods consists of the following steps, although other techniques, such as the overlap-save method [13], may be more efficient for large L.

1. Take L samples $\{x\}$ and form a new periodic sequence $\{x'\}$ of length $2L$ by padding $\{x\}$ with L zeroes.

2. Compute the (aperiodic) correlation function of $\{x'\}$ and its power spectrum by (aperiodic) techniques described above, i.e., compute the DFT of $\{x'\}$. The power spectrum is estimated by L of the values of $|X_k|^2$. The other L values are useless for the aperiodic case. Similarly, the correlation function has L values where are half of the IDFT of the power spectrum; the other half are useless.

In many applications, it is very costly in memory to pad a sequence with zeros in order to estimate the aperiodic transform with a DFT. These costs are higher for two or three dimensional transforms, and it is not always possible to afford this when the only DFT subroutines available assume L is a power of 2. Furthermore, it may be inefficient to pad a series with zeros when N is small, but slightly larger than a power of 2.

One novel idea I have recently developed for use in spectral applications to avoid the DFT of a padded series is to use a quasi-analytic correlation function [22]. The idea is to develop an (aperiodic) quasi-analytic approximation to the signal $x(t)$ from its samples $\{x\}$ and compute the correlation function and power spectrum analytically in terms of that approximation.

For example, if we approximate the (aperiodic) signal $x(t)$ by B-splines of order k at the uniformly spaced N points $j\Delta t$, we have

$$x'(t) = \sum_{j=0}^{N-1} \alpha_j B_j^k(t) \tag{7}$$

where B^k are k-th order splines on this set of points, and the α_j are spline coefficients determined from $\{x_j\}$ with the boundary conditions that $x'(t)$ is 0 for $t \leq 0$ and for $t \geq N\Delta t$ (Ref. [23] has discussed this idea for cubic splines but without proper attention to boundary conditions.) In the case of a periodic signal $x(t)$, periodic boundary conditions can be applied to the B-splines. For homogeneous boundary conditions, it is necessary to invert a tridiagonal matrix of size $N \times N$ to find the α_j. The analytical Fourier transform $X'(f)$ of x' can be computed exactly in MACSYMA and the result is

$$X'(f) = \int \exp(2\pi i f t) x'(t)\, dt = D^k(f) \sum_{k=0}^{N-1} \alpha_k \exp(2\pi i f k \Delta t) \tag{8}$$

where the window function $D^k(f)$ is known analytically and depends on the order of the spline approximation. For cubic splines, $D^k(f)$ is a Parzen window. The continuous power spectrum is then estimated by $|X'(f)|^2$, this method requires the quasi-analytic evaluation of a certain sum. When the sum is computed for a discrete number N of frequencies f, the sum can be evaluated as a DFT of N values $\{\alpha\}$ of the spline coefficients in contrast to the DFT of $2N$ values of a padded signal. Furthermore, this sum can be evaluated directly more efficiently than the DFT method, especially when only a few values, say $N/10$, of the power spectrum are wanted. If N is not a power of 2, an efficient DFT can be used to compute N values of the power spectrum, but the advantage may be minimized if N must be padded with zeros to make a power of 2. A detailed description of this method will discussed elsewhere [22].

Another technique is to use maximum entropy methods or cepstrum methods for power spectrum estimation [18, 24, 25, 26, 27, and 28]. I do not want to describe the maximum entropy or cepstrum methods in any detail here because they can be technically complex. The only comment relevant to this discussion is that these methods typically require evaluation of a variable number M of certain coefficients γ_k. It is then required to evaluate certain sums involving the γ_k that can be transformed into periodic DFTs of length M of $\{\gamma\}$. The calculations proceed on an iterative basis using M. Thus, the maximum entropy or cepstrum methods require frequent evaluation of sums that look like DFTs for variable lengths and they need to be evaluated efficiently.

THE FAST FOURIER TRANSFORM ALGORITHM

The basis of calculating DFTs with fast methods can be illustrated as follows with the Cooley-Tukey algorithm [13, 16, and 12]. Consider a DFT X_k of length N, when N is composite $N = N_1 \times N_2$

$$X_k = \sum_{m=0}^{N-1} W_N^{km} x_m. \tag{9}$$

The indices m and k may be redefined

$$
\begin{aligned}
m &= N_1 m_2 + m_1 & m_1, k_1 &= 0, \ldots, N_1 - 1 \\
k &= N_2 k_1 + k_2 & m_2, k_2 &= 0, \ldots, N_2 - 1.
\end{aligned}
\tag{10}
$$

Note the asymmetry in the transformation of indices. This asymmetry leads to a scrambled order of the DFT. The calculation typically must also be accompanied by some method for permuting the indices back to natural order.

By inserting these transforms into the DFT (9), there results

$$X_{N_2 k_1 + k_2} = \sum_{m_1=0}^{N_1-1} W_{N_1}^{k_1 m_1} W_N^{k_2 m_1} \sum_{m_2=0}^{N_2-1} W_{N_2}^{k_2 m_2} x_{N_1 m_2 + m_1}. \tag{11}$$

This form shows that a one dimensional DFT of length $N = N_1 N_2$ can almost be factored into a two dimensional DFT of size $N_1 \times N_2$ except for a certain number of *twiddle factors* $W_N^{k_2 m_1}$.

The computational complexity of this calculation can be summarized as follows. An unfactored DFT of length N costs N^2 complex multiplications ($\times$) and $N(N-1)$ complex additions (+). Factoring a DFT of length $N = N_1 N_2$ as indicated above costs N_1 DFTs of N_2 terms, N_2 DFTs of N_1 terms, and $N_1 N_2 \times$ for twiddle factors. The result is $N_1 N_2 (N_2 + N_1 + 1) \times$ and $N_1 N_2 (N_1 + N_2 - 2)$ +.

In practice, this algorithm is very potent in reducing costs because the procedure can be applied iteratively when N is highly composite. The desire for homogeneous computational structure motivates applications for N which are powers of an integer, usually taken as 2. However, there are at least two reasons why algorithms with other composite N are to be preferred in specific applications. First, automatic program generating tools for FORTRAN mitigate the need for homogeneous control structures, as will be seen below in the Good algorithm. Second, for spectral applications, small N DFTs are available for direct use that are more efficient in time or storage costs (or both) than radix-2 DFTs. They can be combined to produce convenient composite N DFTs that use less storage when padding is required than the corresponding radix-2 DFT.

Small N DFTs can be made efficient by exploiting certain symmetries in the phase factor W_N^k [12, 13]. For example, for $N = 4$, all the phase factors are ± 1 or $\pm i$. If the FORTRAN implementation explicitly separates real and imaginary parts of the DFT, precisely *no* complex multiplications are needed. Similarly, symmetries in W_8^k can be exploited so as to minimize multiplications.

A practical FORTRAN code based on the composite Cooley-Tukey algorithm utilizing the small N complex transforms explicitly is given in Appendix B . This code extends the algorithm given by Ref. [29] by giving special treatment to the cases $N = 2, 3, 4, 5, 7$, and 8. The times are competitive with other carefully constructed codes.

Other values of N, like $N = 3$ or $N = 6$ also allow the development of DFTs without multiplications. This can be easily seen because, if $\mu = \exp(2\pi i/3) = W_3$, the phase factors W_3^k and W_6^k can be written entirely as ± 1, $\pm \mu$ or $\pm(1 - \mu)$. If $(1, \mu)$, instead of $(1, i)$ are chosen as a basis for complex numbers, the components may be separated explicitly in the $(1, \mu)$ basis, thereby avoiding multiplications. Thus, efficient DFTs can be constructed based on radix-3, radix-4 or radix-6 transforms. Of course, there is a hidden cost in this procedure. It is required to explicitly represent the complex numbers in the $(1, \mu)$ basis. If the result of the transform is to be used for a specific application, like the power spectrum, this cost may be worth absorbing, similar to the cost of bit-reversal. Nevertheless, the fact that radix-3 and radix-6 algorithms can be developed for long signals (with $N = 6^k 3^m$ for some k, m) that minimize mutliplications may overcome this cost for changing basis. The cost of the transformation of basis appears to be too high when this radix-3 or radix-6 algorithm is combined with other N.

To illustrate this algorithm without multiplications, consider the $N = 3$ DFT

$$X_k = \sum_{j=0}^{2} W_3^{jk} x_j. \tag{12}$$

If we explicitly transform the basis of the complex plane from $(1, i)$ to $(1, \mu)$ so that $z = x + iy \rightarrow (a, b) = a + b\mu$, the DFT (12) may be written explicitly as

$$\begin{aligned} X_0 &= x_0 + x_1 + x_2 \\ X_1 &= x_0 + \mu \times x_1 + \mu^2 \times x_2 \\ X_2 &= x_0 + \mu^2 \times x_1 + \mu \times x_2 \end{aligned} \tag{13}$$

in the $(1, i)$ basis and

$$\begin{aligned} (A_0, B_0) &= (a_0 + a_1 + a_2, b_0 + b_1 + b_2) \\ (A_1, B_1) &= (a_0 - b_1 + b_2 - a_2, b_0 + a_1 - b_1 - a_2) \\ (A_2, B_2) &= (a_0 + b_1 - a_1 - b_2, b_0 - a_1 - a_2 + b_2) \end{aligned} \tag{14}$$

in the $(1, \mu)$ basis.

Similarly, the $N = 6$ DFT

$$X_k = \sum_{j=0}^{5} W_6^{jk} x_j = \sum_{j=0}^{5} (1 + \mu)^{kj} x_j \tag{15}$$

may be transformed to minimize multiplications as follows by changing basis and introducing certain temporary variables

$$T1 = (a_0, b_0) + (a_3, b_3); \quad T2 = (a_1, b_1) + (a_4, b_4); \quad T3 = (a_2, b_2) + (a_5, b_5)$$

$$T4 = (a_0, b_0) - (a_3, b_3); \quad T5 = (a_1, b_1) - (a_4, b_4); \quad T6 = (a_2, b_2) - (a_5, b_5)$$

$$T7 = T2 - T3; \quad T8 = T5 + T6$$

$$(A_0, B_0) = T1 + T2 + T3$$

$$(A_1, B_1) = T4 + (1 + \mu)T5 + (1 + \mu)^2 T6 = T4 + T5 + \mu T8 \tag{16}$$

$$(A_2, B_2) = T1 + (1 + \mu)^2 T2 + (1 + \mu)^4 T3 = T1 - T3 + \mu T7$$

$$(A_3, B_3) = T4 + (1 + \mu)^3 T5 + (1 + \mu)^6 T6 = T4 - T5 + T6$$

$$(A_4, B_4) = T1 + (1 + \mu)^4 T2 + (1 + \mu)^8 T3 = T1 - T2 - \mu T7$$

$$(A_5, B_5) = T4 + (1 + \mu)^5 T5 + (1 + \mu)^{10} T6 = T4 - T6 - \mu T8$$

Since $\mu z = \mu(a, b) = (-b, a - b)$, all multiplications can be avoided by explicitly separating the components in the $(1, \mu)$ basis.

The investigation of these properties in MACSYMA is made possible by the pattern matching capabilities expressed in the code fragment in Appendix A.

The results of systematically investigating the DFTs for small N in order to minimize complex operations can be summarized in the following table (c.f. Refs. 12, 13), where I have also included for comparison the operation counts for radix-2 DFT. The counts for the $N = 3$ and $N = 6$ DFTs are in the $(1, \mu)$ basis while the numbers in parentheses are in the $(1, i)$ basis.

Table 1. Operation Counts for FFTs

	FAST DFT			RADIX-2 DFT		
	Complex Multiplications $\times$	Complex Additions $+$	Total Operations	Complex Multiplications $\times$	Complex Additions $+$	Total Operations
N				$(N/2)\log_2 N$	$N\log_2 N$	$(3N/2)\log_2 N$
3	0(2)	6	6 (8)	4	8	12
4	0	8	8	4	8	12
5	5	17	22	12	24	36
6	0(4)	20	20 (24)	12	24	36
7	8	36	44	12	24	36
8	2	25	27	12	24	36
9	10	44	54	32	64	96
16	10	74	84	32	64	96

Using the Good algorithm

840	2,570	9,801	12,371	11,672	17,508	29,180

From this table, one can conclude that small N DFTs can be very fast, and indeed, faster than radix-2 FFTs (except, possibly, for $N = 7$). One can use these DFTs in composite algorithms to perform the DFT for more convenient N in both fewer operations and storage[1]. As a practical example, when $N = 840 = 3 \times 5 \times 7 \times 8$, the total storage is approximately 75% of that of the $N = 1024$ radix-2 DFT needed for computation and the total operation count is less than half that of the $N = 1024$ DFT when using the Good algorithm, described next.

---

[1]When trying to estimte the actual CPU time that result from the operation counts listed here, it is necessary to take into account the costs of array indexing calculations. These costs depend critically on the langauge and hardware details of implementing the algorithm – whether it is in FORTRAN or assmebler language; whether there is specific indexing computing hardware. These associated costs of array indexing are omitted in Table 1.

THE GOOD ALGORITHM

The composite N DFT calculation is somewhat complicated because of the presence of twiddle factors. I now want to turn to an algorithm developed by Good, whose work preceded the Cooley-Tukey algorithm[2], in which the twiddle factors can be eliminated by a new index mapping which results in a more efficient factorization of W_N [17, 16, 14, 18, 12, and 19].

Consider the DFT

$$X_k = \sum_{n=0}^{N-1} W_N^{kn} x_n. \tag{17}$$

Let N have M mutually prime factors $N = N_1 N_2 \cdots N_M$. It is possible to represent uniquely any integer n in the range $[0, N-1]$ by the M-tuple of residues

$$n = (n_1, \ldots, n_M) \quad \text{where} \quad n_j = n \,(\text{mod } N_j). \tag{18}$$

The validity of this representation of n is a consequence of the Chinese Remainder Theorem [30]. Similarly, the index k can be represented as a modular M-tuple.

Another consequence of the Chinese Remainder Thereom is that the modular representation is invertible [30]. There are M numbers μ_j such that $\mu_j \,(\text{mod } N_k) = 1$ if $j = k$ and is 0 if $j \neq k$. Thus, n may be reconstructed from $(n_1, \ldots, n_M)$ by

$$n = \sum_{j=0}^{M-1} n_j \mu_j \,(\text{mod } N). \tag{19}$$

Furthermore, number theoretic arguments show that the μs can be calculated explicitly as

$$\mu_j = \left(\frac{N}{N_j}\right)^{\phi(N_j)} (\text{mod } N) \tag{20}$$

where $\phi(k)$ is Euler's totient function [$\phi(k)$ is the number of integers that have no factors in common with k].

Using the modular representation for the indices, the DFT (17) is transformed into

$$X(k_1, \ldots, k_M) = \sum_{n_1=0}^{N_1-1} \sum_{n_2=0}^{N_2-1} \cdots \sum_{n_M=0}^{N_M-1} x(n_1, n_2, \ldots, n_M) W_N^{kn} \tag{21}$$

---

[2]In their original papers, Cooley and Tukey [16, 14] described their motivation in reference to Good's algorithm [17], but judged the index calculations in factoring kn to be too complex and memory too expensive to precalculate tables [See Eqns. (22) – (24) below]. They considered the computational price of twiddle factors to be worthwhile to avoid the complicated calculations and memory costs involved in factoring and storing permutations of kn. However, the economic situation, especially for modern supercomputers, microcomputers and superminicomputers is quite different in mid-1984 than it was in 1965.

Now, the problem is reduced to factoring the phase W_N. One observation is that $W_N^m = W_N^{[m(\bmod\ N)]}$ so the product kn may be factored modulo N as

$$kn\ (\bmod\ N) = \left(\sum_{j=0}^{M-1} k_j \mu_j\right)\left(\sum_{l=0}^{M-1} n_l \mu_l\right)(\bmod\ N)$$

$$= \left(\sum_{j=0}^{M-1} n_j k_j \mu_j^2\right)(\bmod\ N).$$

(22)

The phase factor W_N may be rewritten and further simplified by taking $R_j = (N/N_j)^{2\phi(N_j)-1}$ $(\bmod\ N_j)$ as

$$W_N^{kn} = W_{N_1}^{R_1 n_1 k_1} \times W_{N_2}^{R_2 n_2 k_2} \cdots \times W_{N_M}^{R_M n_M k_M}.$$

(23)

if this is rewritten with a new index permutation $m_j = R_j n_j\ (\bmod\ N_j)$, we can factor the phase and obtain

$$X(k_1, \ldots, k_M) = \sum_{n_1=0}^{N_1-1} \sum_{n_2=0}^{N_2-1} \cdots \sum_{n_M=0}^{N_M-1} x'(m_1, m_2, \ldots, m_M) W_{N_1}^{R_1 n_1 k_1} \times W_{N_2}^{R_2 n_2 k_2} \cdots \times W_{N_M}^{R_M n_M k_M}$$

(24)

where the sequence x' is just a reindexing of x and all the twiddle factors have been explicitly removed.

In summary, the steps that must be taken to use the Good algorithm in a useful FORTRAN implementation are:

1. Compute the Rs and μs.

2. Use the Rs to compute an M-tuple from the one-dimensional index n (or a one-dimensional FORTRAN index)

$$(m_1, \ldots, m_M) = (R_1 n\ (\bmod\ N_1), \ldots, R_M n\ (\bmod\ N_M))$$

3. Convert the M-tuple (m) to a one-dimensional FORTRAN array index J.

4. Perform the M-dimensional DFT of x' using the one-dimensional FORTRAN indices.

5. Map the one-dimensional FORTRAN index L of X'_L to an M-tuple (k). Map the M-tuple (k) to a one-dimensional index k using the μs

$$k = \sum_{j=0}^{M-1} \mu_j k_j\ (\bmod\ N).$$

(25)

The advantages of the Good algorithm are very simple: a one-dimensional DFT has been mapped into an M-dimensional DFT where each composite DFT is very fast, resulting in a composite DFT without twiddle factors.

The problem areas are also clear:

1. The factors need to be relatively prime.

2. The R_j and μ_j need to be computed.

3. The index transformations need to be converted into FORTRAN notation. Fencepost problems relating FORTRAN indices to the Good indices need to be dealt with explicitly.

4. Modular arithmetic is needed in the index mappings; and

5. There are a variable number of dimensions M to deal with.

These problems can be solved with MACSYMA as follows:

1. It has already been shown that very fast small N DFTs for $N = 2, 3, 4, 5, 6, 7, 8, 9,$ and 16 are available and that it is convenient to build composite N from these factors [12].

2, 3 and 4. MACSYMA has number theoretic tools to precalculate the index mappings with number theoretic tools already present. Further permutations can be made to transform the M-dimensional index into a one-dimensional FORTRAN index. Also, transformations from the FORTRAN index back to the modular representation are required. The fencepost counting problems can be taken into account in the indices; and

5. One can apply symbol manipulation programming techniques such as variable length lists and recursion to deal with variable dimensions.

The following illustrates these calculations for $N = 360$.

MACSYMA Generation of Parts of the Good FORTRAN DFT algorithm

Example: $N = 360 = 5 \times 8 \times 9$. Calculate μ and R

$$\mu_1 = \left(\frac{360}{5}\right)^{\phi(5)} \pmod{360} = 72^4 \pmod{360} = 26,873,856 \pmod{360} = 216$$

$$\mu_2 = \left(\frac{360}{8}\right)^{\phi(8)} \pmod{360} = 45^4 \pmod{360} = 4,100,625 \pmod{360} = 225$$

$$\mu_3 = \left(\frac{360}{9}\right)^{\phi(9)} \pmod{360} = 40^6 \pmod{360} = 4,096,000,000 \pmod{360} = 280$$

$$R_1 = \left(\frac{360}{5}\right)^{2\phi(5)-1} \pmod 5 = 72^7 \pmod 5 = 10,030,613,004,288 \pmod 5 = 3$$

$$R_2 = \left(\frac{360}{8}\right)^{2\phi(8)-1} \pmod 8 = 45^7 \pmod 8 = 373,669,453,125 \pmod 8 = 5$$

$$R_3 = \left(\frac{360}{9}\right)^{2\phi(9)-1} \pmod 9 = 40^{11} \pmod 9 = 41,930,400,000,000,000 \pmod 9 = 7$$

There are two index mappings:

$n \rightarrow (m)$ where the modular index mapping is $m_j = R_j n \pmod{N_j}$ and

modular index $(k) \rightarrow k$ where $k = \sum_{j=0}^{M-1} k_j \mu_j$

Finally, the modular index is converted from a threedimensional FORTRAN index to a one-dimensional FORTRAN array index

$(i_1, i_2, i_3) \rightarrow J$ and in FORTRAN, $x(i_1 + 1, i_2 + 1, i_3 + 1) \rightarrow x(J)$ where $i_1 = 0, \ldots, 4$, $i_2 = 0, \ldots, 7$, $i_3 = 0, \ldots, 8$

and $J = \Phi(i_1, i_2, i_3)$ is the FORTRAN one-dimensional index

$\Phi(i_1, i_2, i_3) = 1 + i_1 + i_2 \times 5 + i_3 \times 40$

There is a similar transformation on output from the FORTRAN index K

to modular index (k) to linear index k

These two mappings can be represented by mappings between FORTRAN indices as

Input: $n \rightarrow (m) \rightarrow J \qquad x(n) \rightarrow x(J)$ and

Output: $K \rightarrow (k) \rightarrow k$.

The indices can be precalculated and stored in tables.

HANDLING A VARIABLE NUMBER OF DIMENSIONS WITH MACSYMA

It is worthwhile to demonstrate one method using MACSYMA code to manipulate a variable number of loops controlling the implementation of the M-dimensional DFT. I want to generate FORTRAN code targeted for the Cray supercomputer for illustrative purposes, although I could have chosen another FORTRAN machine if I had wanted to. In the Cray environment, previous experiences with optimizing FORTRAN code indicates that one should try to compute the one dimension index calculations explicitly, and to write out the innermost part of the nested loops [5, 4]. Thus, the style of the FORTRAN programs generated by MACSYMA is deliberately chosen for a specific FORTRAN environment. It can easily be changed to another style by modifying the inner code generators.

One idea in MACSYMA is use a recursive program to try to generate a variable number of loops and make them somewhat readable for humans. A recursive program to do this needs to pass the following information to the next level of recursion: the current level of loop nesting (J), the permutation of factors N_j indicating the order in which to perform the loops (ORDER, $N[J]$), an offset to control pretty printing (OFF), a set of statement labels (LABEL), and the code kernel or body for the inner loop (BODY). A schematic form of this code looks

```
FFT-GENERATOR(J. LABEL, ORDER, BODY,OFF):= block(
     if J=END then
          (generate-comment("begin....FFT of length n=",N[J]),
           code-generate(BODY),
           perform(FINISH-ACTIONS)),
     else if J=START then
   (perform(START-ACTIONS),
          generate-comment("inner loop is dimension",N[J]),
     FFT-GENERATOR(J-1, LABEL[j-1], ORDER, BODY,OFF[J-1]))
     else
          (code-generate(" do LABEL[J] I[J] = 1, N[J]", OFF[J]),
           FFT-GENERATOR(J-1,LABEL, ORDER, BODY,OFF[J-1]),
           code-generate(" LABEL[J]      continue")))
```

A representative code fragment for $N = 360$ is

```
c    inner loop is dimension    8
     IT2 = - 46 +1
     do  200 I1 = 1, 5
          IT1 = 1 * I1 + IT2
          do  200 I3 = 1, 9
               IT3 = 40 * I3 + IT1
c    begin....FFT of length n=8
                    tr1=xr( IT3 + 5 )+xr( IT3 + 25 )
                    tr2=xr( IT3 + 15 )+xr( IT3 + 35 )
                    tr3=xr( IT3 + 10 )+xr( IT3 + 30 )
                    tr4=xr( IT3 + 10 )-xr( IT3 + 30 )
c.... ETC
```

225

```
200     continue
```

Several features are worth noting. First, the code is indented to make the control structure moderately readable by humans. Next, the index calculations (three-dimensional index to one-dimensional FORTRAN index) are explicitly and systematically done with uppercase dummy variables (e.g., IT1, IT2, and IT3) while the code kernel is done in lowercase. Furthermore, the choice of names, such as IT2, refer to the (second) variable in the input factorization of N. It is also important to observe that the control structure is not homogeneous insofar as the innermost loop is explicitly written out. This potentially results in more FORTRAN code, but faster execution. Without an automatic code generation facility, it could be considered tedious to do this by hand.

An example, showing the output for generating a DFT with $N = 20$ in an intentionally chosen style of FORTRAN is given in Appendix C.

Another important capability for the automatically generated code facility is that code produced this way can be tested and authenticated by a variety of methods. One way is to also produce a test program to check the code numerically in either the code generation environment or the target FORTRAN machine. Another is to check the algorithm symbolically. Yet another way is with a pictorial test rather than a numerical test by producing a graph of known behavior. Each of tests has been performed and the automatically generated code has passed. All tests can be easily and systematically formulated in a symbolic computing environment.

CONCLUSIONS: PEOPLE SHOULD NOT WRITE PROGRAMS; COMPUTERS SHOULD.

I have described several efficient techniques for dealing with the spectral analysis for signals whose length N is not a power of 2. In particular, I have described several MACSYMA investigations of small N DFTs, including the use of $(1, \mu)$ as a basis for complex numbers, and used these DFTs in the Good algorithm to develop arbitrary N DFTs. In doing so, I have described several general and systematic techniques for generating code with MACSYMA that are applicable to other problems. I have described an automatic program generator that produces FORTRAN code optimized for a specific FORTRAN environment. That code generator can be modified to produce code for a different architecture at little cost.

This example has illustrated several important advantages of having an automatic code generation facility. They include:

1, producing portable, mathematically correct FORTRAN code and the ability to reproduce that code;

226

2, minimization of human transcription errors;

3, avoiding mathematical errors in producing constants for the program;

4, automatic checking facilities, including symbolic algorithm testing, test program generations and special case examination;

5, program readability, layout and automatic documentation; and

6, generation of run-time efficient code targeted for, possibly, multiple computer architectures.

All of these features are highly desirable in a complex programming environment. The illustrate the trend that computers can take over certain of the systematic production tasks associated with programming projects. The automatic code generation facility is especially useful in providing linkages among programming and mathematical efficiencies considering symbolic, numeric and FORTRAN aspects of solving problems. The benefits to the user in reducing production costs, program development time and increasing gains in productivity are of obvious value.

ACKNOWLEDGMENTS

It is a pleasure to thank John Aspinall for a number of stimulating discussions, including recursive programming techniques for automatic code generators. This work is supported by the National Science Foundation, Office of Naval Research and the Department of Energy.

APPENDIX A. MACSYMA CODE FRAGMENT FOR MANIPULATING EXPRESSIONS WITH μ

The following code fragment is useful for manipulating expressions involving μ in the investigation of DFTs for $N = 3$ and $N = 6$. The first part of the code fragment modifies the MACSYMA simplifier by inserting the mathematical knowledge needed to transform nonnegative integer powers of μ. It consists of a rule AMU which matches a pattern μ^N and replaces it with the value of a function $UMU(N)$. The second part of the fragment creates a rule to perform the transformation from the $(1, i)$ basis to the $(1, \mu)$ basis.

```
/* transform nonnegative powers of mu to a form linear in mu
   sample usage:    apply1(expand(exp),amu)  */

/* create mathematical information that N is an integer */
matchdeclare(n,integerp);

/* define a rule AMU to replace the pattern mu**N with UMU(N)
          when N is an integer  */
```

```
defrule(amu,mu**n,umu(n));

/* define a function replacement  UMU(n) */
umu(n):=block([rem:abs(remainder(n,3))],
      if rem=0 then 1 else
        if rem=1 then mu else
            if rem=2 then -1-mu);

/* set up pattern variables and match function CMU
    to identify the form A + I*B
    A and B must be expressions independent of I */
  matchdeclare(a,freeof(i),b,freeof(i));
  defmatch(cmu,a+b*i,i);

/* convert x+ i*y --> x + (mu+1/2)*2*y/sqrt(3) = (x+y/sqrt(3), 2y/sqrt(3))
    Sample call:    tomu(x+i*y)                     */
    tomu(exp):=block( [a,b,var,tmp],
        tmp:cmu(exp,i),a:part(tmp,1,2),b:part(tmp,2,2),
  var:part(tmp,3,2),
  if var.ne.i then return(true),
  a + (mu+1/2)*2*b/sqrt(3));
```

APPENDIX B. COMPOSITE FORTRAN FFT FOR ARBITRARY N

The following FORTRAN implementation of the Cooly-Tukey algorithm, motivated by Refs. [29] and [12], explicitly takes into account small N transforms for efficiency.

```
      subroutine zzft(za,zb,n)

c--------------------- fft ---------------------
c
c    transform complex array za with exp(+2*pi*i/N),
c       sum(za(j)*exp(+2*pi*i*(j-1)/N),j,1,n)
c
c    can be used as inverse transform by
c        ift(z) = conjg(fft(conjg(z)))
c
c    normalization is such that   ift(fft(z)) = N *z
c    return result in za
c    zb is always a (complex) work area
c
c    subroutine setdf must be called once to initialize certain
c        constants
c
c--------------------- fft ---------------------

      parameter(npr=12)
```

```
      integer in,n,after,before,next,nextmx,now,prime(npr)
      complex za(n),zb(n)
      data prime/2,3,5,7,11,13,17,19,23,29,31,37/
      after=1
      before=n
      next=1
      in=1
      do 200 k=1,n
200      za(k)=conjg(za(k))
10    if (mod(before,prime(next)).ne.0) then
         next=next+1
         if(next.le.npr) then
              go to 10
         else
              now=before
              before=1
         endif
      else

c   check for power of 2

         if (prime(next).eq.2) then
           if(mod(before,8).eq.0) then
             now=8
             before=before/8
           elseif(mod(before,4).eq.0) then
             now=4
             before=before/4
           elseif(mod(before,2).eq.0) then
             now=2
             before=before/2
           endif
         else
           now=prime(next)
           before=before/prime(next)
         endif
      endif

      if(in.eq.1) then
        call ffstp(za,after,now,before,zb)
      else
        call ffstp(zb,after,now,before,za)
      endif
      in=mod(in,2)+1

c   restore restore to wa

      if(before.eq.1) then
         if (in.eq.2) then
            do 20 k=1,n
20               za(k)=conjg(zb(k))
         else
```

```
      do 201 k=1,n
201        za(k)=conjg(za(k))
      endif

      return
      endif

      after=after*now
      goto 10
      end



      subroutine ffstp(zin,after,now,before,zout)

c   performs 1 step of FFT

      integer after,before,now,ia,ib,in,j
      real angle,ratio,tpi
      complex zin(after,before,now),zout(after,now,before)
      complex arg,omega,value,ang,sqm1,c1,cfac32
      complex cf55,cf53,cf54,cf75,cf76,cf77,cf78,cf82
      complex ang0,ang1,ang2,ang3,bng0,bng1,bng2
      complex t0,t1,t2,t3,t4,t5,t6,t7,t8,t9,t10,t11,t12,t13,
     +        t14,t15,t16,t17,t18,t19,t20,t21,t22,t23,t24,
     +        t25,t26,t27
      complex s0,s1,s2,s3,s4,s5,s6,s7,s8,s9,s10,s11,s12,s13,
     +        s14,s15,s16,s17,s18,s19,s20,s21,s22,s23,s24,
     +        s25,s26,s27
      complex m0,m1,m2,m3,m4,m5,m6,m7,m8,m9,m10,m11,m12,m13,
     +        m14,m15,m16,m17,m18,m19,m20,m21,m22,m23,m24,
     +        m25,m26,m27

      data sqm1,c1/(0.,1.),(1.,0.)/
      data nblz/64/
      data tpi/6.2831853071795864769e0/
      data fac31/-1.5/
      data fac32/8.660254038469492e-1/
      data fac51,fac52,fac53,fac54,fac55/5*0./
      data fac71,fac72,fac73,fac74,fac75,fac76,fac77,
     +     fac78/8*0./
      data fac81,fac82/2*0./
      data fac91,fac92,fac93,fac94,fac95,fac96,fac97
     +     /7*0./
      data fac161,fac162,fac163,fac164,fac165,fac166,
     +     fac167,fac168/8*0./

      angle=tpi/float(now*after)
      omega=cmplx(cos(angle),-sin(angle))
      arg=c1

      if(now.eq.2) then
```

230

```
c     now= 2

      do 190 ia=1,after
         do 180 ib=1,before
            value=zin(ia,ib,2)*arg
            zout(ia,1,ib)=value + zin(ia,ib,1)
            zout(ia,2,ib)=zin(ia,ib,1)-value
180      continue

190      arg=arg*omega

      elseif(now.eq.3) then

c     now= 3

      do 390 ia=1,after
         ang0=arg*cfac32
         do 380 ib=1,before
            t20=zin(ia,ib,3)*arg
            t21=zin(ia,ib,2)
            t1=t21+ t20
            t1=t1*arg
            m0=zin(ia,ib,1)+t1
            m1=fac3i*t1
            m2=(t20-t21)*ang0
            s1=m0+m1
            zout(ia,1,ib)=m0
            zout(ia,2,ib)=s1+m2
            zout(ia,3,ib)=s1-m2
380      continue

390      arg=arg*omega

      elseif(now.eq.4) then

c     now= 4

      do 490 ia=1,after
         ang0=arg
         ang1=arg*arg
         bng0=arg*sqm1
         do 480 ib=1,before
            s1=ang1*zin(ia,ib,3)
            s2=ang1*zin(ia,ib,4)
            s3=zin(ia,ib,1)
            s4=zin(ia,ib,2)
            t1= s3+s1
            t2= s4+s2
            t2=t2*arg
            zout(ia,1,ib)=t1+t2
            zout(ia,3,ib)=t1-t2
```

```
              m2=s3 - s1
              m3=bng0*(s2-s4)
              zout(ia,2,ib)=m2+m3
              zout(ia,4,ib)=m2-m3
480   continue

490   arg=arg*omega


      elseif(now.eq.5) then

c     now= 5

          do 590 ia=1,after
            ang1=arg
            ang2=arg*arg
            ang3=ang2*arg
            do 580 ib=1,before
              t21=ang3*zin(ia,ib,5)
              t22=arg*zin(ia,ib,4)
              t1=zin(ia,ib,2)+t21
              t1=t1*arg
              t2=zin(ia,ib,3)+t22
              t2=t2*ang2
              t3=zin(ia,ib,2)-t21
              t3=t3*arg
              t4=t22-zin(ia,ib,3)
              t4=t4*ang2
              t5=(t1+t2)
              m0=zin(ia,ib,1)+t5
              m1=fac51*t5
              m2=fac52*(t1-t2)
              m3= cf53*(t3+t4)
              m4= cf54*t4
              m5= cf55*t3
              s3=m3-m4
              s5=m3+m5
              s1=m0+m1
              s2=s1+m2
              s4=s1-m2
              zout(ia,1,ib)=m0
              zout(ia,2,ib)=s2+s3
              zout(ia,3,ib)=s4+s5
              zout(ia,4,ib)=s4-s5
              zout(ia,5,ib)=s2-s3
580   continue

590   arg=arg*omega


      elseif(now.eq.7) then

c     now= 7
```

```
        do 790 ia=1,after
c   arg**2 = ang1
        ang1=arg*arg
c   arg**3 = ang2
        ang2=arg*ang1
c   arg**4 = ang3
        ang3=ang1*ang1
c   arg**5 = bng0
        bng0=ang3*arg

        do 780 ib=1,before

            m21=bng0*zin(ia,ib,7)
            m22=ang2*zin(ia,ib,6)
            m23=arg*zin(ia,ib,5)
            t21=zin(ia,ib,2)
            t22=zin(ia,ib,3)
            t23=zin(ia,ib,4)
            t1=arg*(t21 + m21)
            t2=ang1*(t22 +m22)
            t3=ang2*(t23 + m23)
            t4=t1+t2+t3
            t5=arg*(t21 - m21)
            t6=ang1*(t22 - m22)
            t7=ang2*(-t23 + m23)
            t8=t1-t3
            t9=t3-t2
            t10=t5+t6+t7
            t11=t7-t5
            t12=t6-t7
            m0=zin(ia,ib,1)+t4
            m1=fac71*t4
            t13=-t8-t9
            m2=fac72*t8
            m3=fac73*t9
            m4=fac74*t13
            s0=-m2-m3
            s1=-m2-m4
            m5=cf75*t10
            t14=-t11-t12
            m6=cf76*t11
            m7=cf77*t12
            m8=cf78*t14
            s2=-m6-m7
            s3=m6+m8
            s4=m0+m1
            s5=s4-s0
            s6=s4+s1
            s7=s4+s0-s1
            s8=m5-s2
            s9=m5-s3
```

```
            s10=m5+s2+s3
            zout(ia,1,ib)=m0
            zout(ia,2,ib)=s5+s8
            zout(ia,3,ib)=s6+s9
            zout(ia,4,ib)=s7-s10
            zout(ia,5,ib)=s7+s10
            zout(ia,6,ib)=s6-s9
            zout(ia,7,ib)=s5-s8

780    continue

790    arg=arg*omega

       elseif(now.eq.8) then

c    now= 8

          do 890 ia=1,after
            ang1=arg*arg
            t21=sqm1*ang1
            ang2=arg*ang1
            ang3=ang1*ang1
            do 880 ib=1,before
               m21=ang3*zin(ia,ib,5)
               s21=zin(ia,ib,1)
               s22=zin(ia,ib,2)
               s23=zin(ia,ib,3)
               s24=zin(ia,ib,4)
               m22=ang3*zin(ia,ib,6)
               m23=ang3*zin(ia,ib,7)
               m24=ang3*zin(ia,ib,8)
               t1=s21+m21
               t2=ang1 * (s23  + m23)
               t3=arg   * (s22 + m22)
               t4=arg   * (s22 - m22)
               t5=ang2 * (s24 + m24)
               t6=ang2 * (s24 - m24)
               t7=t1+t2
               t8=t3+t5
               m0=t7+t8
               m1=t7-t8
               m2=t1-t2
               m3=s21-m21
               m4=fac81*(t4-t6)
               m5=sqm1*(t5-t3)
               m6=t21*(-s23+m23)
               m7=cf82*(t4+t6)
               s1=m3+m4
               s2=m3-m4
               s3=m6+m7
               s4=m6-m7
               zout(ia,1,ib)=m0
```

```
                zout(ia,2,ib)=s1+s3
                zout(ia,3,ib)=m2+m5
                zout(ia,4,ib)=s2-s4
                zout(ia,5,ib)=m1
                zout(ia,6,ib)=s2+s4
                zout(ia,7,ib)=m2-m5
                zout(ia,8,ib)=s1-s3


880     continue
890     arg=arg*omega

        else

c    otherwise any n will do

        do 100 j=1,now
           do 90 ia=1,after
              do 80 ib=1,before
                 value=zin(ia,ib,now)
                 do 70 in=now-1,1,-1
70                  value=value*arg + zin(ia,ib,in)
80               zout(ia,j,ib)=value
90            arg=arg*omega
100     continue

        endif

        return


        entry setdf
           pi=4.*atan2(1.,1.)
           tpi=2.*pi
c   n=3
              th3= 2.*pi/3.
              fac31= cos(th3)-1.
              fac32= sin(th3)
              cfac32=cmplx(0.,fac32)

c   n=5
              th=2.*pi/5.
              fac51=(cos(th)+cos(2.*th))/2.-1.
              fac52=(cos(th)-cos(2.*th))/2.
              fac53=sin(th)
              fac54=sin(th)+sin(2.*th)
              fac55=sin(th)-sin(2.*th)
              cf53 = cmplx(0.,-fac53)
              cf54 = cmplx(0.,-fac54)
              cf55 = cmplx(0., fac55)
```

```
c   n=7
            th=2.*pi/7.
            fac71=(cos(th)+cos(2.*th)+cos(3.*th))/3.-1.
            fac72=(2.*cos(th)-cos(2.*th)-cos(3.*th))/3.
            fac73=(cos(th)-2.*cos(2.*th)+cos(3.*th))/3.
            fac74=(cos(th)+cos(2.*th)-2.*cos(3.*th))/3.
            fac75=(sin(th)+sin(2.*th)-sin(3.*th))/3.
            fac76=(2.*sin(th)-sin(2.*th)+sin(3.*th))/3.
            fac77=(sin(th)-2.*sin(2.*th)-sin(3.*th))/3.
            fac78=(sin(th)+sin(2.*th)+2.*sin(3.*th))/3.
            cf75=cmplx(0.,-fac75)
            cf76=cmplx(0.,fac76)
            cf77=cmplx(0.,fac77)
            cf78=cmplx(0.,fac78)


c   n=8

            th=2.*pi/8.
            fac81=cos(th)
            fac82=sin(th)
            cf82=cmplx(0.,-fac82)

        return
        end
```

APPENDIX C. A GOOD DFT WITH $N = 20$ GENERATED IN MACSYMA

A full DFT for $N = 20$ is provided in the following code fragment in which the style of FORTRAN is chosen for the Cray environment for illustrative purposes. Other FORTRAN environments could easily have been used.

Note the automatic documentation and layout for readability. The two index mappings for input and output have been precalculated and entered in data statements. The choice of 10 values per data statement is arbitrary and could be assigned differently. Further knowledge about the structure of the code into which this subroutine will be placed permits use of FORTRAN blank COMMON areas for temporary storage.

The calculated loop variables are presented in uppercase while the fourier transform calculations are in lower case. The names of the loops variables (IT1 and IT2) are choosen with reference to the order in which the factors of N appear (here 4 and 5). The first inner loop is a DFT of length 4. Note that there are no multiplications when the real and imaginary parts are explicitly separated. The second inner loop is a DFT of length 5. The numerical constants that occur in the DFT of length 5 are already linear combinations of $\sin(2\pi/5)$, $\cos(2\pi/5)$, $\sin(4\pi/5)$, and $\cos(4\pi/5)$ that result from exploiting symmetries of the trigonometric functions. They are evaluated to 16 decimal places and explicitly written into the code.

```
c... starting to generate N =20
c (C) R.H. Berman, 1984 MIT. all rights reserved
c Number of transform =  20
c Number of factors =  2
c    factors are  (4,5)
c    mu's are  (5,16)
c    R's are  (1,4)


      subroutine  F20  (n,zr,zi)
c
c generation of kin and kout for nc= 20
c nc =  20  =  (4,5)
c    Thursday, Aug 23, 1984  2:06pm
c
      dimension zr(1),zi(1)
      common// xr( 20 ),xi( 20 ),tr( 20 ),ti( 20 )
      integer n1,r1,m1
      dimension n1( 2 ),r1( 2 ),m1( 2 )
c...using 1d index scheme for kin and kout
      dimension kin( 20 ), kout( 20 )
c...generate kin
      data (kin(k),k= 1 , 10 )/
     + 1,18,15,12,5,2,19,16,9,6/
      data (kin(k),k= 11 , 20 )/
     + 3,20,13,10,7,4,17,14,11,8/
c...generate kout
      data (kout(k),k= 1 , 10 )/
     + 1,6,11,16,17,2,7,12,13,18/
      data (kout(k),k= 11 , 20 )/
     + 3,8,9,14,19,4,5,10,15,20/

      n1( 1 )= 4
      n1( 2 )= 5
      r1( 1 )= 1
      r1( 2 )= 4
      m1( 1 )= 5
      m1( 2 )= 16
c    perform transformation k--> (m) -> kin(k)
      do 10 k=1, 20
        kt=kin(k)
        xr(kt)=zr(k)
        xi(kt)=zi(k)

10      continue

      IT1 = - 5 +1
      do  100 I2 = 1, 5
          IT2 = 4 * I2 + IT1
c    begin....FFT of length n=4
```

```
        tr1=xr( IT2 + 1 )+xr( IT2 + 3 )
        ti1=xi( IT2 + 1 )+xi( IT2 + 3 )
        tr2=xr( IT2 + 2 )+xr( IT2 + 4 )
        ti2=xi( IT2 + 2 )+xi( IT2 + 4 )
        xrm0=tr1+tr2
        xim0=ti1+ti2
        xrm1=tr1-tr2
        xim1=ti1-ti2
        xrm2=xr( IT2 + 1 )-xr( IT2 + 3 )
        xim2=xi( IT2 + 1 )-xi( IT2 + 3 )
        xrm3=xi( IT2 + 2 )-xi( IT2 + 4 )
        xim3=xr( IT2 + 4 )-xr( IT2 + 2 )
        xr( IT2 + 1 )=xrm0
        xi( IT2 + 1 )=xim0
        xr( IT2 + 2 )=xrm2+xrm3
        xi( IT2 + 2 )=xim2+xim3
        xr( IT2 + 3 )=xrm1
        xi( IT2 + 3 )=xim1
        xr( IT2 + 4 )=xrm2-xrm3
        xi( IT2 + 4 )=xim2-xim3
c   end  FFT of length n=4

100       continue


        IT2 = - 5 +1
        do  200 I1 = 1, 4
            IT1 = 1 * I1 + IT2
            tr1=xr( IT1 + 8 )+xr( IT1 + 20 )
            tr2=xr( IT1 + 12 )+xr( IT1 + 16 )
            tr3=xr( IT1 + 8 )-xr( IT1 + 20 )
            tr4=xr( IT1 + 16 )-xr( IT1 + 12 )
            tr5=tr1+tr2
            xrm0=xr( IT1 + 4 )+tr5
            xrm1= - 1.25  * tr5
            xrm2= 0.5590169943749474 *(tr1-tr2)
            ti1=xi( IT1 + 8 )+xi( IT1 + 20 )
            ti2=xi( IT1 + 12 )+xi( IT1 + 16 )
            ti3=xi( IT1 + 8 )-xi( IT1 + 20 )
            ti4=xi( IT1 + 16 )-xi( IT1 + 12 )
            ti5=ti1+ti2
            xim0=xi( IT1 + 4 )+ti5
            xim1= - 1.25  * ti5
            xim2= 0.5590169943749474 *(ti1-ti2)
            xrm3=  0.9510565162951535 *(ti3+ti4)
            xim3= - 0.9510565162951535 *(tr3+tr4)
            xrm4=  1.538841768587627 *ti4
            xim4= - 1.538841769587627 *tr4
            xrm5= - 0.3632712640026802  *ti3
            xim5=  0.3632712640026802  *tr3
            sr3=xrm3-xrm4
            sr5=xrm3+xrm5
            si3=xim3-xim4
```

238

```
          si5=xim3+xim5
          sr1=xrm0+xrm1
          sr2=sr1+xrm2
          sr4=sr1-xrm2
          si1=xim0+xim1
          si2=si1+xim2
          si4=si1-xim2
          xr( IT1 + 4 )=xrm0
          xr( IT1 + 8 )=sr2+sr3
          xr( IT1 + 12 )=sr4+sr5
          xr( IT1 + 16 )=sr4-sr5
          xr( IT1 + 20 )=sr2-sr3
          xi( IT1 + 4 )=xim0
          xi( IT1 + 8 )=si2+si3
          xi( IT1 + 12 )=si4+si5
          xi( IT1 + 16 )=si4-si5
          xi( IT1 + 20 )=si2-si3
c     end  FFT of length n=5
200          continue
c     perform transformation k --> (m) --> kout(k)
          do 20 k=1, 20
            kt=kout(k)
            zr(kt)=xr(k)
            zi(kt)=xi(k)
20        continue
        return
        end
```

REFERENCES

[1] G. Cook, Ph. D. thesis, UCRL-53324, U. California (1982).

[2] M.C. Wirth, Ph. D. thesis, UCRL-52996, U. California (1980).

[3] R.W. Hockney, 1978, in *Fast Poisson Solvers and Applications*, ed. U. Schumann, (Advance Publications: London), 78 - 97.

[4] R. H. Berman, 1984, "Measuring the Performance of a Computational Physics Environment" in *Proceedings of the 1984 MACSYMA Users' Conference*, (GE: Schenectaday).

[5] R.H. Berman, "Vectorizing a Particle Push for the Cray-1" *Buffer*, 3, 12 (1979).

[6] D.J. Kuck and M. Wolfe, "A Debate: Retire FORTRAN ? No." *Physics Today*, 37, 67 (1984).

[7] J. R. McGraw, "A Debate: Retire FORTRAN ? Yes." *Physics Today*, 37, 66 (1984).

[8] K.G. Wilson, "Science, Indiustry and the New Japanese Challenge" *Proc. IEEE*, 72, 6 (1984).

[9] R.W. Hockney and J.W. Eastwood, 1981, *Computer Simulation Using Particles*, (McGraw Hill: New York).

[10] R.H. Berman, D.J. Tetreault, and T.H. Dupree, "Observation of Turbulent Self-binding Fluctuations in Simulation Plasma and their Relation to Kinetic Theory of Plasma" *Phys. Fluids*, 26, 2437 (1983).

[11] R.H. Berman, D.J. Tetreault, and T.H. Dupree, "Simulation of Growing Phase Space Holes and the Development of Intermittent Plasma Turbulence" to appear in *Phys. Fluids*, 1984.

[12] H.J. Nussbaumer, 1982, *Fast Fourier Transform and Convolution Algorithms* (New York: Springer-Verlag).

[13] E.O. Bingham, 1974, *The Fast Fourier Transform*, (Prentice-Hall: New York).

[14] J.W. Cooley, P.A.W. Lewis, and P.D. Welch, "Historical Notes on the Fast Fourier Transform Algorithm" *IEEE Trans. Audio Electroacoust.*, AU-15, 76 (1967).

[15] J.W. Cooley, P.A.W. Lewis, and P.D. Welch, "The Fast Fourier Transform Algorithm: Programming Considerations in the Calculation of Sine, Cosine and Laplace Transforms" *J. Sound Vib.* 12, 315 (1970).

[16] J.W. Cooley and J.W. Tukey, "An Algorithm, for the Machine Calculation of Complex Fourier Series" *Math. Comp*, 19, 297 (1965).

[17] I.J. Good, "The Interaction Algorithm and Practical Fourier Series" *J. Roy. Statis. Soc.*, **20**, 361 (1958).

[18] D.E. Dudgeon and R.M. Mersereau, 1984, *Multidimensional Signal Processing*, (Prentice-Hall: New Jersey).

[19] C.M. Rader, "Discrete Fourier Transforms When the Number of Data Samples is Prime" *Proc. IEEE*, **56**, 1107 (1968).

[20] T. G. Stockham, "High Speed Convolution and Correlation" *AFIPS Conf. Proc.*, **28**, 229 (1966).

[21] C.M. Rader, "An Improved Algorithm for High Speed Autocorrelation with Application to Spectral Estimation" *IEEE Trans. Audio Electroacoust.*, **18**, 439 (1970).

[22] R.H. Berman, 1984 in preparation.

[23] H.S. Hou and H.C. Andrews,"Cubic Splines for Image Interpolation and Digital Filtering" *IEEE Trans. Acous., Speech, Signal Processing*, **26**, 508 (1978).

[24] S. Haykin and S. Kesler, 1983, in *Nonlinear Methods of Spectral Analysis*, ed. S. Haykin (New York: Springer-Verlag).

[25] A.V. Oppenheim and R.W. Schafer, 1975, *Digital Signal Processing* (Prentice-Hall: New Jersey).

[26] R.W. Johnson and J.E. Shore "Which is the Better Entropy Expression for Speech processing: $-S \log S$ or $\log S$?" *IEEE Trans. Acoustics, Speech and Signal Proc.*, **32** (1983).

[27] A. Papoulis "Maximum Entropy and Spectral Estimation: A Review" *IEEE Trans. Acoustics, Speech and Signal Proc.* **29**, 1176 (1981).

[28] N-L. Wu "An Explicit Solution and Data Extension method in the Maximum Entropy Method" *IEEE Trans. Acoustics, Speech and Signal Proc.* **486** (1983).

[29] C. de Boor "FFT as Nested Multiplication, with a Twist" *SIAM J. Sci. Stat. Comput.* **1**, 173 (1980).

[30] D. E. Knuth, 1969, *The Art of Computer Programming. Vol. II. Seminumerical Algorithms*, (Addison-Wesley: Reading).

11

COMPUTER ALGEBRA AS A TOOL FOR SOLVING OPTIMAL CONTROL PROBLEMS

C. GOMEZ ; J.P. QUADRAT ; A. SULEM

Institut National de Recherche en Informatique et en Automatique(INRIA)
Domaine de Voluceau, B.P. 105, 78153 Le Chesnay Cedex, FRANCE.

ABSTRACT
    Solving optimal control problems usually consists in modelizing type problem, proving the existence of a solution, choosing a method leadint to a numerical solution and writing a program (generally in FORTRAN). These steps can be automatized using formal calculus and inference techniques. An expert system in stochastic control is presented. It is embedded in the MACSYMA system. After the description of the problem by the user in semi-natural language, the system generates the equations which modelize the problem. Then by using PROLOG (for encoding the used theorems) and symbolic manipulations on the equations, the system can prove the existence of a solution. Finally the system chooses a numerical method to solve it using symbolic differentiations and mtrix inversions and it generates the associated FORTRAN program.

INTRODUCTION

    An expert system in stochastic control problems is presented. It is in fact a decision support system using computer algebra and artificial intelligence techniques. Its purpose is to automatize the studies in the filed of system analysis and optimization. These systems are often dynamic systems, perturbated by random events. Usually the studies made on these systems lead to handwritten reports and FORTRAN software. In these studies we make many trials to test various formulations and algorithms by using scientific libraries.

    The parts of the study we have to automatize are :
- choice of the mathematical model
- choice of the solution algorithm

241

- theoretical analysis of the equations
- algebraic manipulation of the equations
- obtention of the numerical software (FORTRAN)
- report edition.

The practical application fields of such an expert system are :
- stock management (hydraulic dams, energy in the core of nuclear power stations,...)

- modelization of firm development
- portfolio management
- maintenance problems
- stochastic systems regulation

This approach can be extended to over kinds of problems : optimization of networks, discrete event systems etc...

PROBLEM STATEMENT.

A decision study system can be represented by Fig. 1 where the operators and the information transits are showed.

The user specifies the problem and gives the data to a modelizer whose final purpose is to give a report to the user, indicating how to manage the system. Usually this report contains some purely numerical programs used to compute the decision or to give the informations to compute it on line.

The modelizer, with the specifications and the data of the user, chooses a class of mathematical models for the problem in a family which is his internal knowledge data base. In the chosen class of models the system must be able to identify the missing parameters, simulation model.

The model used by the optimization can be different from the one used by the simulation.

Identification, optimization and simulation are different

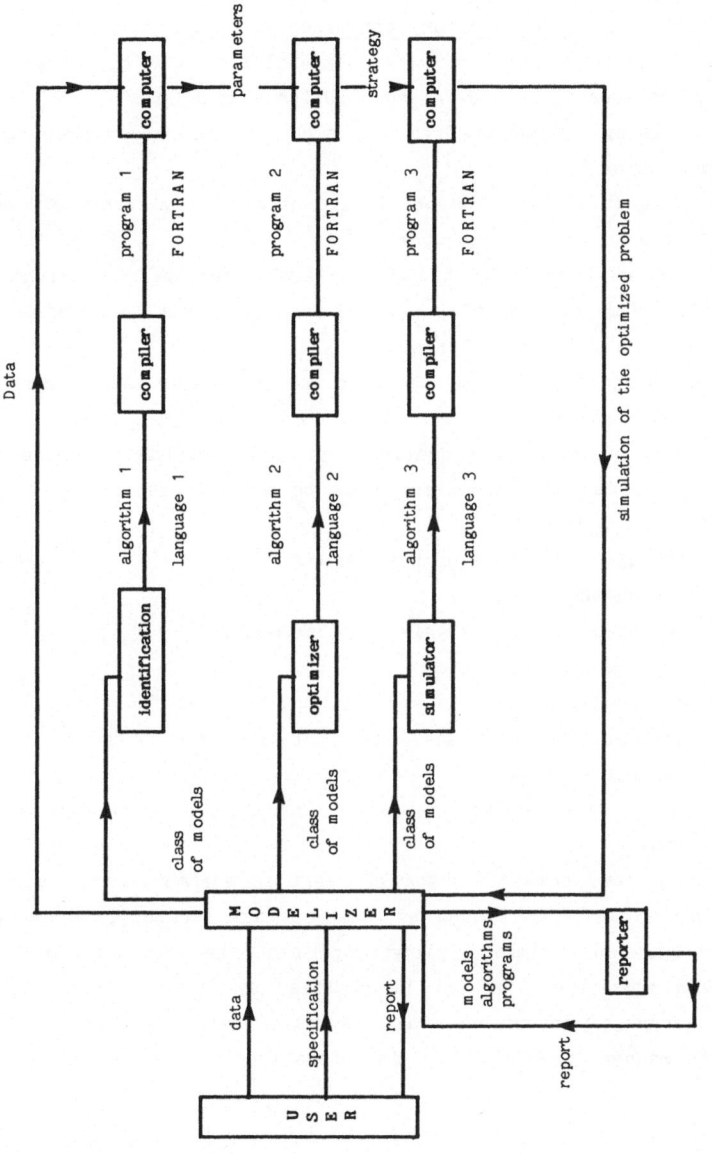

Fig. 1 : Decision study system : operators and information transits.

specialities. Identification is based on statistics, simulation on numerical analysis or probabilities, and optimization is a speciality in itself. We call identificator, optimizer and simulator these three specialists. Having consulted those specialists, the modelizer has consistent and acceptable models and solution algorithms.

Obtaining a program from the algorithm is only a problem of symbolic manipulation :

- formal calculus for obtaining the algebraic formulae needed by the numerical algorithm
- compilation from the high level language in which the algorithm is specified in mathematical terms to a numerical language, usually FORTRAN.

With these programs, it is possible to estimate parameters, to simulate and to optimize.

Having analyzed the qualities of the decision by using the simulation program, the modelizer have to write a report about the study. This report describes :

- the specifications of the problem
- the chosen models
- the theoretical analysis of these models
- the solution algorithms
- the numerical results representing the cost function at the optimum, the decisions, the simulation results of the optimal strategy.

Then the report is given to the user ; he accepts it or change the specifications.

Until now, only the numerical part is automatized (computing the parameters to be estimated and the optimal strategy, simulating the system). But almost everything can be automatized. The conversation with the user can be realized by using a specialized natural language interface which accesses a data base in an intelligent manner. The modelizer is automatized by making inference by using rules in a knowledge data base describing families of models and the conditions of their utilization. The theoretical analysis of the model is made by theorem proving ; one must apply mathematical theorems which are encoded in another data base. The choice of the solution algorithm also is inference ; the conditions for the application of an algorithm are given by theorems or heuristics. Generation of FORTRAN code from an algorithm specification can be automatized by using compilation techniques. And, at each step, computer algebra and algebraic manipulations must be used to handle the equations. Even the report edition can be automatized ; indeed all the knowledge, software and results are included in the system and it is possible to print them, after having classified them, by using a system of automatic text writing.

AN EXPERT SYSTEM IN STOCHASTIC CONTROL.

The decision support system is a kind of expert system in stochastic control. As it was pointed out, the system relies heavily on symbolic and algebraic manipulations ; for this reason, it is entirely embedded in MACSYMA[*] (1), a large computer programming system written in LISP. Fig. 2 shows the parts of the expert system. The specialized natural language interface is used for the conversation with the user in all the parts of the system. The report edition part and the modelization part are not yet available and the algorithm choice part is under development. Each expert part, modelization, theoretical analysis and algorithm choice possesses its own knowledge data base encoded in rules in a PROLOG system written in LISP and called LOGIS (2).

---

* MACSYMA was developed by the Mathlab Group of the MIT Laboratory for Computer Science (formerly Project MAC) and is currently supported and distributed by Symbolics, Inc.

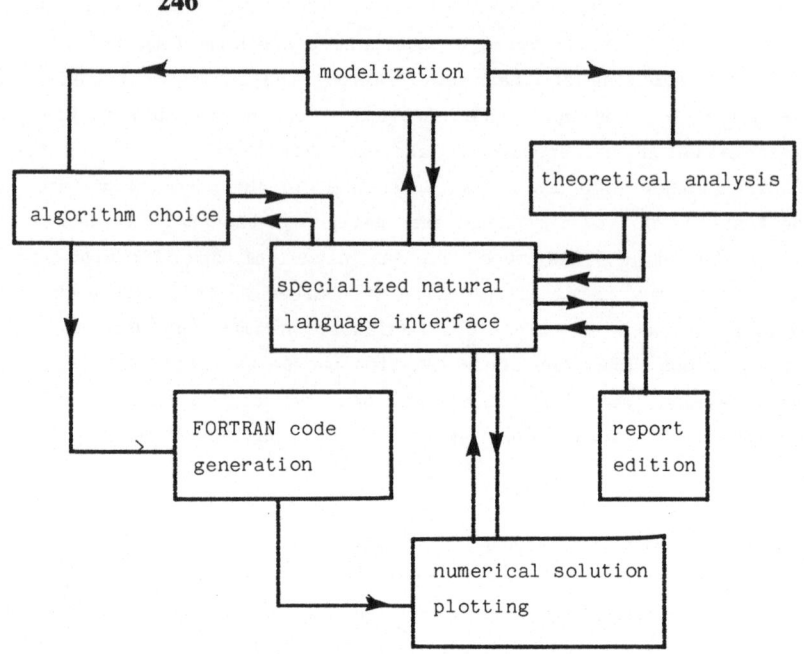

Fig. 2 : Parts of the expert system.

The specialized natural language interface has patterns in order to recognize the sentences entered by the user. All these parts are written in MACLISP (3,4). The FORTRAN code generation part is written in MACSYMA and uses the compilation techniques. The system works now on a HB68 under Multics operating system. We hope it to work soon on a LISP machine.

In the following the parts of the system are presented in a more detailed way. Results of studies also are presented.

SPECIALIZED NATURAL LANGUAGE INTERFACE.

The user talks to the system by using a natural language interface specialized in stochastic control. The sentences he enters are analyzed and matched against patterns of typical sentences. Then according to the type of the sentence, various actions are triggered (answer to the user, access to a data base for putting or retrieving, etc..). A system of this type is described in (5). This interface is written in MACLISP.

Here is an example of a sentence matched by the system ; entering the data of the problem, the user types :

let $[s_1,s_2]$ be a state variable denoting a water level

The corresponding pattern is :

?- let !x !- !- !type ?- denoting !- !physical-meaning

where ?- matches anything, !- matches one word and !x matches one word and binds it to x.

The result of the pattern matching is an association list :

(x . [s1,s2], type . state, physical-meaning . water level)

The associated actions are :

     - to transform the data and to intern them

     - to verify properties and meaning of the data

     - to answer the user.

The answer is :

state : $[s_1,s_2]$

physical meaning of $[s_1,s_2]$: water level

Moreover, the user can delete a variable, ask for the meaning of a variable, ask for all the data and variables (access to the data base).

There are specific actions the user can ask for : the generation of the dynamic programming equation (then the program asks for missing data), the generation of the FORTRAN program associated to a given algorithm, etc...

Here is a sample session for a stochastic control problem

```
solved by the dynamic programmation method :

the sentences typed by the user are preceded by ==>

hello
we shall try to solve your stochastic control problem
please give the statement of the problem
==>  let (s1,s2) be a state variable denoting a water-level
state : (s1, s2)
physical meaning of (s1, s2) : water-level
==>  s1 belongs to (0,1)
domain of s1 : (0, 1)
==>  s2 belongs to (0,1)
domain of s2 : (0, 1)
==>  (v1,v2) is a control variable denoting the turbined-water
control : (v1, v2)
physical meaning of (v1, v2) : turbined-water
==>  v1 belongs to (0,1)
domain of v1 : (0, 1)
==>  v2 belongs to (0,1)
domain of v2 : (0, 1)
==>  d is a drift term equal to v1
drift : d
value of d : v1
the dimensions of the state and the drift must be the same
==>  delete d
drift : d
value of d : v1
deleted
==>  dd is a drift term equal to (v1,v2)
drift : dd
value of dd : (v1, v2)
==>  what do you know
drift :dd
value of dd : (v1,v2);
control : (v1, v2)
physical meaning of (v1, v2) : turbined-water
state : (s1, s2)
physical meaning of (s1, s2) : water-level
==>  l is a diffusion diagonal matrix equal to (1,1)
diffusion : l
value of l : (1, 1)
==>  the problem consists in the minimization of a cost function
the problem leads to a hamilton-jacobi equation
==>  f is the instantaneous-cost equal to
v1**2+v2**2+s1*(1-s1)*s2*(1-s2)+((0.5-s1)*s2*(1-s2))**2+
((0.5-s2)*s1*(1-s1))**2+2*s2*(1-s2)+2*s1*(1-s1)
instantaneous-cost : f
```

value of f : $v2^2 + v1^2 + (0.5 - s1)^2 (1 - s2)^2 s2^2 +$

$(1 - s1) s1 (1 - s2) s2 + 2 (1 - s2) s2 +$

$+ (1 - s1)^2 s1^2 (0.5 - s2)^2 + 2 (1 - s1) s1$
==> what is dd ?
drift : dd
value of dd : (v1, v2)
==> for s1 = 0 the limit condition is of the dirichlet type and
equal to s1*s2
limit condition for s1 = 0 : s1 s2
 dirichlet type
domain of s1 : (0, 1)
==> s2 = 1 stopping (1-s1)*(1-s2)
limit condition for s2 = 1 : (1 - s1) (1 - s2)
 dirichlet type .
domain of s2 : (0, 1)
==> s2 = 0 dirichlet s1*s2
limit condition for s2 = 1 : (1 - s1) (1 - s2)
 dirichlet type .
limit condition for s2 = 0 : s1 s2
 dirichlet type
domain of s2 : (0, 1)
==> stop
you must precise if the problem is static or parabolic
a limit condition is missing for s1 = 1
do you still wish to stop ?
no
==> static
the problem is static
==> s1 = 1 dirichlet (1-s1)*(1-s2)
limit condition for s1 = 1 : (1 - s1) (1 - s2)
 dirichlet type .
limit condition for s1 = 0 : s1 s2
 dirichlet type
domain of s1 : (0, 1)
==> stop
a discount factor is missing
do you still wish to stop ?
no
==> let a be a discount factor equal to 1
discount : a
value of a : 1
==> what is the dynamic programming equation ?
Hamilton-Jacobi equation :

```
        dv      dv    2    2
- v + min (v1 -- + v2 -- + v2  + v1  +
        ds1     ds2


      2       2   2
 (0.5 - s1)  (1 - s2)  s2  + (1 - s1) s1 (1 - s2) s2


                    2   2              2
+ 2 (1 - s2) s2 + (1 - s1)  s1  (0.5 - s2)  + 2 (1 - s1) s1 +


   2       2
  d v     d v
+ ---- + ----)
   2       2
  ds1     ds2
```

==> please generate the fortran program
done
you will find the subroutine in the segment belman.fortran
==> give the principal program
drift is of the order 0.5
diffusion is of the order 1
you will find the principal program in the segment pp.fortran
==> do a numerical test
you will find the the fortran program in the segment num.fortran
==> stop
good bye

THEORETICAL ANALYSIS.

The problem consists in finding automatically theoretical results on partial differential equations with boundary values (regularity, existence, uniqueness of a solution). Indeed, most stochastic control problems lead to such a type of equations.

The abstract problem is :

(P)

$$Au = f \qquad \text{in } \Omega \text{ bounded open of } R^n$$

$$B_j u = g_j, \; j=0,\ldots,m-1 \text{ on the boundary } \Gamma \text{ of } \Omega$$

A, B_j are differential operators.

One of the purposes of the study is to find the functional space to which the solution belongs. The Sobolev spaces (6) are used for these types of problems. $L^P(\Omega)$ denoting the space of these functions f such that f^P is summable, the Sobolev space $W^{m,P}(\Omega)$ is the space of these functions belonging to $L^P(\Omega)$ with the m first derivatives (in the weak sense) belonging to $L^P(\Omega)$ and $W_o^{m,P}(\Omega)$ is the closure of $W^{m,P}(\Omega)$ in $D(\Omega)$ (indefinitely differentiable functions with compact support). $H^m(\Omega)$ is only but a simplified notation for $W^{m,2}(\Omega)$.

The mathematical objects in the equations are put in a relational data base managed by the system of property lists of LISP. For instance, for the equation :

$$- \sum_{i=1}^{n} \frac{\partial}{\partial x_i} \left(a \frac{\partial u}{\partial x_i} \right) + |u|^{P-2} u = f,$$

we have the following table :

OBJECT	NATURE	SPACE	PROPERTIES
u	solution	unknown	nil
a	coefficient	L^q	a > 0
q	constant	R	nil
p	constant	R	p ≥ 2
f	smember	unknown	nil

The internal representation of the functional spaces shows a way to use internal MACSYMA. For instance the space $W_o^{m,P}(\Omega)$ is handled in the following way :
it is typed by the user in the MACSYMA syntax as :
space (w,m,p,o) ;

space is a MACSYMA function that generates the list

((space simp) ((p m o)) nil) which is used as the internal representation of the space

and when the display function of MACSYMA sees this list, it prints W(m,p,o).

It is an easy way to deal with the spaces in the MACSYMA environment.

To find results on the equation, we use theorems of functional analysis (7,8). They are encoded in LOGIS (PROLOG written in LISP). For instance, the Lax Milgram theorem is encoded in the following clause :

((solution space)   ← conclusion

 (well defined)

 (linear $principal-op $solution)

 (elliptic $principal-op space)   hypothesis

 (continuous $principal-op space))

the meaning of which is : the problem (P) has a solution in a space V if (P) is well defined, if the principal operator A is linear with respect to the solution u, if A is elliptic in V and if A is continuous in V.

Each clause triggers the execution of another clause, a LISP function, or a MACSYMA function.

In the demonstrations, an important part is taken by symbolic manipulations. For instance, we have to obtain the Green formula of the problem (P). It consists in multiplying the equation Au = f by a function v and to integrate the whole over Ω. Then successive by parts integrations are used. We wrote this module in LISP and manipulate the internal form of the equation in internal MACSYMA syntax. Another sample example is given by the study of the coercivity :

The purpose is to find α positive such that $a(u) \geq \alpha \| u \|_V^2$ where $\| u \|_V$ is the norm of u in V.

The steps are the following :

1 : choice of the norm of V among the equivalent norms (LOGIS)

2 : obtention of a canonical form for a(u) :

$$a(u) = \sum_i a_i N_i \quad \text{with } N_i = \text{norm } (u, p_i, m_i)$$

and $\text{norm } (u,p,m) = \sum_{i=1}^{n} \int_\Omega \left| \frac{\partial^m u}{\partial x_i^m} \right|^p d\omega$

this part is written in LISP and uses the MACSYMA internal relational data base in order to find minimum and maximum of functions

3 : minimization of the canonical form :

for instance $a(u) = a_1 N_1 + a_2 N_2 + a_3 N_3$

$$\| u \|_v^2 = N_1 + N_2 \qquad \Rightarrow \alpha = \text{Min}(a_1, a_2)$$
$$a_3 \geq 0$$

In the parts written in LISP, the programming technique of data driven programming (9) has been largely used (rules for Green formula obtention, inequalities solution).

A sample session of this module is given below. The problem is :

$$-\sum_{i=1}^{n} \frac{\partial}{\partial x_i} (a_i \frac{\partial u}{\partial x_i}) + u |u|^{p-2} + u = f \text{ in } \Omega$$

$$u = 0 \text{ on } \Gamma$$

The system shows that the problem is non linear, obtains the variational formulation in the space $H_0^1(\Omega) \sqcap L^p(\Omega)$ and shows that the problem is coercive. The following steps (monotony, continuity) leading to the existence of a solution are under development.

```
                           n
dimension of the space R   :
==> 2;
the dimension of the space is 2
ok ?
==> yes;
equation to be solved in omega of the type "A(u)=f" :
```

```
==> u+abs(u)(p-2)*u-sum(diff(a(i)*diff(u,x(i)),x(i)),i,1,n)=f;
conditions on the boundary gamma of the type
"[B[1](u)=0,...,B[m](u)=0]" :
==> [u=0];
the problem is :
   n
  ====
  \        d       du                 p - 2
- >      --- (a  ---)  + u abs(u)         + u = f in omega
  /       dx    i dx
  ====     i       i
  i = 1
with on gamma :
u = 0
ok ?
==> yes;
nature of u :
==> solution;
function : u
*********
    nature : solution
    space : unknown
    properties : nil
ok ?
==> yes;
nature of p :
==> constant;
properties of p :
==> p>2;
function : p
*********
    nature : constant
    space : R
    properties : p > 2
ok ?
==> yes;
nature of a :
==> coefficient;
space of a :
==> space(1,inf);
properties of a :
==> a>1;
function : a
*********
    nature : coefficient
    space : L(inf)
    properties : a > 1
```

```
ok ?
==> yes;
properties of f :
==> nil;
function : f
*********
    nature : smember
    space : unknown
    properties : nil
ok ?
==> yes;
==>  ((solution space))
**** well-defined ****
**** non-linear ****
principal-op
solution
**** variational-formulation ****
principal-op
 2
====  /                /                      /
\     (     du  dv    (         p - 2        (
 >    I (a  --- ---) + I (u abs(u)      v) + I (u v)
/     )  i dx  dx     )                      )
====  /       i  i    /                      /
i = 1
/
(
I (f v)
)
/
(H(1, 0), L(p))
**** coercive ****
principal-op
(H(1, 0), L(p))
1
==> end
```

ALGORITHM CHOICE.

According to the type of stochastic control problem, various methods are used (10) : dynamic programming method (dynamic programming equation solution), decoupling techniques (optimization in the class of local feedbacks), stochastic gradient method (optimization in a parametrized class of feedbacks), perturbation method (small noises) etc.. The good method and the associated algorithm will be choiced by the means of rules encoded in LOGIS. This part is under development.

A part of the system also deals with signal processing (11).

FORTRAN CODE GENERATION.

Having chosen the algorithm for the solution of the stochastic control problem (at the moment it is done by hand), the problem and the algorithm are specified by a sentence of a specific language.

For instance, suppose we solve the dynamic programming equation :

$$
\begin{cases}
-\lambda v + \underset{u}{\text{Min}} \left[b(x,u) \frac{\partial v}{\partial x} + c(x,u) \right] + \frac{1}{2} \frac{\partial^2 v}{\partial x^2} = 0 \text{ for } x \in [0,1] \\
v(0) = v(1) = 0
\end{cases}
$$

We must find v as a solution.

In order to minimize $b(x,u) \frac{\partial v}{\partial x} + c(x,u)$, we have to specify the algorithm, for instance the Newton method. It leads to the following sentence of a specific language :
[1, stat, condlim, [[dirich,0]], [[dirich,0]], elp, λ, belm, 1, newton, au, $b(x,u)p + c(x,u) + \frac{1}{2} q_1$] which is a MACSYMA list describing the problem and the algorithm solution.

From this list, a compiler generates a subroutine in FORTRAN code using finite difference methods to discretize the equation.

Here is the subroutine in FORTRAN code generated after the conversation with the user showed above :

```
      subroutine prodyn(n1,n2,epsimp,impmax,v,ro,u,eps,nmax)
      dimension v(n1,n2),u(2,n1,n2)
c     Resolution de 1 equation de Bellman dans le cas ou:
c         Les parametres sont
```

```
c              L etats-temps est:  x1 x2
c              La dynamique du systeme est decrite par 1 operateur
c                         2          2  2
c      Minu( (0.5 - x1)  (1 - x2)   x2  + (1 - x1) x1 (1 - x2) x2
c
c                             2  2              2                        2
c + 2 (1 - x2) x2 + (1 - x1)  x1  (0.5 - x2)  + 2 (1 - x1) x1 + u2
c                                                        + p2 u2
c
c        2
c + u1  + p1 u1 + q2 + q1 )
c              ou v designe le cout optimal
c              ou pi designe sa derivee premiere par rapport a xi
c              ou qi designe sa derivee seconde par rapport a xi
c              Le probleme est statique
c              Les conditions aux limites sont:
c                   x2 = 0 v= x1 x2
c                   x2 = 1 v= (1 - x1) (1 - x2)
c                   x1 = 0 v= x1 x2
c                   x1 = 1 v= (1 - x1) (1 - x2)
c       Les nombres de points de discretisation sont: n1 n2
c                   x2 = 1 correspond a i2 = n2
c                   x2 = 0 correspond a i2 = 1
c                   x1 = 1 correspond a i1 = n1
c                   x1 = 0 correspond a i1 = 1
c       Le taux d actualisation vaut: 1
c       impmax designe le nbre maxi d iterations du systeme implicite
c       epsimp designe 1 erreur de convergence du systeme implicite
c       ro designe le pas de la resolution du systeme implicite
c                                     par une methode iterative
c       Minimisation par la methode de Newton de l'Hamiltonien
c       L inversion de la Hessienne est faite formellement
c       nmax designe le nombre maxi d iteration de la methode de Newton
c       eps designe 1 erreur de convergence de la methode de Newton
      h2 = float(1)/(n2-1)
      h1 = float(1)/(n1-1)
      u2 = u(2,1,1)
      u1 = u(1,1,1)
      hih2 = h2**2
      hih1 = h1**2
      h22 = 2*h2
      h21 = 2*h1
      nm2 = n2-1
      nm1 = n1-1
      do  119  i2 = 1 , n2 , 1
      do  119  i1 = 1 , n1 , 1
      v(i1,i2) = 0.0
```

```
119 continue
    imiter = 1
113 continue
    erimp = 0
    do 111    i1 = 1 , n1 , 1
    x1 = h1*(i1-1)
    v(i1,n2) = 0
    v(i1,1) = 0
111 continue
    do 109    i2 = 2 , nm2 , 1
    x2 = h2*(i2-1)
    v(n1,i2) = 0
    v(1,i2) = 0
110 continue
    do 109    i1 = 2 , nm1 , 1
    x1 = h1*(i1-1)
    q2 = (v(i1,i2+1)-2*v(i1,i2)+v(i1,i2-1))/hih2
    q1 = (v(i1+1,i2)-2*v(i1,i2)+v(i1-1,i2))/hih1
    p2 = (v(i1,i2+1)-v(i1,i2-1))/h22
    p1 = (v(i1+1,i2)-v(i1-1,i2))/h21
    niter = 0
    w0 = -1.0e+20
101 continue
    niter = niter+1
    if ( niter - nmax )  102 , 102 , 103
103 continue
    write(8,901)i1,i2
901 format(' newton n a pas converge' , 2 i3)
    goto  104
102 continue
    u1 = -p1/2.0
    u2 = -p2/2.0
    ww = (0.5-x1)**2*(1-x2)**2*x2**2+(1-x1)*x1*(1-x2)*x2+2*(1-x2)*x2+(
   1   1-x1)**2*x1**2*(0.5-x2)**2+2*(1-x1)*x1+u2**2+p2*u2+u1**2+p1*u1+
   2   q2+q1
    er = abs(ww-w0)
    if ( er - eps )  104 , 104 , 105
105 continue
    w0 = ww
    goto  101
104 continue
    u(1,i1,i2) = u1
    u(2,i1,i2) = u2
    w0 = ww
    w0 = w0-v(i1,i2)
    vnew = ro*w0+v(i1,i2)
    v(i1,i2) = vnew
```

```
      erimp = abs(w0)+erimp
109 continue
      imiter = imiter+1
      if ( imiter - impmax )  116 , 115 , 115
116 continue
      if ( epsimp - erimp )  113 , 112 , 112
115 continue
      write(8,907)
907 format(' schema implicite n a pas converge')
112 continue
      do  117    i1 = 1 , n1 , 1
      do  117    i2 = 1 , n2 , 1
      write(8,900)i1,i2,v(i1,i2)
900 format(' v(', (i3,','), i3,'):', e14.7,'$')
      write(8,902)i1,i2,u(1,i1,i2),u(2,i1,i2)
902 format(' u(', (i3,','), i3,'):(', (e14.7,','), e14.7,')$')
117 continue
      return
      end
```

For the obtention of the FORTRAN code, many symbolic manipulations are made using MACSYSMA ; in particular, the derivatives necessary for the Newton method are computed by MACSYMA.

For more details on the way the system handles the various methods, see (12), (13).

FUTURE DEVELOPMENTS AND CONCLUSION.

The system has been already used for generating programs in economics applications (management of dams, optimization of strategy for insurance companies, optimum advertising, quality production).
It is being developed in order to make the missing parts, automatic modelization and report edition, to introduce other methods in the algorithm choice part and to extend the system to a larger class of stochastic processes.

Moreover, frames and object oriented languages (14,15) will be used for a better structure of the whole system. We will use a PROLOG recognizing objects (16) and the inference will be much more important.

Another approach for solving dynamic programming equations based on the formal calculus in the (R ,max,+) algebra is in project.

REFERENCES.

1.   MACSYMA Reference Manual. Version 10. The Mathlab Group, MIT, Laboratory for Computer Science. MIT. December 1983.

2.   Gloess P., Logis User's Manual. Second Edition. UTC/GI, BP 233, 60206 Compiègne Cedex, France. June 1984.

3.   Davis J. and Moon D., Multics Maclisp Reference Manual, Revision 1, MIT, Laboratory for Computer Science, 1980.

4.   Pitman K.M., The revised MACLISP manual, MIT, Laboratory for Computer Science, 1983.

5.   Queinnec C., LISP : Langage d'un autre type, Edition Eyrolles, 1983.

6.   Adams R.A., Sobolev spaces. Academic Press. 1975.

7.   Lions J.L. and Magenes E., Problèmes aux limites non homogènes et applications, Tome I, Dunod, Paris, 1968.

8.   Lions J.L., Quelques méthodes de résolution des problèmes aux limites non linéaires. Dunod, Gauthier-Villars, Paris, 1969.

9.   Charniak E., MacDermott D. and Riesbeck C., Artificial Intelligence Programming. Lawrence Erlbaum Associates, Hillsdale, New Jersey. 1979.

10.   Bensoussan A. et al., Commande optimale de systèmes stochastiques. RAIRO, Systems Analysis and Control, Vol. 18, n°2. 1984.

11.   Blankenship G.L. et al., An expert system for control and signal processing with automatic FORTRAN code generation. CDC 1984, Las Vegas, to appear.

12.   Gomez C., Quadrat J.P. and Sulem A., Towards an expert system in stochastic control : the Hamilton-Jacobi equation part. Nice, Juin 1984. Springer Verlag, Lecture Notes in Control and Information Sciences, 62.

13.   Gomez C., Quadrat J.P. and Sulem A., Towards an expert system

in stochastic control : optimization in the class of local
feedbacks. Rome, 1984. To appear.

14.   Gloess P., Understanding Expert systems. Alfred Handy Guide.
       1984. To appear.

15.   Feigenbaum E.A. and Barr A., The Handbook of Artificial
       Intelligence, Heuristics Press, Stanford, California, W.
       Kaufmann Inc., Los Altos, California. 1982.

16.   Gloess P. and Nguyen D., OBLOGIS, an object oriented
       implementation of PROLOG in a LISP environment. Colloque
       International d'Intelligence Artificielle, Marseille, France,
       October 1984.

12

APPLICATION OF MACSYMA
TO KINEMATICS AND MECHANICAL SYSTEMS

M.A. Hussain
General Electric Company
Corporate Research and Development

B. Noble
Mathematics Research Center
University of Wisconsin

ABSTRACT

The objective of this paper is to illustrate that symbol manipulation systems can readily handle many of the typical symbolic calculations arising in the formulation of problems in kinematics and mechanical systems.

The paper consists of two parts. First, we discuss the use of MACSYMA in connection with the algebraic manipulations involved in transferring a body from one position to another in space, with particular reference to Rodrigues and Euler parameters and successive rotations, and an example involving quaternions. Second, we indicate how MACSYMA can be used to set up dynamical equations for the Stanford manipulator arm, and a spacecraft problem.

INTRODUCTION

Kinematics is a basic tool for the analysis of mechanisms and mechanical systems. Until recently, the most common approach has been to use vectors and Euler angles. More recently, other approaches have been gaining in popularity because of computers. We illustrate by several examples that these approaches are particularly amenable to symbolic manipulation. The immediate objective is limited, namely to indicate that several methods of representing rotations including Rodrigues and Euler parameters, and quaternions can be handled by MACSYMA by a unified approach that would seem to have some elements of novelty. But also it should be clear that our examples suggest a different approach to dynamical problems such as those considered by Branets and Shmyglevskiy [3] using quaternions and Dimentberg [4] using the screw calculus. The nearest connected account of the type of approach we have in mind is the mss. [12] by Nikravesh et al., but a systematic use of computer symbolic manipulation would certainly affect the detailed treatment. This is the first part of the paper.

It is clear that the complexity of mechanical systems is increasing to the point where symbol manipulation must play an important part in their formulation and solution. We illustrate by two dynamical examples, one involving a robot arm, the other a spacecraft problem. The main reason for choosing these particular examples is that

---

This paper will also be published in the Proceedings of the Third MACSYMA Users' Conference 1984, supported by DOD.

the equations have been formulated and published in quite a detailed form already. By comparing our treatment with those already published, the reader will be able to make a judgment for himself concerning the usefulness of MACSYMA, and also how thinking in terms of symbol manipulation does change one's approach to the formulation of the equations. This is the second part of the paper.

We give the MACSYMA programs in detail in Appendices for two reasons. Experienced MACSYMA users may be able to suggest improvements. Readers familiar with other symbol manipulation systems may care to write programs for the same problems, and compare their programs with those in the Appendices. For the benefit of new and old users we encourage authors to publish their programs in detail, as done here.

KINEMATICS EXAMPLES

1. The Representation of Rotation by Orthogonal Matrices

We remind the reader of some standard results. We work in terms of matrices (this can be converted into vector interpretations as appropriate) using upper case for matrices.

A rotation of a body with a fixed point by an angle ϕ around an axis defined by the unit column matrix $\mathbf{n} = [n_1, n_2, n_3]^T$ transfers a point $\mathbf{r} = [x, y, z]^T$ into a point $\mathbf{r}' = [x', y', z']^T$ by (cf. Bottema and Roth [1] p. 59) $\mathbf{r}' = \mathbf{Ar}$ where (see Figure 1)

$$\mathbf{A} = [\cos\phi\,\mathbf{I} + (1 - \cos\phi)\mathbf{nn}^T + \sin\phi\,\mathbf{N}] \ , \tag{1}$$

$$\mathbf{N} = \begin{bmatrix} 0 & -n_3 & n_2 \\ n_3 & 0 & -n_1 \\ -n_2 & n_1 & 0 \end{bmatrix} \tag{2}$$

(Note that $\mathbf{Nr}$ corresponds to the vector $\mathbf{n} \times \mathbf{r}$, and $\mathbf{I}$ is the identity matrix).

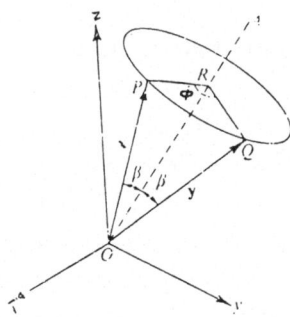

Figure 1. Rotation of a body with a fixed point.

The matrix $\mathbf{A}$ is orthogonal. We discuss three different ways of proving this using MACSYMA:

a) The simplest and most direct way is to express (1) in component form and simply check by brute force that $\mathbf{A}^T\mathbf{A} = \mathbf{I}$, using TELLSIMP to impose side-conditions.

b) Alternatively we could use MACSYMA interactively as follows. It is easily checked that

$$\mathbf{n}^T\mathbf{n} = 1, \quad \mathbf{N}^T = -\mathbf{N}, \quad \mathbf{n}^T\mathbf{N} = 0, \quad \mathbf{N}\mathbf{n} = 0, \quad \mathbf{N}^2 = \mathbf{n}\mathbf{n}^T - \mathbf{I} \tag{3}$$

We use MACSYMA to form $\mathbf{A}^T\mathbf{A}$, which will give nine terms involving $\mathbf{n}^T\mathbf{N}$, $\mathbf{n}\mathbf{n}^T\mathbf{n}\mathbf{n}^T$, $\mathbf{N}^2$ etc., and we use SUBST to simplify and finally derive $\mathbf{A}^T\mathbf{A} = \mathbf{I}$.

c) We can use TELLSIMP to build the rules (3) into MACSYMA. Then a MACSYMA program can be written to produce the result $\mathbf{I}$ for $\mathbf{A}^T\mathbf{A}$.

Method a) is clearly simplest. Method c) is surprisingly tricky in MACSYMA because in addition to (3) we have to distinguish between scalars and matrices, and set proper switches. For verifying that $\mathbf{A}\mathbf{A}^T = \mathbf{I}$, the simplest method is to use a) not c), but for more complicated problems, method a) soon produces algebraic expressions of horrendous complexity. As problem size increases, method c) will become preferable. In this paper, we have used the component form but further developments may require the more abstract approach.

2. Rodrigues Parameters

We introduce these by stating the result that any 3×3 orthogonal matrix A can be expressed in the following product form by the Cayley-Klein decomposition which says that there exists a skew-symmetric 3×3 matrix $\mathbf{B}$ such that (cf. Bottema and Roth [1], p. 10):

$$\mathbf{A} = (\mathbf{I} - \mathbf{B})^{-1} (\mathbf{I} + \mathbf{B}) \tag{4}$$

This tells us immediately that $\mathbf{B} = (\mathbf{A} - \mathbf{I})(\mathbf{A} + \mathbf{I})^{-1}$. MACSYMA gives us directly (Appendix I):

$$b_i = n_i \tan \frac{1}{2} \phi \qquad i = 1,2,3 \tag{5}$$

The b_i $(i = 1,2,3)$ are the Rodrigues parameters.

We first express $\mathbf{A}$ in terms of the Rodrigues parameters. We find (Appendix II) cf. Bottema and Roth [1] p. 148:

$$\mathbf{A} = \frac{1}{\Delta} \begin{bmatrix} 1 + b_1^2 - b_2^2 - b_3^2 & 2(b_1b_2 - b_3) & 2(b_1b_3 + b_2) \\ 2(b_2b_1 + b_3) & 1 - b_1^2 + b_2^2 - b_3^2 & 2(b_2b_3 - b_1) \\ 2(b_3b_1 - b_2) & 2(b_3b_2 + b_1) & 1 - b_1^2 - b_2^2 + b_3^2 \end{bmatrix} \tag{6}$$

where $\Delta = 1 + b_1^2 + b_2^2 + b_3^2$. Using the notation $\mathbf{A} = [a_{ij}]$, it is clear from this result that:

$$b_1 = (a_{32} - a_{23})/d$$
$$b_2 = (a_{13} - a_{31})/d \tag{7}$$
$$b_3 = (a_{21} - a_{12})/d$$

with $d = 1 + a_{11} + a_{22} + a_{33}$. Having established the necessary background, we derive typical basic results by means of MACSYMA. The reader should compare our derivation with those of, for example, Bottema and Roth [1], Gibbs [5], and Dimentberg [4].

Consider the result of first rotating a body round an axis $\mathbf{n}$ by angle ϕ, then around a second axis $\mathbf{n}'$ by an angle ϕ'. Euler's theorem tells us that the result is equivalent to a rotation by some angle ϕ'' round some axis $\mathbf{n}''$. In matrices, if the matrices corresponding to these three rotations are $\mathbf{A}$, $\mathbf{A}'$, $\mathbf{A}''$ and we start with a point $\mathbf{r}$, this is first transformed into $\mathbf{r}' = \mathbf{A}\mathbf{r}$, and then $\mathbf{r}'$ is transformed into $\mathbf{r}'' = \mathbf{A}'\mathbf{r}'$. We also have $\mathbf{r}'' = \mathbf{A}''\mathbf{r}$ so that

$$\mathbf{A}'' = \mathbf{A}'\mathbf{A} \quad \cdot$$

The Rodrigues parameters corresponding to $\mathbf{n}''$, ϕ'' are given by (7) where a_{ij} are the elements of $\mathbf{A}''$. But these are given in terms of the first two rotations by the the corresponding elemnts of $\mathbf{A}'\mathbf{A}$ These matrix relations are carried out by MACSYMA in Appendix III, giving the result:

$$\mathbf{b}'' = \frac{\mathbf{b} + \mathbf{b}' - \mathbf{B}\mathbf{b}'}{1 - \mathbf{b}^T\mathbf{b}'} \tag{8}$$

where $\mathbf{B}$ is related to $\mathbf{b}$ as $\mathbf{N}$ was to $\mathbf{n}$ in (2).

Note that this is a straightforward derivation that would be laborious to carry out by hand, as compared with derivations carried out in the literature designed for the ease of hand computation.

3. Euler Parameters

Instead of using Rodrigues parameter b_i, it is often convenient to use Euler parameters c_i related to b_i by (Bottema and Roth [1] p. 150)

$$b_i = c_i/c_0 \ , \quad c_0^2 + c_1^2 + c_2^2 + c_3^2 = 1 \tag{9}$$

The relation (5) then gives

$$\mathbf{A} = \begin{bmatrix} c_0^2 + c_1^2 - c_2^2 - c_3^2 & 2(-c_0c_3 + c_1c_2) & 2(c_0c_2 + c_1c_3) \\ 2(c_0c_3 + c_2c_1) & c_0^2 - c_1^2 + c_2^2 - c_3^2 & 2(-c_0c_1 + c_2c_3) \\ 2(-c_0c_2 + c_3c_1) & 2(c_0c_1 + c_3c_2) & c_0^2 - c_1^2 - c_2^2 + c_3^2 \end{bmatrix} \tag{10}$$

Although it would seem that the Euler parameters are straightforward homogeneous forms of the Rodrigues parameters, it turns out that some relations are expressed much more simply in terms of the Euler parameters.

One example is the Euler parameter analog of (8) for two successive rotations. To derive this, substitute $\mathbf{b} = \mathbf{c}/c_0$, $\mathbf{b}' = \mathbf{c}'/c_0$ in (8) which gives:

$$\mathbf{b}'' = \frac{c_0'\mathbf{c} + c_0\mathbf{c}' - \mathbf{C}\mathbf{c}}{c_0c_0' - \mathbf{c}^T\mathbf{c}'} \tag{11}$$

When this is written out in detail we find that by introducing

$$c_0'' = c_0'c_0 - c_1'c_1 - c_2'c_2 - c_3'c_3 \tag{12}$$

$$c_0'' = c_0'c_0 - c_1'c_1 - c_2'c_2 - c_3'c_3$$

$$c_1'' = c_1'c_0 + c_0'c_1 - c_3'c_2 + c_2'c_3$$

$$c_2'' = c_2'c_0 + c_3'c_1 + c_0'c_2 - c_1'c_3 \tag{12}$$

$$c_3'' = c_3'c_0 - c_2'c_1 + c_1'c_2 + c_0'c_3$$

equation (11) can be written in the simple form

$$\mathbf{b}'' = \mathbf{c}''/c_0'' \tag{13}$$

In Appendix IV we check by MACSYMA that if $c_0^2 + \mathbf{c}^T\mathbf{c} = 1$, $(c_0')^2 + (\mathbf{c}')^T\mathbf{c}' = 1$, then $(c_0'')^2 + (\mathbf{c}'')^T\mathbf{c}'' = 1$, which is a well-known result due to Euler. This result and (12) mean that c_0'', c_1'', c_2'', c_3'' are the Euler parameters corresponding to the total rotation.

In the literature, the result (12) is often derived via quaternions (e.g. Bottema and Roth [1], p. 518,520). It is of some interest to express this approach in the present context of Euler parameters and matrices which can be done without mentioning quaternions explicitly. Introduce γ and Γ defined as follows:

$$\gamma = \begin{bmatrix} c_0 \\ c_1 \\ c_2 \\ c_3 \end{bmatrix}, \quad \Gamma = \begin{bmatrix} c_0 & -c_1 & -c_2 & -c_3 \\ c_1 & c_0 & -c_3 & c_2 \\ c_2 & c_3 & c_0 & -c_1 \\ c_3 & -c_2 & c_1 & c_0 \end{bmatrix}$$

If γ', Γ' are the corresponding matrices with c' in place of c, and similarly for γ'', Γ'', we define the product $\gamma'\gamma$ by (compare the remark following (2)):

$$\gamma'' = \gamma'\gamma = \Gamma'\gamma \tag{14a}$$

which says exactly the same as (12). We first note that if we define $\gamma^{-1} = [c_0, -c_1, -c_2, -c_3]^T$ then $\gamma\gamma^{-1} = \gamma^{-1}\gamma = [1,0,0,0]^T$. It can be verified (e.g. by the MACSYMA program in Appendix V) that introducing $\rho = [r_0, r_1, r_2, r_3]^T$, $\mathbf{r} = [r_1, r_2, r_3]^T$ and $\rho', \mathbf{r}'$ correspondingly, then if we form $\gamma\rho\gamma^{-1}$, and denote the result by ρ', then

$$\begin{bmatrix} r_0' \\ \mathbf{r}' \end{bmatrix} = \begin{bmatrix} 1 & 0 \\ 0 & \mathbf{A} \end{bmatrix} \begin{bmatrix} r_0 \\ \mathbf{r} \end{bmatrix}$$

Where $\mathbf{A}$ is precisely the matrix that appeared in (10), i.e., $\gamma\rho\gamma^{-1}$ represents a rotation. This is our version of the standard quaternion theorem on rotation, derived of course from a completely different point of view (cf. Brand [2], p. 417). A second rotation would give $\rho'' = \gamma'\rho'(\gamma')^{-1}$, and combining the rotations leads to $\rho'' = \gamma'\gamma\rho(\gamma')^{-1}\gamma^{-1}$, i.e., if γ'' represents the combined rotation then $\gamma'' = \gamma'\gamma$, which is identical with (12).

Still another way of obtaining (12) is suggested by the discussion of Cayley-Klein parameters in Bottema and Roth ([1] p. 529), namely that a result corresponding to equation (9.8) in that reference should hold for Euler parameters. We introduce the notations:

$$\mathbf{V} = c_0\mathbf{I} + \mathbf{S}$$

$$\mathbf{V}^{-1} = c_0\mathbf{I} - \mathbf{S}$$

$$S = \begin{bmatrix} 0 & -c_1 & -c_2 & -c_3 \\ c_1 & 0 & -c_3 & c_2 \\ c_2 & c_2 & 0 & -c_1 \\ c_3 & -c_2 & c_1 & 0 \end{bmatrix} , \quad q = \begin{bmatrix} y & z & 0 & -x \\ z & -y & x & 0 \\ 0 & x & y & z \\ -x & 0 & z & -y \end{bmatrix}$$

$$Q = \begin{bmatrix} Y & Z & 0 & -X \\ Z & -Y & X & 0 \\ 0 & X & Y & Z \\ -X & 0 & Z & -Y \end{bmatrix}$$

The MACSYMA program in Appendix VI does the following. We form VqV^{-1} and equate this to Q. This gives 16 equations. However, it is easily checked by MACSYMA that, in fact, there are only *three* independent relations involving x,y,z and X,Y,Z which can be written in the form

$$Aq = Q$$

where A is exactly the A given in (10). The implication of this, in connection with repeated rotations, is that if q corresponds to r and Q to r' defined in the second paragraph of Section 1 and the corresponding A is denoted by V, then

$$VrV^{-1} = r'$$

Similarly, the second rotation gives $V'r'V'^{-1} = r''$ and the rotation from the initial position to the final position gives $V''rV''^{-1} = r''$. Eliminating r' we have $V''r(V'')^{-1} = V'VrV^{-1}V'^{-1}$ so that finally

$$V'' = V'V \tag{14b}$$

and this is precisely equation (12) cf. (14a).

4. An Example Involving Dual Quaternions

The discussion in the last two sections was concerned with the rotation of a body with a fixed point and involved only three independent parameters. The general motion of a body involves displacement, as well as rotation, and requires six independent parameters. Rather than extending the methods of the last two sections, we illustrate how MACSYMA deals with a rather different approach to kinematics, namely via quaternions, by considering a calculation in a classic paper by Yang and Freudenstein ([14], 1964) dealing with a spatial four-bar mechanism.

In Figure 2, MA and NB are two nonparallel and nonintersecting lines. MN is the common perpendicular. Let a,b denote unit vectors in the direction of MA, NB respectively, and let r_a, r_b denote the vectors $\vec{OM}$, $\vec{ON}$. We introduce the quaternions

$$\hat{a} = a + \epsilon(r_a \times a) \quad , \quad \hat{b} = b + \epsilon(r_b \times b)$$

where ϵ is a symbol with the property that $\epsilon^2 = 0$. Note that this implies, for example, that if $\hat{\theta} = \theta + \epsilon s$ then

$$\sin\hat{\theta} = \sin\theta + \epsilon s\cos\theta \quad , \quad \cos\hat{\theta} = \cos\theta - \epsilon s\sin\theta \tag{15}$$

As discussed by Yang et al. [14], the relative shift between $\hat{a}$ and $\hat{b}$ can be expressed as

$$\hat{b} = Q\hat{a} \quad , \quad \hat{a} = \hat{b}Q$$

where Q is a dual quaternion (see [14], (22, 23)). Successive application of formulae of this type gives rise to a loop closure equation for the mechanism of the form:

$$A(\hat{\theta}_1)\sin\hat{\theta}_4 + B(\hat{\theta}_1)\cos\hat{\theta}_4 = C(\hat{\theta}_1) \tag{16}$$

where

$$A(\hat{\theta}_1) = \sin\hat{\alpha}_{12}\sin\hat{\alpha}_{34}\sin\hat{\theta}_1$$

$$B(\hat{\theta}_1) = -\sin\hat{\alpha}_{34}\,(\sin\hat{\alpha}_{41}\cos\hat{\alpha}_{12} + \cos\hat{\alpha}_{41}\sin\hat{\alpha}_{12}\cos\hat{\theta}_1)$$

$$C(\hat{\theta}_1) = \cos\hat{\alpha}_{23} - \cos\alpha_{34}\,(\cos\hat{\alpha}_{41}\,\cos\hat{\alpha}_{12} - \sin\hat{\alpha}_{41}\sin\hat{\alpha}_{12}\cos\hat{\theta}_1)$$

Here

$$\hat{\alpha}_{12} = \alpha_{12} + \epsilon a_{12} \quad , \quad \hat{\theta}_1 = \theta_1 + \epsilon s_{11}$$

$$\hat{\alpha}_{23} = \alpha_{23} + \epsilon a_{23} \quad , \quad \hat{\theta}_2 = \theta_2 + \epsilon s_2$$

$$\hat{\alpha}_{34} = \alpha_{34} + \epsilon a_{34} \quad , \quad \hat{\theta}_3 = \theta_3 + \epsilon s_3$$

$$\hat{\alpha}_{41} = \alpha_{41} + \epsilon a_{41} \quad , \quad \hat{\theta}_4 = \theta_4 + \epsilon s_4$$

It is then clear that (15) can be reduced to the form

$$\cdot P + \epsilon Q = R + \epsilon S$$

where P, Q, R, and S are independent of ϵ. It is required to find the explicit form of P, Q, R, and S. To calculate this by hand is extremely laborious, but straightforward in MACSYMA. The program is given in Appendix VII.

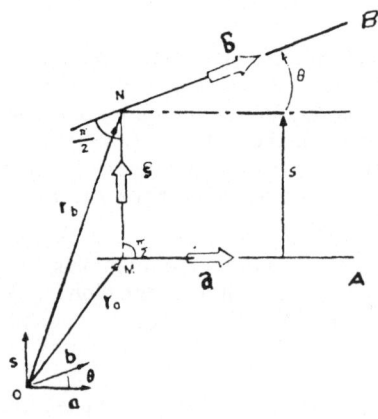

Figure 2. Relative position of two line vectors.

TWO EXAMPLES IN DYNAMICS

5. Equations of Motion for the Stanford Manipulator Arm

There are a number of ways to set up dynamical equations for robot manipulator arms (see Paul [13]). Kane-Levinson [9] have given an example of setting up dynamical equations for the Stanford manipulator. Our objective is to reproduce these equations from an algorithmic point of view, without having to do by hand the kind of extensive manipulation given in that paper. The method can help us to set up similar sets of equations for any manipulator automatically, thereby reducing the labor. We also show that MACSYMA can simplify the Kane-Levinson end-result, reducing the numbers of arithmetic operations required to obtain numerical results.

We consider the Stanford manipulator arm (Paul [13]), a six-element, six-degree-of-freedom manipulator. A schematic representation of this arm is given in Figure 3, from Kane-Levinson [9], where more details can be found. The six bodies are designated A, ..., F. Body A can be rotated about a vertical axis fixed in space. A supports B which can be rotated about a horizontal axis fixed relative to A. The figure should now be self-explanatory, the joint connecting B and C being translational, and the remaining joints rotational.

q_1, ..., q_6 are generalized coordinates characterizing the instantaneous configuration of the arms, the first five being rotational and q_6 translational. For the plane configuration of the arms as drawn in Figure 3, it is assumed that $q_1, \cdots q_5$ are zero.

We choose coordinate axes as follows. n_1, n_2, n_3 are unit vectors fixed in space as indicated in Figure 3, n_1, n_2 lying in the plane of the paper. a_1, a_2, a_3 are unit vectors fixed in the arm A which coincide with n_1, n_2, n_3 when the arm is in the configuration of Figure 3. Similarly, b_1, b_2, b_3 are unit vectors attached to the arm B and similarly for C, D, E, and F.

We give a mathematical description of an algorithm for setting up the dynamical equations. This is essentially the algorithm described by Kane-Levinson [9], but organized in a somewhat different way in order to facilitate implementation on MACSYMA. The stages and details of the MACSYMA program which are in Appendix VIII, parallel the mathematical description that follows:

Stage 1: Set up angular velocities:

Rotations about x,y,z axes can be described by orthogonal matrices of simple form as discussed in detail by Paul [13], Chapter 1. For instance, rotation by an angle θ about the x-axis involves ([13], p. 15)

$$\text{Rot}_x(\theta) = \begin{bmatrix} 1 & 0 & 0 \\ 0 & \cos\theta & -\sin\theta \\ 0 & \sin\theta & \cos\theta \end{bmatrix}$$

Let R_1, ... R_5 denote matrices corresponding to rotations θ_1, ..., θ_5 about axes, y,x,y,x,y respectively in the local coordinates fixed relative to arms A, B, D, E, F. Let $\dot{q}_1$, ..., $\dot{q}_5$ denote angular velocities around y,x,y,x,y axes respectively. These are vector quantities represented by matrices that we denote by ω_1, ..., ω_5. For instance, $\omega_1 = [0,\dot{q}_1,0]$ etc. Similarly, for the linear velocity $\dot{q}_6$.

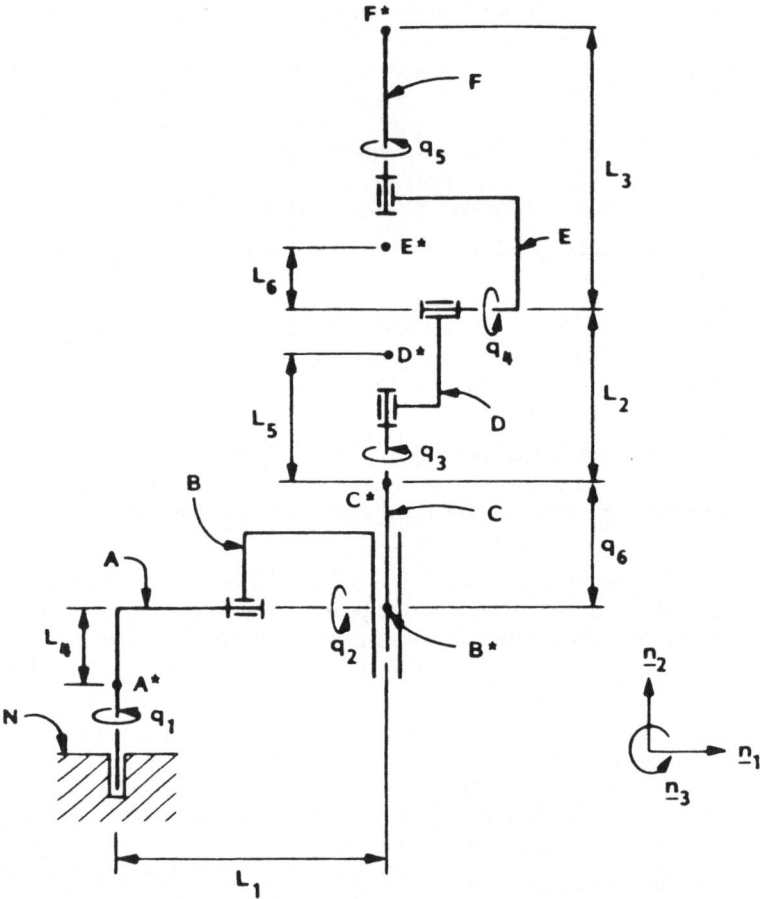

Figure 3. A schematic representation of Stanford manipulator arm.

Next introduce ω^A, ..., ω^F, the angular velocities of A, ..., F in our Newtonian frame of reference, but with components expressed in the local coordinate frame of reference. For example:

$$\omega^D = [u_1,u_2,u_3] \quad \text{means:} \quad \omega^D = u_1\underline{d}_1 + u_2\underline{d}_2 + u_3\underline{d}_3 \qquad (17)$$

The algorithm for computing ω^A, ..., ω^F is given by:

$$\omega^A = \omega_1 R_1$$

$$\omega^B = (\omega^A + \omega_2)R_2$$

$$\omega^C = \omega^B$$

$$\omega^D = (\omega^C + \omega_3)R_3$$

$$\omega^E = (\omega^D + \omega_4)R_4$$

$$\omega^F = (\omega^E + \omega_5)R_5$$

If these formulae are used as they stand, the expression for ω^F in terms of $\dot{q}_i$ will be complicated. The complexity can be reduced using a method due to Kane-Levinson [9]. The u_i that occur in (16) can be expressed in terms of $\dot{q}_1$, $\dot{q}_2$, $\dot{q}_3$ as follows

$$u_1 = \dot{q}_1 \sin q_2 \sin q_3 + \dot{q}_2 \cos q_3$$

$$u_2 = \dot{q}_1 \cos q_2 + \dot{q}_3$$

$$u_3 = -\dot{q}_1 \sin q_2 \cos q_3 + q_2 \sin q_1$$

$$u_i = \dot{q}_i \qquad i = 4,5,6$$

Stage 2: Set up linear velocities:

In stage 1, the angular velocities were always expressed in local coordinates corresponding to the arm being considered. This is not necessarily the case for the way in which Kane-Levinson [9] formulate the linear velocities (see paragraph preceeding (28) in the paper). Because we wish our results to be comparable to those in [9], we state the formulae we use, which will lead to results that are the same as those in equations (28-43) in [9]. (Note that the stars in the following refer to the velocities of the centers of mass of the corresponding arms.)

$$v^{A^*} = 0$$

$$v^{B^*} = \omega^A \times R^B$$

$$v^{C^*} = \omega^B \times R^C + \bar{\dot{q}}_6$$

$$v^{D^*} = \omega^B \times R^D + \bar{\dot{q}}_6$$

The expressions for v^{E^*}, v^{F^*} correspond to those in equation (40) and (42) in the Kane-Levinson paper [9]. The exact form we use can be found from the expressions for VE and VF in the MACSYMA program given in Appendix VIII.

The remaining stages are relatively straightforward.

Stage 3: Find the partial angular velocities.

Stage 4: Find the partial linear velocities.

These are explained in the Kane-Levinson paper [9] and the MACSYMA implementation in Appendix VIII is self-explanatory.

Stage 5: Find the angular accelerations.

Stage 6: Find the linear accelerations.

These are obtained by simple differentiation of the corresponding angular and linear velocities as given in the MACSYMA program in Appendix VIII.

Stage 7: Define moments of inertia.

We next have to consider forces.

Stage 8: Define torques.

Stage 9: Set up generalized forces.

Stage 10: Set up active forces.

Stage 11: Set up Kane's equations.

These steps are straightforward; the MACSYMA program is given in Appendix VIII.

Finally, Figure 4 gives a comparison of some numerical results obtained from MACSYMA and Kane-Levinson [9].

It is of some interest to compare the mathematical equations in the Kane-Levinson paper with the corresponding MACSYMA expressions. For example, consider:

Kane-Levinson [9]	MACSYMA
(underlined quantities are vectors)	

$$\underline{\omega}^A = \dot{q}_1\underline{a}_2 \qquad (13)$$

WA: EXPAND(W1.R1)

But

$\underline{\omega}^A \equiv WA$, $\dot{q}_1\underline{a}_2 \equiv W1.R1$

$$\dot{q}_1 = \frac{u_1 s_3 - u_3 c_3}{s_2} \qquad (8)$$

Here ω^A, $\dot{q}_1$, $\underline{a}_2$ are vectors;

WA, W1.R1 are matrices

Introduce

$$Z_4 = \frac{s_3}{s_2}, \quad Z_5 = -\frac{c_3}{c_2}$$

Then (13) becomes:

$$\underline{\omega}^A = (Z_4 u_1 + Z_5 u_3)\underline{a}_2 \qquad (15)$$

Similarly,

$$\underline{\omega}^B = Z_2\underline{b}_1 + Z_{10}\underline{b}_2 + Z_{11}\underline{b}_3 \qquad (16)$$

WB: EXPAND ((WA+W2).R2)

where

$$Z_{10} = Z_6 u_1 + Z_7 u_3, \quad Z_{11} = Z_8 u_1 + Z_9 u_3$$

One point here is that because Kane-Levinson [9] are carrying out the algebra by hand, it is convenient for them to introduce intermediate symbols Z_1, Z_2 $\cdots$ going up to Z_{196}, and similarly, 36 X's and 31 W's. MACSYMA has no difficulty in generating the end result in explicit form. These end results are no more complex than the complexity of the equations given in [9]. At the time of writing this paper a preliminary number count on additions and multiplication for X_{ij}, the coefficients of equations of motion, obtained by MACSYMA, as compared to those in [9], shows a reduction by approximately a factor of two.

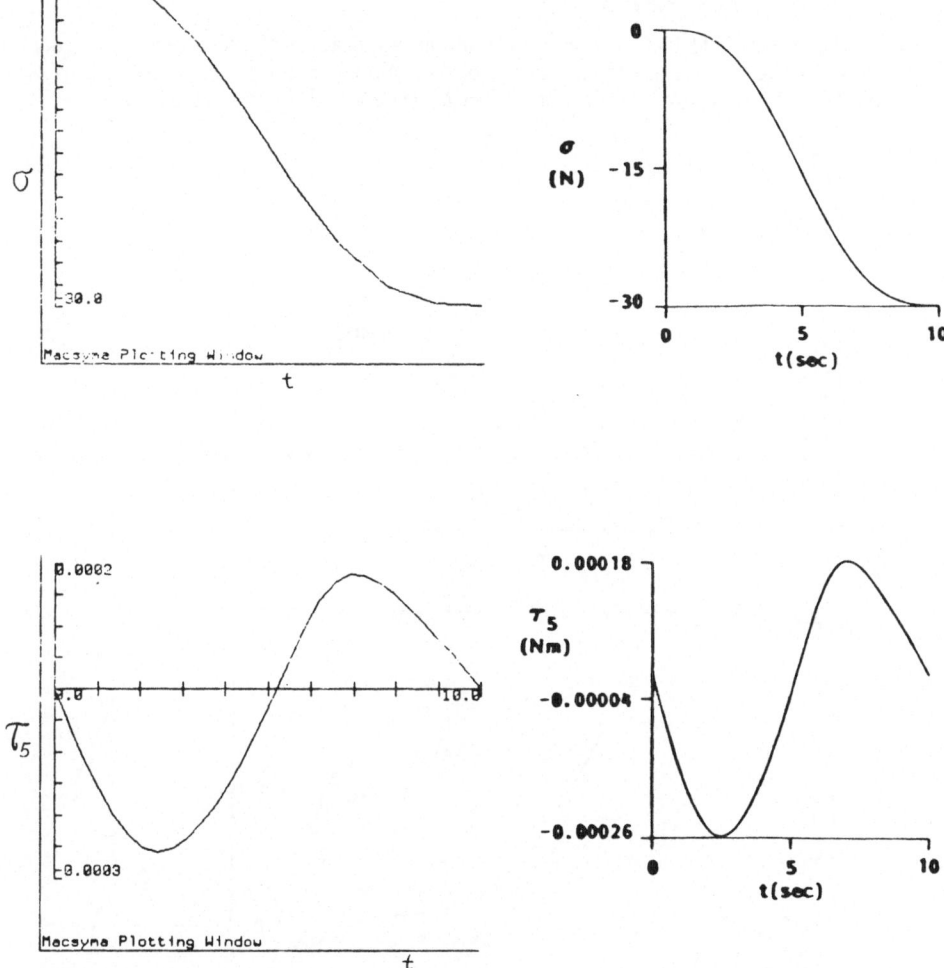

Figure 4a. Comparison of numerical results for σ, τ_5 obtained by MACSYMA and Reference 9.

In conclusion, we note that Paul [13] sets up the dynamical equation of the Stanford manipulator arm using the Lagrangian equation approach. See also [6]. Some applications of the Lagrange method using MACSYMA are discussed in [9].

Various methods of setting up dynamical equations that could be carried out by MACSYMA are illustrated in [8].

6. A Spacecraft Problem

Levinson [11] has described in detail an application of the symbolic language FORMAC to formulate the spacecraft problem shown in Figure 5, consisting of two rigid bodies with a common axis of rotation b. (See also [10], pp.279-285).

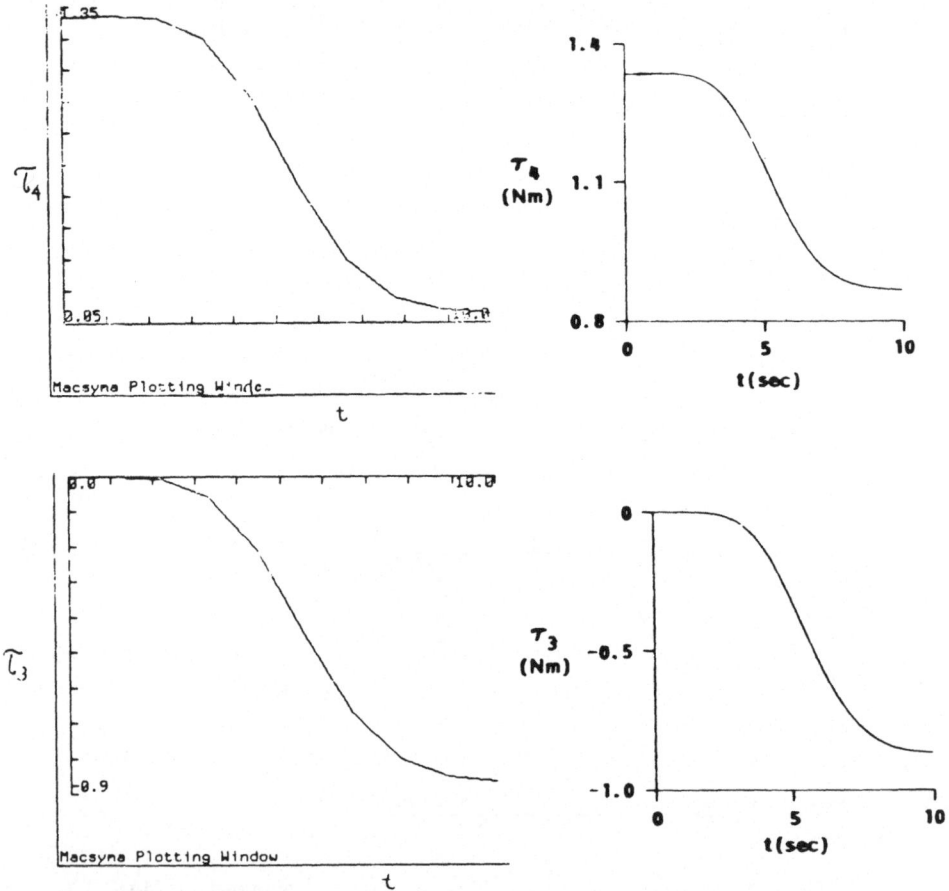

Figure 4b. Comparison of numerical results for τ_4 and τ_3 obtained by MACSYMA and Reference 9.

275

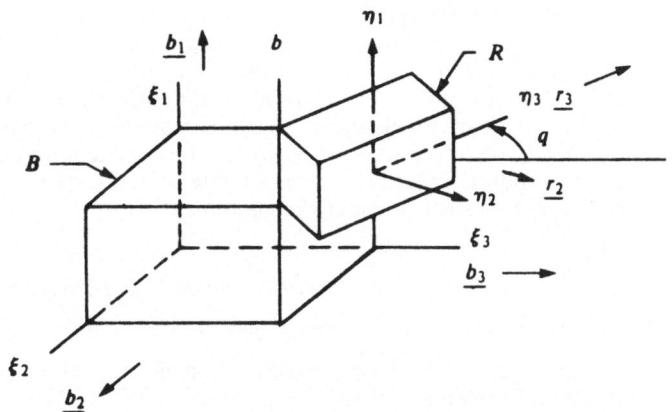

Figure 5. Two rigid bodies with a common axis of rotation.

The equations are given in complete detail in Ref. [11], and translated into MACSYMA in Appendix IX. In the example in the last section, we wrote the MACSYMA program in terms of matrices. In Appendix IX, the present example is written in terms of vectors, by writing BLOCK functions to perform the dot and cross products. To illustrate the comparison of the vector equation with the corresponding MACSYMA expressions:

Equations from Ref. [11]		MACSYMA
$\underline{r}_2 = \cos q\ \underline{b}_2 + \sin q\ \underline{b}_3$	(1)	R[2]:COS(Q)*B[2]+SIN(Q)*B[3];
$\omega^B = u_1 \underline{b}_1 + u_2\ \underline{b}_2 + u_3\ \underline{b}_3$	(3)	WB:U[1]*B[1]+U[2]*B[2]+U[3]*B[3];
$\mu_4 = \dot{q}$		U[4]:DIFF(Q,T);
$\alpha^R = \dfrac{d}{dt}(\omega^R) + \omega^B \times \omega^R$	(7)	ALPR:DIFF(WR,T)+CROSS(WB,WR);

We discuss only one other correspondence. Equation (27) in Ref. [11] is

$$F_r = \frac{\partial \nu^{B^*}}{\partial \mu_r} \cdot (F)_B + \frac{\partial \omega^B}{\partial \mu_r} \cdot (T)_B \quad (r=1, \cdots, 7)$$

which becomes in MACSYMA

F[R]:=DOT(DIFF(VBS,U[R]),FB) + DOT(DIFF(WB,U[R]),TB);

The complete set of equations given in Ref. [11] is generated by Appendix IX. The reader should compare the corresponding FORMAC program given in Levinson [11].

CONCLUDING REMARKS

It should be clear from the examples given that symbolic manipulation by computer can carry out many of the laborious and routine calculations involved in the analysis of mechanical systems. But, potentially even more important, is the influence that symbolic manipulation is likely to have on the methods used to formulate problems. The reader should compare, for example, the algorithmic approach we have adopted to the Stanford manipulator arm problem with the approach in [9]. As another example, if symbolic manipulation methods are used, this will influence whether we formulate problems in terms of Euler angles, Euler parameters, Rodrigues parameters, or quaternions, etc. In addition, one can visualize the production of standard MACSYMA software to produce equations corresponding to those of Kane-Levinson [9] for *any* given combination of rotating and sliding joints.

REFERENCES

1. O. Bottema and B. Roth, *Theoretical Kinematics*, North-Holland, 1979.

2. L. Brand, *Vector and Tensor Analysis*, Wiley, 1957.

3. V. N. Branets and I. P. Shmyglevskiy, "Application of Quarternions to Rigid Body Rotation Problems," NASA Tech. Transl. TTF-15, 414, 1974 (1973 Russian original).

4. F. M. Dimentberg, "The Screw Calculus and its Applications in Mechanics," NTIS Transl. FTD-HT-23-1632-67, 1968 (1965 Russian original).

5. J. W. Gibbs, *Vector Analysis*, Dover reprint, 1960 (original published in 1909).

6. J. M. Hollerbach, "A Recursive Lagrangian Formulation of Manipulator Dynamics and a Comparative Study of Dynamics Formulation Complexity," *IEEE Trans. on Syst., Man and Cyb. SMC-10*, 1980, 730-736.

7. M. A. Hussain and B. Noble, "Application of Symbolic Computation to the Analysis of Mechanical Systems, Including Robot Arms," General Electric Technical Report 84CRD062, 1984. Also to be published in the Proceedings of the NATO Conference on Mechanisms, E. Haug, ed., University of Iowa, 1984.

8. T. R. Kane and D. A. Levinson, "Formulation of Equations of Motion for Complex Spacecraft," *J. Guidance and Control*, 1980, 99-112.

9. T. R. Kane and D. A. Levinson, "The Use of Kane's Dynamical Equations in Robotics", *Int. J. Robotics Research 2*, 1983, 3-21.

10. T. R. Kane, P. W. Likins, and D. A. Levinson, *Spacecraft Dynamics*, McGraw-Hill, 1983.

11. D. A. Levinson, "Equations of Motion for Multiple-Rigid-Body Systems via Symbol Manipulation," *J. Spacecraft and Rockets 14*, 1977, 479-487.

12. P. E. Nikravesh, R. A. Wehage, and E. J. Haug, *Computer-Aided Analysis of Mechanical Systems*, to be published.

13. R. P. Paul, *Robot Manipulators - Mathematics, Programming, and Control*, M. I. T. Press, 1981.

14. A. T. Yang and F. Freudenstein, "Application of Dual-Number Quaternion Algebra in the Analysis of Spatial Mechanisms," *Trans ASME J. Appl. Mech.*, 1966, 300-308.

APPENDIX I

```
/*PROVE THE IDENTITY OF EQUATION 5 */
/* TRIGNOMETRIC SIMPLIFICATION */
MATCHDECLARE(A,TRUE);
TELLSIMP(SIN(A)^2,1-COS(A)^2);
/* DEFINE CROSS PRODUCT MATRIX OR ALTERNATING TENSOR */
ALT(N):=MATRIX([0,-N[3,1],N[2,1]],[N[3,1],0,-N[1,1]],
[-N[2,1],N[1,1],0]);
N:MATRIX([N1],[N2],[N3]);
NN:ALT(N);
I:IDENT(3);
AA:COS(ALPHA)*I+(1-COS(ALPHA))*(N . TRANSPOSE(N))+NN*SIN(ALPHA);
AAP:I+AA;
/* WORK WITH HALF ANGLES */
ALPHA:BTA*2;
EV(AA);
TRIGEXPAND(%);
AA:%$
/* ADD IDENTITY MATRIX AND INVERT */
AAP:AA+I$
IAAP:AAP^^(-1)$
/*SUBTRACT IDENTITY MATRIX AND FORM MATRIX PRODUCT AS ANSWER*/
AAM:AA-I$
ANSWER:AAM . IAAP$
/*USE IDENTITY THAT N1^2+N2^2+N3^2=1 */
NN3:1-N1**2-N2**2;
ANSWER:RATSUBST(NN3,N3^2,ANSWER);
ANSWER:RATSIMP(%);
```

APPENDIX II

```
/* CAYLEY'S DECOMPOSITION OF ORTHOGONAL MATRIX
A =(I-B)^-1(I+B),WHERE B1,B2,B3 ARE RODRIGUES PARAMETERS*/
/* DEFINE CROSS PRODUCT OR ALTERNATING TENSOR MATRIX*/
ALT(N):=MATRIX([0,-N[3,1],N[2,1]],[N[3,1],0,-N[1,1]],
[-N[2,1],N[1,1],0]);
B:MATRIX([B1],[B2],[B3]);
BB:ALT(B);
I:IDENT(3);
INBB:(I-BB)^^-1;
A:INBB.(I+BB);
ANSWER:RATSIMP(%);
/*SOLVE ABOVE FOR B1 B2 B3,FOLLOWIN IS A CROSS CHECK */
DEL:RATSIMP(1+A[1,1]+A[2,2]+A[3,3]);
BB1:RATSIMP(1/DEL*(A[3,2]-A[2,3]));
BB2:RATSIMP(1/DEL*(A[1,3]-A[3,1]));
BB3:RATSIMP(1/DEL*(A[2,1]-A[1,2]));
```

APPENDIX III

```
/*TWO SUCCESIVE ROTATIONS IN TERMS OF RODRIGUES PARAMETER*/
ALT(N):=MATRIX([0,-N[3,1],N[2,1]],[N[3,1],0,-N[1,1]],[-N[2,1],N[1,1],0]);
B:MATRIX([B1],[B2],[B3]);
BB:ALT(B);
I:IDENT(3);
INBB:(I-BB)^^-1;
A:INBB.(I+BB)$
A:RATSIMP(%);
BP:MATRIX([BP1],[BP2],[BP3]);
BBP:ALT(BP);
INBBP:(I-BBP)^^-1;
AP:INBBP.(I+BBP)$
AP:RATSIMP(%);
APP:AP.A;
/*SOLVE ABOVE FOR BPP1 BPP2 BPP3 */
DEL:RATSIMP(1+APP[1,1]+APP[2,2]+APP[3,3]);
BPP1:RATSIMP(1/DEL*(APP[3,2]-APP[2,3]));
BPP2:RATSIMP(1/DEL*(APP[1,3]-APP[3,1]));
BPP3:RATSIMP(1/DEL*(APP[2,1]-APP[1,2]));
/*THE ABOVE RESULTS ARE SAME AS EQUATION (11) */
```

APPENDIX IV

```
/* DERIVE EULER IDENTITY SEE ALSO  BRAND REF. [2] P.408*/
S:MATRIX(
[0,-CC1,-CC2,-CC3],
[CC1,0,-CC3,CC2],
[CC2,CC3,0,-CC1],
[CC3,-CC2,CC1,0]);
SP:MATRIX(
[0,-CP1,-CP2,-CP3],
[CP1,0,-CP3,CP2],
[CP2,CP3,0,-CP1],
[CP3,-CP2,CP1,0]);
I:IDENT(4);
V:CC0*I+S;
VP:CP0*I+SP;
MAT1:V.VP;
/* NOW TAKE THE FIRST COLUMN OF THE ABOVE MATRIX AND SQUARE IT*/
MAT2:SUBMATRIX(%,2,3,4);
ANSWER:%.%;
ANSWER:FACTOR(ANSWER);
/* NOTE ABOVE IS A COMPLETE SQUARE */
```

APPENDIX V

```
/*QUATERNION MULTIPLICATION EXAMPLE */
/*ANALOG OF CAYLEY-KLEIN RESULT */
I:IDENT(4);
/* NOW WE DEFINE AN OPERATION SS ON A COLUMN MATIX BASED ON ANALOG
OF CAYLEY KLEIN DECOMPOSITION */
SS(CC):=MATRIX([CC[1,1],-CC[2,1],-CC[3,1],-CC[4,1]],
[CC[2,1],CC[1,1],-CC[4,1],CC[3,1]],
[CC[3,1],CC[4,1],CC[1,1],-CC[2,1]],
[CC[4,1],-CC[3,1],CC[2,1],CC[1,1]]);
/* DEFINE AN INVERSE OPERATION */
INV(CC):=1/(CC.CC)*MATRIX([CC[1,1]],[-CC[2,1]],[-CC[3,1]],[-CC[4,1]]);
/* NOW THE BRANDS'S THEOREM ON QUATERNION FORMULATED IN MATRIX FORM */
RHO:MATRIX([R0],[R1],[R2],[R3]);
GAM:MATRIX([Q0],[Q1],[Q2],[Q3]);
/*NOW DEFINE QUATERNION PRODUCT */
APROD(R,Q):=SS(R).Q;
A:MATRIX([A0],[A1],[A2],[A3]);
RATSIMP(APROD(INV(A),A));
ANSWER:RATSIMP(APROD(GAM,APROD(RHO,INV(GAM))));
EQ1:ANSWER[1,1];
EQ2:ANSWER[2,1];
EQ3:ANSWER[3,1];
EQ4:ANSWER[4,1];
/* NOW GENERATE COEFFICIENT MATRIX FOR RHO */
COEFMATRIX([EQ1,EQ2,EQ3,EQ4],[R0,R1,R2,R3]);
/* THE ABOVE IS SAME AS EXTENDED EULER PARAMETER MATRIX */
```

APPENDIX VI

```
/* THE BASIC DECOMPOSITION FOR EULER PARAMETER */
/*TEST OUT (C0*I+S)X(C0*I-S)     */
I:IDENT(4);
SS:MATRIX([0,-CC1,-CC2,-CC3],
[CC1,0,-CC3,CC2],
[CC2,CC3,0,-CC1],
[CC3,-CC2,CC1,0]);
Q:MATRIX([Y,Z,0,-X],
[Z,-Y,X,0],
[0,X,Y,Z],
[-X,0,Z,-Y]);
EQ1:EXPAND((CC0*I+SS).Q.(CC0*I-SS));
T1:EQ1[2,3];
T2:EQ1[1,1];
T3:EQ1[4,3];
ANSWER:COEFMATRIX([T1,T2,T3],[X,Y,Z]);
/*ABOVE IS SAME A S EQUATION 11 */
```

APPENDIX VII

```
/*........ALGEBRA FOR QUATERNIONS FROM YANG'S PAPER....*/
NNPRED(N):=IS(N>=2);
MATCHDECLARE(NN,NNPRED);
TELLSIMPAFTER(EP^NN,0);
/*ABOVE WILL ELIMINATE EP**2 TERMS */
AL12H:AL12+EP*A12;
AL23H:AL23+EP*A23;
AL34H:AL34+EP*A34;
AL41H:AL41+EP*A41;
 TH1H:TH1+EP*S11;
 TH2H:TH2+EP*S2;
 TH3H:TH3+EP*S3;
 TH4H:TH4+EP*S4;
SAL12H: EXPAND(TAYLOR(SIN(AL12H),EP,0,1));
SAL23H: EXPAND(TAYLOR(SIN(AL23H),EP,0,1));
SAL34H: EXPAND(TAYLOR(SIN(AL34H),EP,0,1));
SAL41H: EXPAND(TAYLOR(SIN(AL41H),EP,0,1));
STH1H:  EXPAND(TAYLOR(SIN(TH1H),EP,0,1));
STH2H:  EXPAND(TAYLOR(SIN(TH2H),EP,0,1));
STH3H:  EXPAND(TAYLOR(SIN(TH3H),EP,0,1));
STH4H:  EXPAND(TAYLOR(SIN(TH4H),EP,0,1));
CAL12H:EXPAND(TAYLOR(COS(AL12H),EP,0,1));
CAL23H:EXPAND(TAYLOR(COS(AL23H),EP,0,1));
CAL34H:EXPAND(TAYLOR(COS(AL34H),EP,0,1));
CAL41H:EXPAND(TAYLOR(COS(AL41H),EP,0,1));
CTH1H: EXPAND(TAYLOR(COS(TH1H),EP,0,1));
CTH2H: EXPAND(TAYLOR(COS(TH2H),EP,0,1));
CTH3H: EXPAND(TAYLOR(COS(TH3H),EP,0,1));
CTH4H: EXPAND(TAYLOR(COS(TH4H),EP,0,1));
AATH1H:SAL12H*SAL34H*STH1H;
BBTH1H:-SAL34H*(SAL41H*CAL12H+CAL41H*SAL12H*CTH1H);
CCTH1H:CAL23H-CAL34H*(CAL41H*CAL12H-SAL41H*SAL12H*CTH1H);
EQ1:AATH1H*STH4H+BBTH1H*CTH4H-CCTH1H;
PRIMARY:EV(EQ1,EP=0);
DUAL:RATCOEFF(EQ1,EP);
A:RATCOEFF(PRIMARY,SIN(TH4));
B:RATCOEFF(PRIMARY,COS(TH4));
C:EXPAND(PRIMARY-A*SIN(TH4)-B*COS(TH4));
DUAL1:DUAL-S4*(A*COS(TH4)-B*SIN(TH4));
A0:RATCOEFF(DUAL1,SIN(TH4));
B0:RATCOEFF(DUAL1,COS(TH4));
CC0:EXPAND(DUAL1-A0*SIN(TH4)-B0*COS(TH4));
CC0:RATSIMP(CC0);
```

APPENDIX VIII

```
/*DYNAMICAL EQUATIONS FOR STANFORD MANIPULATOR*/
MATCHDECLARE(A,TRUE);
DEPENDS([Q1,Q2,Q3,Q4,Q5,Q6],T);
DEPENDS(U,T);
/*TRIGNOMETRIC SIMPLIFICATIONS */
TELLSIMP(SIN(A)^2,1-COS(A)^2);
S1:SIN(Q1);
CC1:COS(Q1);
S2:SIN(Q2);
CC2:COS(Q2);
S3:SIN(Q3);
CC3:COS(Q3);
/* EXPRESS LOCAL ANGULAR VELOCITIES IN TERMS OF GENERALIZED ONES*/
QD1:1/S2*(U[1]*S3-U[3]*CC3);
QD2:U[1]*CC3+U[3]*S3;
QD3:U[2]+(U[3]*CC3-U[1]*S3)*CC2/S2;
QD4:U[4];
QD5:U[5];
QD6:U[6];
GRADEF(Q1,T,QD1);
GRADEF(Q2,T,QD2);
GRADEF(Q3,T,QD3);
GRADEF(Q4,T,QD4);
GRADEF(Q5,T,QD5);
GRADEF(Q6,T,QD6);
/*DEFINE ROTATIONS */
ROTX(Q):=MATRIX([1,0,0],[0,COS(Q),-SIN(Q)],[0,SIN(Q),COS(Q)]);
ROTY(Q):=MATRIX([COS(Q),0,SIN(Q)],[0,1,0],[-SIN(Q),0,COS(Q)]);
ROTZ(Q):=MATRIX([COS(Q),-SIN(Q),0],[SIN(Q),COS(Q),0],[0,0,1]);
W1:MATRIX([0,QD1,0]);
```

```
W2:MATRIX([QD2,0,0]);
W3:MATRIX([0,QD3,0]);
W4:MATRIX([QD4,0,0]);
W5:MATRIX([0,QD5,0]);
W6:MATRIX([0,QD6,0]);
/*SET UP ROTATION MATRICES */
R1:ROTY(Q1);
R2:ROTX(Q2);
R3:ROTY(Q3);
R4:ROTX(Q4);
R5:ROTY(Q5);
/*STAGE 1. SET UP ANGULAR VELOCITIES */
WA:EXPAND(W1 . R1);
WB:EXPAND(W1 . R1 . R2+W2 . R2);
WC:WB;
WD:EXPAND(W1 . R1 . R2 . R3+W2 . R2 . R3+W3 . R3);
WE:EXPAND(W1 . R1 . R2 . R3 . R4+W2 . R2 . R3 . R4+W3 . R3 . R4+W4 . R4);
WF:EXPAND(W1.R1.R2.R3.R4. R5+W2.R2.R3.R4.R5+W3.R3.R4.R5+W4.R4.R5+W5.R5);
/* SET UP BASE VECTORS AND CROSS PRODUCT */
AA:MATRIX([AA1,AA2,AA3]);
BB:MATRIX([BB1,BB2,BB3]);
CC:MATRIX([CC1,CC2,CC3]);
DD:MATRIX([DD1,DD2,DD3]);
EE:MATRIX([EE1,EE2,EE3]);
FF:MATRIX([FF1,FF2,FF3]);
CROSS(A,B,BASE):=BLOCK([],MATRIX([A[1,2]*B[1,3]-A[1,3]*B[1,2],
-(A[1,1]*B[1,3]-A[1,3]*B[1,1]),A[1,1]*B[1,2]-A[1,2]*B[1,1]]));
/* LENTHE VECTORS FOR VELOCITIES */
VECL1:MATRIX([L1,0,0]);
VECQ6:MATRIX([0,Q6,0]);
VECL5:MATRIX([0,L5,0]);
VECL2:MATRIX([0,L2,0]);
VECL6:MATRIX([0,L6,0]);
VECL3:MATRIX([0,L3,0]);
VECL4:MATRIX([0,L4,0]);
/*STAGE 2. SET UP LINEAR VELOCITIES */
VA:MATRIX([0,0,0]);
RB:MATRIX([L1,L4,0]);
VB:CROSS(WA,RB,AA);
RC:VECQ6+VECL1 . R2;
VC:CROSS(WB,RC,CC);
/*ADD LINEAR COMPONENT */
VC:VC+W6;
RD:VECL1 . R2+VECQ6+VECL5;
VD:CROSS(WB,RD,CC);
/*ADD LINEAR COMPONENT */
VD:VD+W6;
/*FOR VF START WITH VELOCITY OF C */
VE:EXPAND(VC . R3 . R4)+CROSS(WE-W4,VECL2 . R3 . R4,EE)+CROSS(WE,VECL6,EE)
/*REPLACE L6 BY L3 IN ABOVE FOR VELOCITY OF F*/
VF:RATSUBST(L3,L6,%);
/*STAGE 3. SET UP PARTIAL ANGULAR VELOCITIES */
FOR I THRU 6 DO LDISPLAY(WAR[I]:RATCOEF(WA,U[I]));
FOR I THRU 6 DO WBR[I]:RATCOEF(WB,U[I]);
FOR I THRU 6 DO WCR[I]:RATCOEF(WC,U[I]);
FOR I THRU 6 DO WDR[I]:RATCOEF(WD,U[I]);
FOR I THRU 6 DO WER[I]:RATCOEF(WE,U[I]);
FOR I THRU 6 DO WFR[I]:RATCOEF(WF,U[I]);
/*STAGE 4. SET UP PARTIAL LINEAR VELOCITIES */
FOR I THRU 6 DO VAR[I]:RATCOEF(VA,U[I]);
FOR I THRU 6 DO VBR[I]:RATCOEF(VB,U[I]);
FOR I THRU 6 DO VCR[I]:RATCOEF(VC,U[I]);
FOR I THRU 6 DO VDR[I]:RATCOEF(VD,U[I]);
FOR I THRU 6 DO VER[I]:RATCOEF(VE,U[I]);
FOR I THRU 6 DO VFR[I]:RATCOEF(VF,U[I]);
/*STAGE 5. FIND THE ANGULAR ACCELERATIONS */
ALPHAA:DIFF(WA,T)$
ALPHAB:DIFF(WB,T)$
ALPHAC:DIFF(WC,T)$
ALPHAD:DIFF(WD,T)$
ALPHAE:DIFF(WE,T)$
ALPHAF:DIFF(WF,T)$
/*STAGE 6. FIND THE LINEAR ACCELERATION */
ACCA:DIFF(VA,T);
ACCB:DIFF(VB,T)+CROSS(WA,VB,AA)$
ACCC:DIFF(VC,T)+CROSS(WB,VC,BB)$
ACCD:DIFF(VD,T)+CROSS(WC,VD,CC)$
ACCE:DIFF(VE,T)+CROSS(WE,VE,EE)$
ACCF:DIFF(VF,T)+CROSS(WE,VF,EE)$
/*STAGE 7. MOMENTS OF INERTIA */
```

```
IA:MATRIX([IA1,IA2,IA3]);
IB:MATRIX([IB1,IB2,IB3]);
IC:MATRIX([IC1,IC2,IC3]);
ID:MATRIX([ID1,ID2,ID3]);
IE:MATRIX([IE1,IE2,IE3]);
IIF:MATRIX([IIF1,IIF2,IIF3]);
/*STAGE 8,9,10. DEFINE TORQUES,REACTIONS,AND GENERALIZED FORCES FOR A,B,C,D,E,F*/
TAS:-ALPHAA*IA-CROSS(WA,IA*WA,AA)$
RAS:-MA*ACCA$
FOR I THRU 6 DO LDISPLAY(KAS[I]:WAR[I] . TAS+VAR[I] . RAS)$
TBS:-ALPHAB*IB-CROSS(WB,IB*WB,BB)$
RBS:-MB*ACCB$
FOR I THRU 6 DO KBS[I]:WBR[I] . TBS+VBR[I] . RBS$
TCS:-ALPHAC*IC-CROSS(WC,IC*WC,CC)$
RCS:-MC*ACCC$
FOR I THRU 6 DO KCS[I]:WCR[I] . TCS+VCR[I] . RCS$
TDS:-ALPHAD*ID-CROSS(WD,ID*WD,DD)$
RDS:-MD*ACCD$
FOR I THRU 6 DO KDS[I]:WDR[I] . TDS+VDR[I] . RDS$
TES:-ALPHAE*IE-CROSS(WE,IE*WE,EE)$
RES:-ME*ACCE$
FOR I THRU 6 DO KES[I]:WER[I] . TES+VER[I] . RES$
TFS:-ALPHAF*IIF-CROSS(WF,IIF*WF,FF)$
RFS:-MF*ACCF$
FOR I THRU 6 DO KFS[I]:WFR[I] . TFS+VFR[I] . RFS$
/*SUM ALL CORRESPONDING GENERALIZED FORCES */
KK1:KAS[1]+KBS[1]+KCS[1]+KDS[1]+KES[1]+KFS[1]$
KK2:KAS[2]+KBS[2]+KCS[2]+KDS[2]+KES[2]+KFS[2]$
KK3:KAS[3]+KBS[3]+KCS[3]+KDS[3]+KES[3]+KFS[3]$
KK4:KAS[4]+KBS[4]+KCS[4]+KDS[4]+KES[4]+KFS[4]$
KK5:KAS[5]+KBS[5]+KCS[5]+KDS[5]+KES[5]+KFS[5]$
KK6:KAS[6]+KBS[6]+KCS[6]+KDS[6]+KES[6]+KFS[6]$
/*STAGE 10. SET UP ACTIVE FORCES */
GA:MATRIX([0,-G*MA,0]);
GB:MATRIX([0,-G*MB,0]);
GC:-G*MC*MATRIX([0,CC2,-S2]);
GD:-G*MD*MATRIX([0,CC2,-S2]);
GE:-G*ME*MATRIX([0,1,0]) . R1 . R2 . R3 . R4;
GF:-G*MF*MATRIX([0,1,0]) . R1 . R2 . R3 . R4;
TNA:MATRIX([0,TAU1,0]);
TBA:MATRIX([TAU2,0,0]);
TCB:MATRIX([0,-SIGMA,0]);
TDC:MATRIX([0,TAU3,0]);
TED:MATRIX([TAU4,0,0]);
TFE:MATRIX([0,TAU5,0]);
RNA:MATRIX([0,0,0]);
/*SET UP GENERALIZED ACTIVE FORCES */
SPECIAL2[R]:=BLOCK(IF R = 6 THEN -SIGMA ELSE 0);
KTOTALR[R]:=SPECIAL2[R]+WAR[R] . TNA+(WAR[R] . R2-WBR[R]) . TBA
+(WCR[R] . R3-WDR[k]) . TDC . R3+(WDR[R] . R4-WER[R]) . TED . R4
+(WER[R] . R5-WFR[R]) . TFE . R5+VBR[R] . GB+VCR[R] . GC+VDR[R] . GD+VER[R]
KEEPFLOAT:TRUE;
/*NUMERICAL EXAMPLE WITH VALUES GIVEN IN REF. [9]*/
G:9.8; L1:.1; L2:.6; L3:.2; L4:.1; L5:0.7; L6:0.06; MA:9; MB:6; MC:4; MD:1;
ME:0.6; MF:0.5; IA1:0.01; IA2:0.02; IA3:0.01; IB1:0.06; IB2:0.01; IB3:0.05;
IC1:0.4; IC2:0.01; IC3:0.4; ID1:0.0005; ID2:0.001; ID3:0.001; IE1:0.0005;
IE2:0.0002; IE3:0.0005; IIF1:0.001; IIF2:0.002; IIF3:0.003;
EV(FT:(T-10/(2*%PI)*SIN(2*%PI*T/10))*(%PI/180),NUMER);
TQ1:60/10*FT;
TQ2:%PI/2+(60-90)/10*FT;
TQ3:TQ1;
TQ4:TQ1;
TQ5:TQ1;
TQ6:1/10;
U[1]:DIFF(TQ1,T)*SIN(TQ2)*SIN(TQ3)+DIFF(TQ2,T)*COS(TQ3);
U[2]:DIFF(TQ1,T)*COS(TQ2)+DIFF(TQ3,T);
U[3]:-DIFF(TQ1,T)*SIN(TQ2)*COS(TQ3)+DIFF(TQ2,T)*SIN(TQ3);
U[4]:DIFF(TQ4,T);
U[5]:DIFF(TQ5,T);
U[6]:DIFF(TQ6,T);
Q1:TQ1;
Q2:TQ2;
Q3:TQ3;
Q4:TQ4;
Q5:TQ5;
Q6:TQ6;
/* NOW WE PLOT RESULTS AND COMPARE WITH REF. [9] */
FINAL6:KTOTALR[6]+SIGMA+KK6;
FINAL6:EV(FINAL6,DIFF)$
EQUALSCALE:FALSE;
```

```
PLOTNUM:10;
PLOT(FINAL6,T,0,10,"PLOT OF SIGMA ");
FINAL5:EV(KK5,DIFF)$
PLOT(FINAL5,T,0,10,"PLOT OF TAU 5");
FINAL4:EV(KK4+KTOTALR[4]+TAU4,DIFF)$
PLOT(FINAL4,T,0,10,"PLOT OF TAU 4");
FINAL3:EV(KK2+KTOTALR[2]+TAU3,DIFF)$
PLOT(FINAL3,T,0,10,"PLOT OF TAU3 ");
/*TRY TO SIMPLIFY AND COLLECT TERMS  X I J IN EQUATION OF MOTION*/
/*FIRST DELETE NUMERICAL VALUES */
FOR I:1 QRU 6 DO ( FOR J:1 QRU 6 DO XXA[I,J]:RATCOEFF(KAS[I],DIFF(U[J],T)));
FOR I:1 QRU 6 DO ( FOR J:1 QRU 6 DO XXB[I,J]:RATCOEFF(KBS[I],DIFF(U[J],T)));
FOR I:1 QRU 6 DO ( FOR J:1 QRU 6 DO XXC[I,J]:RATCOEFF(KCS[I],DIFF(U[J],T)));
FOR I:1 QRU 6 DO ( FOR J:1 QRU 6 DO XXD[I,J]:RATCOEFF(KDS[I],DIFF(U[J],T)));
FOR I:1 QRU 6 DO ( FOR J:1 QRU 6 DO XXE[I,J]:RATCOEFF(KES[I],DIFF(U[J],T)));
FOR I:1 QRU 6 DO ( FOR J:1 QRU 6 DO XXF[I,J]:RATCOEFF(KFS[I],DIFF(U[J],T)));
FOR I:1 QRU 6 DO (FOR J:1 QRU 6 DO XXX[I,J]:RATSIMP(XXA[I,J]+XXB[I,J]+
XXC[I,J]+XXD[I,J]+XXE[I,J]+XXF[I,J]));
FOR I:1 THRU 6 DO ( FOR J:1 THRU 6 DO LDISPLAY (XXX[I,J]));
/*COMPARE ABOVE X,ij WITH THOSE  OF REF. [9]*/
```

APPENDIX IX
———————

```
/*SPACECRAFT EXAMPLE*/
        /*.......CARTESIAN DIV AND CURL DEFINITION.........
        ...........UNIT VECTORS ARE B1 B2 B3...... SEE LEVINSON    */
        /*..............DEFINE DOT AND CROSS PRODUCTS....*/
        DOT(V1,V2):=BLOCK([P,PP],
        FOR I:1 THRU 3 DO P[I]:RATCOEFF(V1,B[I]),
        FOR I:1 THRU 3 DO PP[I]:RATCOEFF(V2,B[I]),
        P[4]:SUM(P[I]*PP[I],I,1,3),
        RETURN(P[4]))$
        CROSS(V1,V2):=BLOCK([P,PP,PPP],
        FOR I:1 THRU 3 DO P[I]:RATCOEFF(V1,B[I]),
        FOR I:1 THRU 3 DO PP[I]:RATCOEFF(V2,B[I]),
        PPP[1]:(P[2]*PP[3]-P[3]*PP[2]),
        PPP[2]:(-P[1]*PP[3]+P[3]*PP[1]),
        PPP[3]:(P[1]*PP[2]-P[2]*PP[1]),
        PPP[4]:B[1]*PPP[1]+B[2]*PPP[2]+B[3]*PPP[3],
        RETURN(PPP[4]))$
        /*...........NOW WE INPUT EQUATIONS FROM LEVINSON'S PAPER */
        DEPENDS(U,T);
        DEPENDS(Q,T);
        R[2]:COS(Q)*B[2]+SIN(Q)*B[3];
        R[3]:-SIN(Q)*B[2]+COS(Q)*B[3];
        WB:U[1]*B[1]+U[2]*B[2]+U[3]*B[3];
        DERIVABBREV:TRUE;
        U[4]:DIFF(Q,T);
        WR:(U[1]+U[4])*B[1]+U[2]*B[2]+U[3]*B[3];
        ALPB:DIFF(U[1],T)*B[1]+DIFF(U[2],T)*B[2]+DIFF(U[3],T)*B[3];
        ALPR:DIFF(WR,T)+CROSS(WB,WR);
        PPBS:B1*B[1]+B2*B[2]+B3*B[3];
        PPRS:R1*B[1]+R2*R[2]+R3*R[3];
        PRSBS:PPBS-PPRS;
        VBS:U[5]*B[1]+U[6]*B[2]+U[7]*B[3];
        VRS:VBS +DIFF(PRSBS,T)+CROSS(WB,PRSBS);
        ABS:DIFF(VBS,T)+CROSS(WB,VBS);
        ARS:DIFF(VRS,T)+CROSS(WB,VRS);
        IBBSWB:BET1*B[1]*DOT(B[1],WB)+BET2*B[2]*DOT(B[2],WB)+BET3*B[3]*DOT(B[3],WB);
        IRRSWR:RHO1*B[1]*DOT(B[1],WR)+RHO2*B[2]*DOT(B[2],WR)+RHO3*B[3]*DOT(B[3],WR);
        IBBSALPB:BET1*B[1]*DOT(B[1],ALPB)+BET2*B[2]*DOT(B[2],ALPB)+BET3*B[3]*DOT(B[3],ALPB);
        IRRSALPR:RHO1*B[1]*DOT(B[1],ALPR)+RHO2*B[2]*DOT(B[2],ALPR)+RHO3*B[3]*DOT(B[3],ALPR);
        FSB:-MB*ABS;
        FSR:-MR*ARS;
        TSB:CROSS(IBBSWB,WB)-IBBSALPB;
        TSR:CROSS(IRRSWR,WR)-IRRSALPR;
        FB:F1*B[1]+F2*B[2]+F3*B[3];
        TB:T1*B[1]+T2*B[2]+T3*B[3];
        F[R]:=DOT(DIFF(VBS,U[R]),FB)+DOT(DIFF(WB,U[R]),TB);
        FS[R]:=DOT(DIFF(VBS,U[R]),FSB)+DOT(DIFF(VRS,U[R]),FSR)
        +DOT(DIFF(WB,U[R]),TSB)+DOT(DIFF(WR,U[R]),TSR);
        EQ[R]:=F[R]+FS[R];
        EQ[1];
        FOR I:1 THRU 7 DO LDISPLAY ( X[1,I]:RATCOEFF(EQ[I],DIFF(U[I],T)));
```

13

STABILITY ANALYSIS OF A ROBOTIC MECHANISM USING COMPUTER ALBEGRA

M. GOLNARAGHI, W. KEITH, AND F. C. MOON
Theoretical and Applied Mechanics, Cornell University, Ithaca, N.Y.

ABSTRACT

This paper concerns the stability analysis of a two degree of freedom lumped mass model of a robotic mechanism. The equations of motion as well as the Hamiltonian for the system are derived using the symbolic manipulation system MACSYMA. The second order, nonlinear, coupled, ordinary differential equations which govern the system are linearized about a particular operating point. We emphasize the advantage of using MACSYMA to derive this linear system. The regions of flutter and divergence instabilities in a two dimensional parameter space are determined from a quartic characteristic equation. A MACSYMA program is used to perform a root locus study of this equation. A discussion of the application of the results to the control of a mechanical robot is presented.

INTRODUCTION

Since the early 1960's, the robotic industry has received a great deal of attention from researchers in engineering. Most of the work has been done in the areas of software, vision, and artificial intelligence. However, one of the most important areas, the dynamics of robotic manipulators, has received relatively little attention. A detailed study of the motion and dynamics of robotic systems could well lead to advances in the design of robots. Smaller and lighter robots with better performance capability and lower cost might replace the current large and expensive ones.

In this paper we perform a stability analysis of a two degree of freedom lumped mass model of a robotic mechanism. Because of the complicated dynamics of the system, the symbolic manipulation system MACSYMA was used extensively (see Appendix 1).

DESCRIPTION OF THE MODEL

We propose the lumped mass model shown in Figure (1). The X, Y, Z reference frame is stationary. Mass M_1 represents the mass of the body transported by the robot, as well as the mass of link L_1. Mass M_2 represents a counterweight mass, as well as the mass of link L_2. Links L_1, L_2, and L_3 are connected at point P. Link L_3 makes an angle ϕ with respect to the Z axis, such that $L_4 = L_3 \sin\phi$ = constant for all time. Link L_3 rotates about the Z axis only with angular velocity $\underline{\Omega}$, which is constant in magnitude and direction with time.

The x, y, z reference frame is centered at point P and rotates with constant angular velocity $\underline{\Omega}$. Note that this reference frame is not taken to be rigidly attached to the linkage assembly. The system has two degrees of freedom which reflect the flexibility of the linkages. We choose the generalized coordinates θ_1 and θ_2, both measured relative to the x, y, z reference frame, as shown in Figure (1). Associated with θ_1 and θ_2 are the spring stiffnesses K_1 and K_2, respectively. K_2 represents the torsional stiffness of link L_3, and K_1 the stiffness of the connection joint at point P.

A control torque $\underline{T}$, whose direction coincides with the y axis is applied at point P. This torque allows any static equilibrium configuration to be attained, independent of the spring stiffness K_1. We take $\underline{T}$ to remain constant in magnitude and direction with respect to the x, y, z reference frame for all time.

For a given set of parameters, we are interested in determining the stability of an operating position defined by $\theta_1 = \theta_1^o$ and $\theta_2 = 0$. In particular, we will determine regions in a parameter space in which flutter and divergence instabilities exist.

DERIVATION OF THE EQUATIONS OF MOTION

In this section we derive the dynamical equations of motion of the model shown in Figure (1), using the Lagrangian formulation. These equations were derived analytically (by hand), and verified using the symbolic manipulation system MACSYMA.

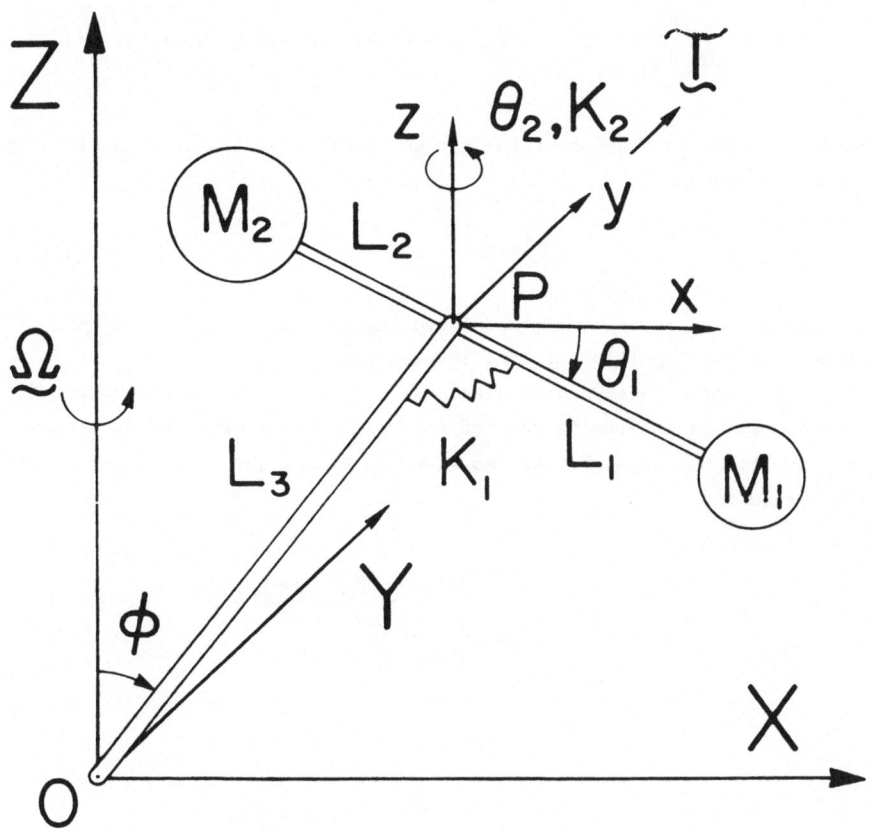

Figure 1.   The Lumped Mass Robotic Model

Figure (2) shows a general n degree of freedom linkage system with a mass M attached at the end. Lagrange's equations of motion for the system are:

$$\frac{d}{dt} \frac{\partial L}{\partial \dot{q}_i} - \frac{\partial L}{\partial q_i} = F_i, \quad i = 1, 2, \ldots, n \qquad (1)$$

where L is the Lagrangian of the system, defined as the kinetic minus the potential energy.

$$L = KE - PE \qquad (2)$$

The q_i, $\dot{q}_i$, and F_i are the generalized angular coordinates, angular velocities, and applied torques, respectively.

The kinetic energy is obtained using a method involving coordinate transformations similar to that of ref. (1). A set of coordinate axes is regidly attached to each link, as shown in Figure (2). The kinetic energy of the system is:

$$KE = \frac{1}{2} M_1 (\underline{V}_{M_1} \cdot \underline{V}_{M_1})$$

$$+ \frac{1}{2} M (\underline{V}_{M_2} \cdot \underline{V}_{M_2}) \qquad (3)$$

where, $\underline{V}_{M_n}$ is the velocity vector for link n;

$$\underline{V}_{M_n} = \sum_{i=1}^{n} \omega \; i \; T \; i \; \underline{L} \; i, \quad i = 1, 2 \qquad (4)$$

$\omega_{(i)}$ is a skew symmetric matrix defining the angular velocity of link i and can be written as

285

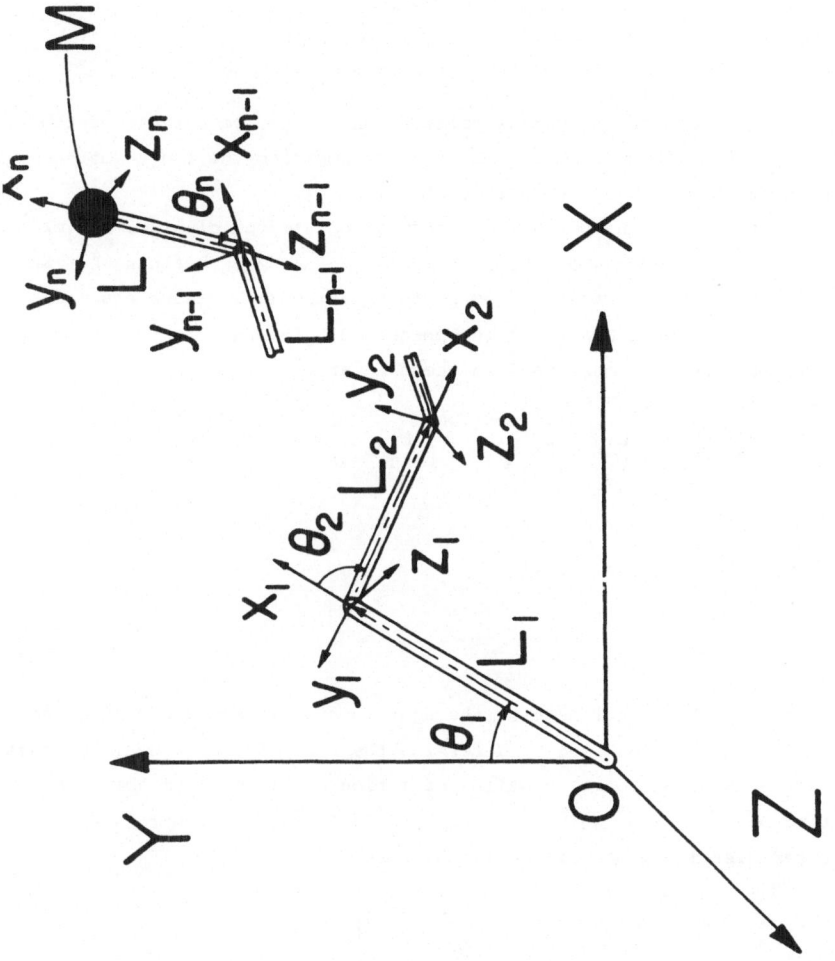

Figure 2. A General N-Degree of Freedom Linkage System

$$\omega \, [\underline{i}] = \begin{bmatrix} 0 & -\omega_z & \omega_x \\ +\omega_z & 0 & -\omega_y \\ -\omega_x & +\omega_y & 0 \end{bmatrix} \quad , \tag{5}$$

$\underline{T}_{(i)}$ is the transformation matrix relating the $x_i y_i z_i$ coordinate system to the base coordinate system XYZ, $\underline{L}_{(i)}$ is a vector defining the length of each link with respect to the ith coordinate system.

For our system, the potential energy is due to the displacement of springs K_1 and K_2 and the position of masses M_1 and M_2. Defining L = KE - PE, the equations of motion are derived using equation (1). These equations obtained using MACSYMA are given in Appendix 2. In order to reduce the number of parameters, we nondimensionalize these equations using the following:

$$M^* = M_2/M_1, \quad L_2^* = L_2/L_1, \quad L_4^* = L_4/L_1, \quad \Omega^* = \Omega / \left\{ \frac{K_1}{M_1 L_1^2} \right\}^{\frac{1}{2}},$$

$$g^* = g M_1 l_1/K_1, \quad K^* = K_2/K_1.$$

STABILITY ANALYSIS

In this section we linearize the equations of motion and derive the characteristic equation used to determine the stability regions in parameter space. The nondimensional equations of motion are linearized about the operating point $\theta_1 = \theta_1^o$ and $\theta_2 = 0$, where $0 < \theta_1 < \pi/2$. The resulting linearized system may be written in the form;

$$\begin{bmatrix} a_{11} & 0 \\ 0 & a_{22} \end{bmatrix} \begin{bmatrix} \ddot{\theta}_1 \\ \ddot{\theta}_2 \end{bmatrix} + \begin{bmatrix} 0 & b_{12} \\ -b_{12} & 0 \end{bmatrix} \begin{bmatrix} \dot{\theta}_1 \\ \dot{\theta}_2 \end{bmatrix}$$

$$+ \begin{bmatrix} c_{11} & 0 \\ 0 & c_{22} \end{bmatrix} \begin{bmatrix} \theta_1 \\ \theta_2 \end{bmatrix} = \begin{bmatrix} \tau \\ 0 \end{bmatrix} \tag{6}$$

The elements of the matrices are as follows;

$$a_{11} = (1 + M^* l_2^{*2})$$

$$a_{22} = (1 + M^* l_2^{*2}) \cos^2\theta_1^o$$

$$b_{21} = \Omega^* \left(3 \cos\theta_1^o + l_4^* + M^* (3l_2^{*2} \cos\theta_1^o - l_2^* l_4^*)\right) \sin\theta_1^o$$

$$c_{11} = g^* \sin\theta_1^o + \Omega^{*2} l_4^* \cos\theta_1^o - \Omega^{*2} (\sin^2\theta_1^o - \cos^2\theta_1^o)$$

$$\qquad - M^* \left[\Omega^{*2} l_2^{*2} (\sin\theta_1^2 - \cos^2\theta_1^o) + \Omega^{*2} l_2^* l_4^* \cos\theta_1^o + g^* l_2^* \sin\theta_I^o\right] + 1$$

$$c_{22} = \Omega^{*2} l_4^* \cos\theta_1^o - M^* \Omega^{*2} l_4^* l_2^* \cos\theta_1^o + K^*$$

We note that the matrices associated with the acceleration and displacement vectors are diagonal, and that which is associated with the velocity vector is skew symmetric. This is the appropriate form for a <u>gyroscopic system</u> (2). The a_{ij}, b_{ij}, and c_{ij} coefficients were derived analytically (by hand) and checked using MACSYMA (3, 4).

We assume a solution to this linear system of the form

$$\begin{bmatrix} \theta_1 \\ \theta_2 \end{bmatrix} = \begin{bmatrix} V_1 \\ V_2 \end{bmatrix} e^{i\rho t}$$

which implies

$$\begin{bmatrix} \theta_1 \\ \theta_2 \end{bmatrix} = \begin{bmatrix} V_1 \\ V_2 \end{bmatrix} (i\rho e^{i\rho t}) \qquad \begin{bmatrix} \theta_1 \\ \theta_2 \end{bmatrix} = - \begin{bmatrix} V_1 \\ V_2 \end{bmatrix} (\rho^2 e^{i\rho t}) \qquad (7)$$

Substituting these expressions into (6), we obtain

$$
\begin{bmatrix}
c_{11} - \rho^2 a_{11} & b_{12} i \rho \\
-b_{12} i \rho & c_{22} - \rho^2 a_{22}
\end{bmatrix}
\begin{bmatrix}
V_1 \\
V_2
\end{bmatrix}
e^{i \rho t}
=
\begin{bmatrix}
0 \\
0
\end{bmatrix}
\qquad (8)
$$

For nontrivial solutions we set the determinant of the matrix of coefficients in (8) equal to zero and obtain the characteristic equation

$$
a_{11} a_{22} \rho^4 - (a_{11} c_{22} + a_{22} c_{11} - b_{12}{}^2) \rho^2 + c_{11} c_{22} = 0 \qquad (10)
$$

Equation (10) is quadratic in ρ. Below are listed the possible solutions and the corresponding stability types where $\gamma, \beta > 0$.

Solution	Stability type
1. $\underline{V} e^{\pm i \gamma t}$	Stable
2. $\underline{V} e^{\pm (i \gamma + \beta t)}$	Flutter instability
3. $\underline{V} e^{\gamma t}$	Divergence instability

The regions in Ω^{*2}, K^* parameter space corresponding to the above stability types are shown in Fig (3), where $M^* = 5$, $L_2^* = L_4^* = g^* = 1$, and $\Theta_1^\circ = 3^\pi/20$. For these parameter values, increasing the stiffness ratio K^* may stabilize or destabilize the system, depending on the value of Ω^{*2}.

SUMMARY

A two degree of freedom dynamical system governed by coupled highly nonlinear equations of motion has been analyzed. The symbolic manipulation system MACSYMA has been shown to be extremely useful in this application. Calculations which are extremely lengthy and difficult to perform by hand without error, may be performed quickly and correctly using computer algebra (3).

REFERENCES

1. Paul, R.P. Robot manipulators: Mathematics, programming, and control. MIT Press, Cambridge, 4th ed., 1982.
2. Huseyin, K. Vibrations and stability of multiple parameter systems. Sijthoff & Noordhoff International Publishers, Alphen aan den Rijn, The Netherlands, 1978.
3. Rand, R.H. Computer algebra in applied mathematics: An introduction to MACSYMA. Pitman Publishing, Inc., Marshfield, Mass., 1st ed., 1984.
4. MATHLAB Group Laboratories for Computer Science MIT. MACSYMA reference manual. MIT, Cambridge, Version 10, Vol. I, 1st printing, 1982.

APPENDIX 1

The following outline shows the particular analyses which were performed
using MACSYMA.

1. Derivation of the equations of motion:
    a) Derive the kinetic and potential energies.
    b) Substitute these into Lagrange's equations, and compute
        the associated partial and total derivatives.

2. Nondimensionalizing the equations of motion:
    a) Define nondimensional parameters.
    b) Substitute these into the full nonlinear equations.
    c) Manipulate these into the desired form.

3. Linearization of the nondimensional equations:
    a) Expand the full nonlinear equations in Taylor series in
        the appropriate variables, retaining only the linear
        terms.
    b) Manipulate these into the form given by equation (6).

4. Stability analysis of the linear system:
    a) Compute the characteristic equation (see eqn (10)).
    b) Solve for the roots of this quartic equation.
    c) Determine the nature of the roots as particular parameters
        are varied.

APPENDIX 2

The full nonlinear coupled equations of motion for this system are given below, where θ_1 = T1, θ_2 = T2, $|\Omega|$ = OM,

EQ1;

$$(L2^2M2 + L1^2M1) \ COS(T1) \ SIN(T1) \ SIN(T2) \ \frac{d^2T2}{dT^2}$$

$$+ \ ((L2^2M2 + L1^2M1) \ COS(T1) \ SIN(T1) \ SIN^2(T2) + (L2^2M2 + L1^2M1)$$

$$COS(T1) \ SIN(T1) \ COS(T2)) \ (\frac{dT2}{dT})^2 + ((-2L2^2M2 - 2L1^2M1) \ OM \ COS(T1)$$

$$SIN(T1) \ SIN^2(T2) + (-2L2^2M2 - 2L1^2M1) \ COS^2(T1)\frac{dT1}{dT} \ COS(T2) \ SIN(T2)$$

$$+ \ (-2L2^2M2 - 2L1^2M1)OM \ COS(T1) \ SIN(T1) \ COS^2(T2)$$

$$+ \ ((-L2^2M2 - L1^2M1)OM \ COS(T1)$$

$$+ \ (L2L3M2 - L1L3M1)OM \ SIN(PHI)) \ SIN(T1) \ COS(T2))\frac{dT2}{dT}$$

$$+ \ (L2^2M2 + L1^2M1)OM \ COS(T1) \ SIN(T1) \ SIN^2(T2)$$

$$+ \ ((L2^2M2 + L1^2M1) \ COS^2(T1)\frac{d^2T1}{dT^2}$$

$$+ \ (L2^2M2 - L1^2M1) \ COS(T1) \ SIN(T1) \ (\frac{dT1}{dT})$$

$$+ \ (L2^2M2 + L1^2M1)OM^2 \ COS(T1) \ SIN(T1) \ COS^2(T2)$$

$$+ \ (L1L3M1 - L2L3M2)OM^2 \ SIN(PHI) \ SIN(T1) \ COS(T2)$$

$$+ \ (L2^2 M2 + L1^2 M1) \ SIN^2 (T1)\frac{d^2T1}{dT^2}$$

$$+ \ (L2^2M2 + L1^2M1) \ COS(T1) \ SIN(T1) \ (\frac{dT1}{dT})^2$$

$$+ \ (GL2M2 - GL1M1) \ COS(T1) + K1T1 = \tau$$

EQ2;

$$((L2^2M2 + L1^2M1) \cos^2(T1) \sin^2(T2)$$

$$+ (L2^2M2 + L1^2M1) \cos^2(T1) \cos^2(T2)) \frac{d^2T2}{dT^2}$$

$$+ ((-2L2^2M2 - 2L1^2M1) \cos(T1) \sin(T1) \frac{dT1}{dT} \sin^2(T2)$$

$$+ (-2L2^2M2 - 2L1^2M1) \cos(T1) \sin(T1) \frac{dT1}{dT} \cos^2(T2)) \frac{dT2}{dT}$$

$$+ (2L2^2M2 + 2L1^2M1)OM \cos(T1) \sin(T1) \frac{dT1}{dT} \sin^2(T2)$$

$$+ ((L2^2M2 + L1^2M1) \cos^2(T1) (\frac{dT1}{dT})^2 \cos(T2)$$

$$+ (L2^2M2 + L1^2M1) \cos(T1) \sin(T1) \frac{d^2T1}{dT^2}$$

$$+ ((-L2^2M2 - L1^2M1) \sin^2(T1)$$

$$+ (L2^2M2 + L1^2M1) \cos^2(T1)) (\frac{dT1}{dT})^2$$

$$+ (L1L3M1 - L2L3M2)OM^2 \sin(PHI) \cos(T1)) \sin(T2)$$

$$+ (2L2^2M2 + 2L1^2M1)OM \cos(T1) \sin(T1) \frac{dT1}{dT} \cos^2(T2)$$

$$+ ((L2^2M2 + L1^2M1)OM \cos(T1)$$

$$+ (L1L3M1 - L2L3M2)OM \sin(PHI)) \sin(T1) \frac{dT1}{dT} \cos(T2)$$

$$+ K2T2 = 0$$

14

DERIVATION OF THE HOPF BIFURCATION FORMULA USING
LINDSTEDT'S PERTURBATION METHOD AND MACSYMA.

RICHARD H. RAND

Department of Theoretical and Applied Mechanics
Cornell University, Ithaca, New York 14853

ABSTRACT

This paper involves the sudden appearance and growth
of a periodic motion called a limit cycle in an autonomous
system of two nonlinear first order ordinary differential
equations. The bifurcation occurs as a parameter is tuned
so that an equilibrium point goes from a stable focus to an
unstable focus. The resulting limit cycle will generically
occur either i) when the equilibrium is stable (in which
case the limit cycle is unstable), or ii) when the
equilibrium is unstable (in which case the limit cycle is
stable). The Hopf bifurcation formula determines which of
these two cases occurs in a given system, and depends in a
complicated way on the second and third derivatives of the
right-hand sides of the differential equations.

While the Hopf formula itself is well-known to many
users, the usual derivations are complicated and less
accessible. In this paper the Hopf formula is derived in a
straightforward fashion using Lindstedt's well-known
perturbation method in conjunction with MACSYMA.

INTRODUCTION

This paper involves the dynamics of a system of two
ordinary differential equations:

(1) $x' = a(u)\ x + b(u)\ y + f(x,y;u)$

(2) $y' = c(u)\ x + d(u)\ y + g(x,y;u)$

where f and g are strictly nonlinear in x and y (i.e.,
their Taylor expansions about x=y=0 have no constant or

linear terms), and where u is a parameter. Associated with
(1),(2) is the corresponding linearized system:

(3)   x' = a(u) x + b(u) y
(4)   y' = c(u) x + d(u) y

We assume that for u=0 the system (3),(4) has purely
imaginary eigenvalues, i.e. there is a center at the
origin. Moreover, we assume that for u small and negative
(positive), the eigenvalues of (3),(4) have negative
(positive) real parts, i.e. there is a stable (unstable)
focus at the origin.

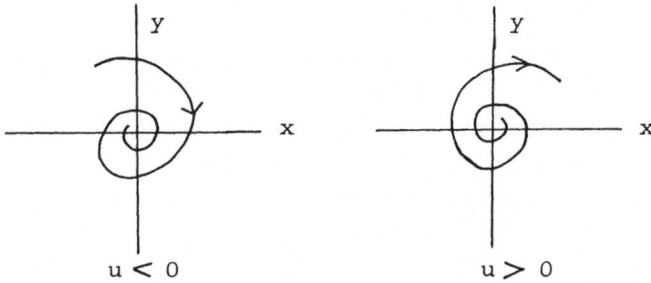

Fig. 1. Phase portraits for the linearized system.

    In such a case, the nonlinear system (1),(2)
generically (i.e. typically, but not always) undergoes the
birth of a limit cycle, a process called a Hopf bifurcation
(1,2,3). (The birth of a limit cycle can be guaranteed
under certain additional conditions, namely (i) that the
derivative of the real part of the eigenvalues with respect
to u be non-zero at u=0, and (ii) that the origin of
(1),(2) be asymptotically stable or unstable at u=0.)
    There are two generic possibilities, as follows. The
limit cycle may occur for u > 0, in which case it is stable
(the "supercritical" case), or the limit cycle may occur
for u < 0, in which case it is unstable (the "subcritical"
case):

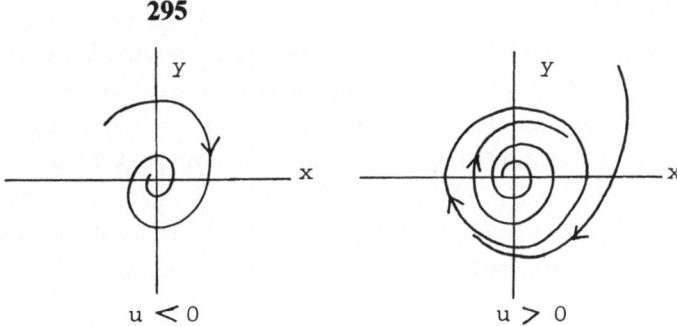

Fig. 2. Phase portraits for the supercritical case.

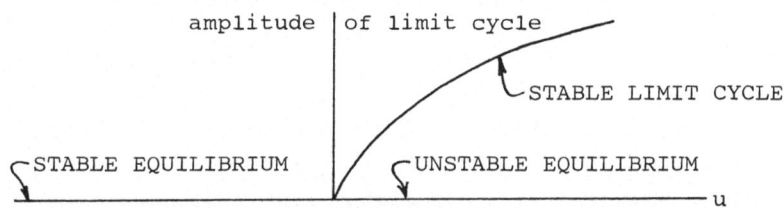

Fig. 3. Bifurcation diagram for the supercritical case.

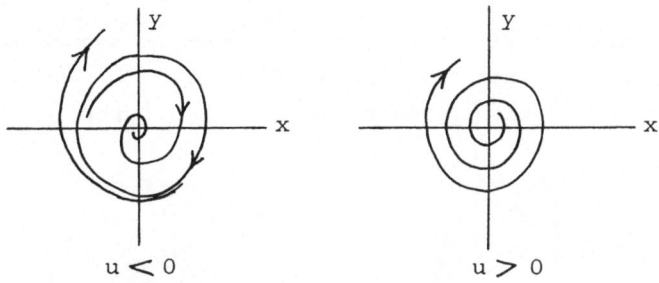

Fig. 4. Phase portraits for the subcritical case.

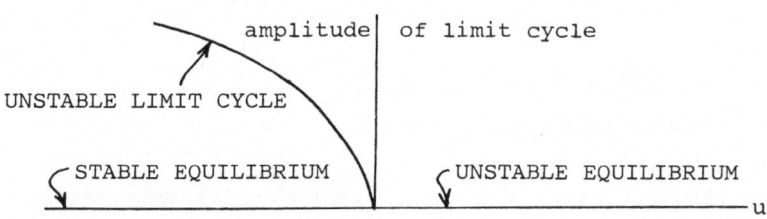

Fig. 5. Bifurcation diagram for the subcritical case.

The Hopf bifurcation formula determines which of these two generic cases occurs. The purpose of this paper is to offer a straightforward derivation of this formula using Lindstedt's perturbation method (see (4)) and MACSYMA.

Example: The following equation is a variant of Van der Pol's equation (ref. (4)):

(5) $$x'' + x - 2\ u\ x' + x^2\ x' = 0,$$

When written as a first order system, this becomes:

(6) $$x' = y$$

(7) $$y' = -x + 2\ u\ y - x^2\ y$$

The eigenvalues of the associated linear system are:

(8) $$u \pm i\ (1 - u^2)^{1/2}$$

Note that eq.(8) satisfies the foregoing assumptions, and hence we may expect a Hopf bifurcation. In fact numerical integration reveals there to be a supercritical Hopf bifurcation:

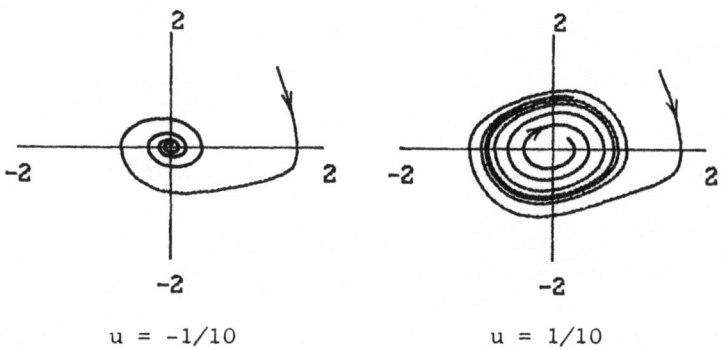

u = -1/10   u = 1/10

Fig. 6. Phase portraits obtained by numerical integration of eq.(5).

CANONICAL FORM

We will begin the computation by assuming that the system (1),(2) is already in the following canonical form:

$$(9) \qquad x' = u\,x - w(u)\,y + f(x,y;u)$$
$$(10) \qquad y' = w(u)\,x + u\,y + g(x,y;u)$$

where again f and g are strictly nonlinear in x and y.

Some elementary linear algebra shows that an arbitrary linear system (3),(4) with complex eigenvalues can be transformed to the linearization of (9),(10) by setting:

$$(11) \qquad xold = b\ xnew$$
$$(12) \qquad yold = (d - a)/2\ xnew - w\ ynew$$

where $w = (-(d-a)^2/4 - b\,c)^{1/2}$, and where it turns out that $u = (a+d)/2$.

Example: Continuing the previous example given by eqs.(6),(7), we set

$$(13) \qquad xold = xnew$$
$$(14) \qquad yold = u\ xnew - w\ ynew$$

where $w = (1 - u^2)^{1/2}$. This gives:

$$(15) \quad xnew' = u\ xnew - w\ ynew$$
$$(16) \quad ynew' = w\ xnew + u\ ynew + (u/w)\ xnew^3 - xnew^2\ ynew$$

which is of the form (9),(10).

LINDSTEDT'S METHOD

Lindstedt's perturbation method (4) is a well known
procedure for obtaining approximate solutions to
differential equations which involve a small parameter. We
will introduce a small perturbation parameter e into
eqs.(9),(10) by scaling variables as follows:

$$(17) \qquad x = e\ X, \quad y = e\ Y, \quad u = e^2\ M$$

Next we expand the function w(u) in a power series
valid for small u:

$$(18) \qquad w = w(u) = w0 + w1\ u + \ldots$$

$$= w0 + e^2\ w1\ M + O(4)$$

where w0 and w1 are given constants, and where O(n) means
terms of order of e raised to the nth power throughout.

We also expand the functions f and g in power series
in x and y:

$$(19) \qquad f(x,y,;u) = \frac{fxx}{2}\ x^2 + \ldots + \frac{fxxx}{6}\ x^3 + \ldots$$

$$= \frac{fxx}{2}\ e^2 X^2 + \ldots + \frac{fxxx}{6}\ e^3 X^3 + \ldots + O(4)$$

and a similar expression for g(x,y;u). Note that the
partial derivatives fij are evaluated at x=y=0, but in
general will depend on u (see e.g. eq.(16)). However,
since u = O(2), we may take the fij to also be evaluated at
u = 0 and still maintain accuracy to O(4) in eq.(19).

As usual in Lindstedt's method (4), we stretch time t
to accomodate the possibility of a dependence of frequency
on amplitude in this nonlinear system. We replace t as

independent variable by z, where

$$(20) \qquad z = (w0 + K e^2 + O(3)) \, t$$

where K is a constant whose value is to be determined.

Finally we expand the scaled dependent variables X and Y in power series in e:

$$(21) \qquad X(z) = X0(z) + e \, X1(z) + e^2 \, X2(z) + O(3)$$

$$(22) \qquad Y(z) = Y0(z) + e \, Y1(z) + e^2 \, Y2(z) + O(3)$$

We substitute eqs.(17)-(22) into (9),(10), collect terms and equate to zero the coefficient of e raised to the power n, for n=1,2,3,... . This yields a sequence of linear eqs. on Xn(z), Yn(z) which may be solved recursively.

The zero order terms satisfy

$$(23) \qquad X0(z)' = - Y0(z)$$

$$(24) \qquad Y0(z)' = \quad X0(z)$$

which has the solution

$$(25) \qquad X0(z) = A \sin z + B \cos z$$

$$(26) \qquad Y0(z) = B \sin z - A \cos z$$

Since the original problem is autonomous, we may without loss of generality select the initial condition y=0 when t=0, or in the new variables, Y=0 when z=0, which, from (22), gives

$$(27) \qquad Y0(0) = Y1(0) = Y2(0) = \ldots = 0$$

Eqs.(26),(27) require that A=0 so that

(28) $X0(z) = B \cos z, \quad Y0(z) = B \sin z$

Here the amplitude B, like K in eq.(20), is a constant to
be determined.

The MACSYMA session which follows may be outlined
thus: We substitute eq.(28) into the differential
equations on X1 and Y1 and solve for X1(z), Y1(z).
Lindstedt's method generally requires the removal of all
resonant (secular, unbounded) terms at each stage of the
process, but it turns out that there are no resonant terms
in the X1, Y1 equations. Next these results are
substituted into the X2, Y2 equations and resonant terms
are removed giving two equations for the undetermined
coefficients B and K. Solving for the amplitude B and
requiring B to be real will yield the Hopf bifurcation
formula.

Before beginning, a word about removal of resonant
terms in the system:

(29) $x(z)' = - y(z) + a \sin z + b \cos z$
(30) $y(z)' = \quad x(z) + c \sin z + d \cos z$

A particular solution to (29),(30) is

(31) $2 x(z) = (a-d) z \sin z + (b+c) z \cos z + (b-c) \sin z$
(32) $2 y(z) = (b+c) z \sin z + (d-a) z \cos z + (a+d) \sin z$

and therefore for no resonant terms we require

(33) $a - d = 0 \quad \text{and} \quad b + c = 0$

MACSYMA SESSION

Schemes for handling Lindstedt's method in MACSYMA
have been extensively treated in (5). (For an introduction
to MACSYMA, see (5) or (6).)

We begin by defining the differential equations
(9),(10) using stretched time z, eq.(20). Note the use of

variables LCX and LCY to represent lower case x and y
respectively:

(C1) STRETCH:WO+K*E**2;

(D1)
$$WO + E^2 K$$

(C2) 'DIFF(LCX,Z)*STRETCH=U*LCX-W*LCY+F;

(D2)
$$LCX_Z (WO + E^2 K) = - LCY W + LCX U + F$$

(C3) 'DIFF(LCY,Z)*STRETCH=W*LCX+U*LCY+G;

(D3)
$$LCY_Z (WO + E^2 K) = LCX W + LCY U + G$$

Next we define the functions f and g (cf. eq.(19)):

(C4) F:FXX/2*LCX**2+FXY*LCX*LCY+FYY/2*LCY**2
+FXXX/6*LCX**3+FXXY/2*LCX**2*LCY+FXYY/2*LCX*LCY**2
+FYYY/6*LCY**3;

(D4)
$$\frac{FYYY\ LCY^3}{6} + \frac{FXYY\ LCX\ LCY^2}{2} + \frac{FYY\ LCY^2}{2} + \frac{FXXY\ LCX^2\ LCY}{2}$$

$$+ FXY\ LCX\ LCY + \frac{FXXX\ LCX^3}{6} + \frac{FXX\ LCX^2}{2}$$

(C5) G:GXX/2*LCX**2+GXY*LCX*LCY+GYY/2*LCY**2
+GXXX/6*LCX**3+GXXY/2*LCX**2*LCY+GXYY/2*LCX*LCY**2
+GYYY/6*LCY**3;

(D5)
$$\frac{GYYY\ LCY^3}{6} + \frac{GXYY\ LCX\ LCY^2}{2} + \frac{GYY\ LCY^2}{2} + \frac{GXXY\ LCX^2\ LCY}{2}$$

$$+ GXY\ LCX\ LCY + \frac{GXXX\ LCX^3}{6} + \frac{GXX\ LCX^2}{2}$$

We complete the specification of the perturbation scheme by
scaling u and w, and then expanding the scaled variables X
and Y (see eqs.(21),(22)):

(C6) U:E**2*M;

(D6)
$$E^2\ M$$

```
(C7) W:WO+W1*U;
```

$$(D7) \qquad\qquad E^2 \ M \ W1 \ + \ WO$$

```
(C8) LCX:E*(XO(Z)+E*X1(Z)+E**2*X2(Z));
```

$$(D8) \qquad\qquad E \ (E^2 \ X2(Z) \ + \ E \ X1(Z) \ + \ XO(Z))$$

```
(C9) LCY:E*(YO(Z)+E*Y1(Z)+E**2*Y2(Z));
```

$$(D9) \qquad\qquad E \ (E^2 \ Y2(Z) \ + \ E \ Y1(Z) \ + \ YO(Z))$$

Now all the previous expansions are substituted into the differential equations labeled D2 and D3:

```
(C10) [D2,D3],DIFF$
```

Here and elsewhere we use the $ terminator to supress the display of the resulting expression. Next we Taylor expand and collect terms:

```
(C11) TAYLOR(EV(%),E,0,3)$
```

```
(C12) FOR I THRU 3 DO EQ[I-1]:COEFF(%,E,I)$
```

As a check we display the zeroth order equations (cf. eqs.(23),(24)):

```
(C13) EQ[0]/WO;
```

$$(D13)/R/ \qquad [XO(Z)_Z \ = \ - \ YO(Z), \ YO(Z)_Z \ = \ XO(Z)]$$

As usual in Lindstedt's method, the first order equations are nonhomogenous versions of the zeroth order:

```
(C14) EQ[1];
```

$$(D14)/R/ \ [WO \ X1(Z)_Z \ = \ - \ (2 \ Y1(Z) \ WO \ - \ FXX \ XO^2(Z)$$

$$- \ 2 \ FXY \ YO(Z) \ XO(Z) \ - \ FYY \ YO^2(Z))/2,$$

$$Y1(Z)_Z \ WO \ = \ (2 \ X1(Z) \ WO \ + \ GXX \ XO^2(Z) \ + \ 2 \ GXY \ YO(Z) \ XO(Z)$$

$$+ \ GYY \ YO^2(Z))/2]$$

We define the zeroth order solution, eq.(28), and substitute into the first order equations:

```
(C15) [XO(Z)=B*COS(Z),YO(Z)=B*SIN(Z)];
(D15)      [XO(Z) = B COS(Z),  YO(Z) = B SIN(Z)]
```

```
(C16) EQ[1],%,DIFF$
```

We clean up the trig terms with TRIGREDUCE before attempting to solve the first order equations:

```
(C17) EXPAND(TRIGREDUCE(EXPAND(%)));
                      2                    2
                     B  FXY SIN(2 Z)      B  FYY COS(2 Z)
(D17) [WO X1(Z)  =  ---------------  -  ---------------
                 Z          2                   4


   2
  B  FXX COS(2 Z)                   2         2
+ ---------------  - WO Y1(Z) +    B  FYY    B  FXX
         4                         ------  + ------,
                                     4         4


                      2                    2
                     B  GXY SIN(2 Z)      B  GYY COS(2 Z)
WO Y1(Z)  =         ---------------  -  ---------------
          Z                  2                   4


   2
  B  GXX COS(2 Z)                   2         2
+ ---------------  + WO X1(Z) +    B  GYY    B  GXX
         4                         ------  + ------]
                                     4         4
```

The first order equations are solved using DESOLVE. The MACSYMA function ATVALUE is used to specify the initial condition eq.(27):

```
(C18) LOAD([DESOLN,MACSYMA,SHARE])$
```

```
(C19) ATVALUE(Y1(Z),Z=0,0)$
```

```
(C20) DESOLVE(D17,[X1(Z),Y1(Z)]);
                   2         2         2
                  (B  GXY - B  FYY + B  FXX) SIN(2 Z)
(D20) [X1(Z) =  -----------------------------------
                              6 WO


     2         2          2
    (B  GYY - B  GXX + 4 B  FXY) COS(2 Z)
  - -------------------------------------
                 12 WO
```

$$- \frac{(2 B^2 GXY - 2 B^2 FYY - B^2 FXX) SIN(Z)}{6 WO}$$

$$+ \frac{(6 X1(0) WO + 2 B^2 GYY + B^2 GXX + 2 B^2 FXY) COS(Z)}{6 WO}$$

$$- \frac{B^2 GYY + B^2 GXX}{4 WO},$$

$$Y1(Z) = - \frac{(B^2 GYY - B^2 GXX + B^2 FXY) SIN(2 Z)}{6 WO}$$

$$- \frac{(4 B^2 GXY - B^2 FYY + B^2 FXX) COS(2 Z)}{12 WO}$$

$$+ \frac{(6 X1(0) WO + 2 B^2 GYY + B^2 GXX + 2 B^2 FXY) SIN(Z)}{6 WO}$$

$$+ \frac{(2 B^2 GXY - 2 B^2 FYY - B^2 FXX) COS(Z)}{6 WO} + \frac{B^2 FYY + B^2 FXX}{4 WO}]$$

Next the zeroth and first order solutions are substituted into the second order equations, and TRIGREDUCE is again used to tidy up:

(C21) EQ[2],D15,%,DIFF$

(C22) EXPAND(TRIGREDUCE(EXPAND(%)))$

Finally we isolate the coefficients of sin z and cos z in order to remove the resonant terms in the second order equations:

(C23) COEFF(D22,SIN(Z));

$$(D23) [- B K = - B M W1 - \frac{7 B^3 FXY GYY}{24 WO} + \frac{B^3 FYY GXY}{6 WO}$$

$$+ \frac{B^3\ FXX\ GXY}{12\ WO} - \frac{5\ B^3\ FXY\ GXX}{24\ WO} + \frac{5\ B^3\ FYY^2}{24\ WO} + \frac{5\ B^3\ FXX\ FYY}{24\ WO}$$

$$+ \frac{B^3\ FXY^2}{12\ WO} + \frac{B^3\ FXX^2}{12\ WO} + \frac{B^3\ FYYY}{8} + \frac{B^3\ FXXY}{8},$$

$$O = -\frac{B^3\ GXY\ GYY}{8\ WO} + \frac{5\ B^3\ FYY\ GYY}{24\ WO} + \frac{7\ B^3\ FXX\ GYY}{24\ WO} - \frac{B^3\ GXX\ GXY}{8\ WO}$$

$$+ \frac{B^3\ FXY\ GXY}{12\ WO} - \frac{B^3\ FYY\ GXX}{12\ WO} + \frac{B^3\ FXX\ GXX}{12\ WO} + B\ M + \frac{B^3\ GYYY}{8}$$

$$+ \frac{B^3\ GXXY}{8}]$$

(C24) COEFF(D22,COS(Z));

$$(D24)\ [O = -\frac{B^3\ FYY\ GYY}{12\ WO} - \frac{7\ B^3\ FXX\ GYY}{24\ WO} - \frac{B^3\ FXY\ GXY}{12\ WO}$$

$$+ \frac{B^3\ FYY\ GXX}{12\ WO} - \frac{5\ B^3\ FXX\ GXX}{24\ WO} + \frac{B^3\ FXY\ FYY}{8\ WO} + \frac{B^3\ FXX\ FXY}{8\ WO}$$

$$+ B\ M + \frac{B^3\ FXYY}{8} + \frac{B^3\ FXXX}{8},$$

$$B\ K = B\ M\ W1 - \frac{B^3\ GYY^2}{12\ WO} - \frac{5\ B^3\ GXX\ GYY}{24\ WO} - \frac{B^3\ FXY\ GYY}{12\ WO} - \frac{B^3\ GXY^2}{12\ WO}$$

$$+ \frac{5\ B^3\ FYY\ GXY}{24\ WO} + \frac{7\ B^3\ FXX\ GXY}{24\ WO} - \frac{5\ B^3\ GXX^2}{24\ WO} - \frac{B^3\ FXY\ GXX}{6\ WO}$$

$$+ \frac{B^3\ GXYY}{8} + \frac{B^3\ GXXX}{8}]$$

Comparison of expressions D23 and D24 with eqs.(29),(30) shows that D23 is of the form [a,c] while D24 represents [b,d]. The conditions (33) for removal of resonant terms involve two equations for B and K. However since we are only interested in B for the Hopf formula, we form only one of the conditions (33), namely b+c = 0:

(C25) PART(D24,1)+PART(D23,2)$

(C26) SOLVE(%,B);
(D26) [B = - 4 SQRT(- M WO/(GYYY WO + GXXY WO + FXYY WO

+ FXXX WO - GXY GYY + FYY GYY - GXX GXY - FXX GXX

+ FXY FYY + FXX FXY)),

B = 4 SQRT(- M WO/(GYYY WO + GXXY WO + FXYY WO + FXXX WO

- GXY GYY + FYY GYY - GXX GXY - FXX GXX + FXY FYY

+ FXX FXY)),

B = 0]

SOLVE returns three values for B. We choose the positive value:

(C27) PART(%,2);
(D27) B = 4 SQRT(- M WO/(GYYY WO + GXXY WO + FXYY WO

+ FXXX WO - GXY GYY + FYY GYY - GXX GXY - FXX GXX

+ FXY FYY + FXX FXY))

The result of the preceding calculation may be expressed thus: For small e, the amplitude of the limit cycle is approximately given by

$$(34) \qquad \text{Amplitude} = e\, B = 4\,(\,-u\,w0\,/\,S\,)^{1/2}$$

where the quantity S is defined by the formula

$$(35) \qquad S = w0\,(\,gyyy + gxxy + fxyy + fxxx\,)$$
$$- gxy\,gyy + fyy\,gyy - gxx\,gxy$$
$$- fxx\,gxx + fxy\,fyy + fxx\,fxy$$

in which all the partial derivatives are evaluated when
x=y=u=0.

In order for the limit cycle to exist, the amplitude
(34) must be real. Thus if $S > 0$, the limit cycle occurs
when $u < 0$ (the subcritical case), while if $S < 0$, the
limit cycle occurs when $u > 0$ (the supercritical case). If
$S = 0$, no conclusion may be drawn (the nongeneric case).

Example: For the Van der Pol example of
eqs.(15),(16), we find

$$(36) \qquad f = 0, \quad g = (u/w) x^3 - x^2 y$$

where $w = (1 - u^2)^{1/2}$. The only nonvanishing derivative
at x=y=u=0 is

$$(37) \qquad gxxy = -2$$

We also see that (cf. eq.(18))

$$(38) \qquad w0 = 1$$

Eqs.(35),(37),(38) give

$$(39) \qquad S = -2 < 0$$

Thus the Hopf theory predicts that we have a
supercritical bifurcation in which a stable limit cycle
emerges for $u > 0$. The approximate expression (34) for the
limit cycle's amplitude becomes

$$(40) \qquad \text{Amplitude} = (8 u)^{1/2}$$

valid for small u. For $u = 1/10$, eq.(40) predicts an
amplitude of about 0.89, which approximately agrees with
the result of numerical integration shown in Fig. 6.

CONCLUSION

This work involves the use of computer algebra to derive a formula for which other derivations have been given (1,2). All such derivations include a vexatious quantity of algebra, making the use of computer algebra more attractive.

This kind of application of computer algebra is distinctly different from traditional computations in which one seeks the answer to a particular problem. Rather, here we see the computer algebra system as functioning as a theorem-prover. We expect to see the increasing appearance of computer algebra proofs and derivations replacing traditional tedious hand calculations in courses in mathematics and engineering.

ACKNOWLEDGEMENT

This work was partially supported by Air Force grant no. AFOSR-84-0051.

REFERENCES

1. Marsden, J.E. and McCracken, M., The Hopf Bifurcation and Its Applications, Springer-Verlag (1976)
2. Guckenheimer, J. and Holmes, P., Nonlinear Oscillations, Dynamical Systems and Bifurcations of Vector Fields, Springer-Verlag (1983)
3. Hassard, B.D., Kazarinoff, N.D. and Wan, Y.H., Theory and Applications of Hopf Bifurcation, Cambridge University Press (1981)
4. Stoker, J.J., Nonlinear Vibrations, Interscience Publishers (1950)
5. Rand, R.H., Computer Algebra in Applied Mathematics: An Introduction to MACSYMA, Pitman Publishing (1984)
6. MACSYMA Reference Manual, 2 vols., The Mathlab Group, Laboratory for Computer Science, MIT, 545 Technology Square, Cambridge, MA 02139, Version 10 (1983)

15

NORMAL FORM AND CENTER MANIFOLD CALCULATIONS ON MACSYMA

R.H. RAND and W.L. KEITH

Department of Theoretical and Applied Mechanics,
Cornell University, Ithaca NY 14853

ABSTRACT

This paper describes the use of the symbolic
manipulation system MACSYMA to facilitate normal form and
center manifold computations, which arise in nonlinear
dynamics problems. These computations represent a
relatively new approach towards obtaining approximate
solutions to systems of ordinary differential equations. A
near-identity coordinate transformation in the form of a
power series is used to locally transform a given system
into a normal (i.e., simple or canonical) form. Center
manifold calculations are a related technique for reducing
the number of dimensions in a given system of ordinary
differential equations by restricting attention to the flow
on an invariant subspace. A MACSYMA program developed to
perform such computations is presented and described in
detail. The program is illustrated by applying it to a
sample nonlinear dynamics problem.

NORMAL FORMS

There exist a variety of methods for obtaining
approximate solutions to ordinary differential equation
problems (e.g., perturbations, averaging, harmonic balance;
see (1)). Most of these share a common approach in which
the solution of the problem is written in the form of an
infinite series, and the method involves a scheme for
obtaining an approximation based on an n term truncation of
the series.

In contrast to these methods, the method of normal forms does not involve expanding the solution in an infinite series. Rather, it is based on the idea of approximately transforming the differential equations themselves into a form which is easily solved. The method involves generating a transformation of dependent variables in the form of an infinite series, and computing the coefficients of the series so that the resulting transformed differential equations are in a normal (simple, canonical) form.

The method can be illustrated by considering a 2 dimensional system (although the same process applies to an n dimensional system):

(1) $x1' = f(x1,x2)$

(2) $x2' = g(x1,x2)$

We assume that the coordinates xi have been chosen such that the origin $x1=x2=0$ is an equilibrium point:

(3) $f(0,0) = g(0,0) = 0.$

We Taylor expand f and g about the origin,

(4) $x1' = a\ x1 + b\ x2 + F(x1,x2)$

(5) $x2' = c\ x1 + d\ x2 + G(x1,x2)$

where F and G are strictly nonlinear (i.e., contain terms of order 2 and higher).

We further assume that the coordinates xi have been chosen such that the linear system associated with (4),(5) is in canonical form (a problem in elementary linear algebra). In the case of real eigenvalues this will be diagonalized (or more generally, in Jordan canonical form), while for complex roots $u \pm iw$ we choose the real (but nondiagonal) form

(6) $\begin{bmatrix} u & -w \\ w & u \end{bmatrix}$

(In the case of an n dimensional system a corresponding form is chosen, built up from Jordan blocks, see e.g. (2).)

The linear transformation which puts the linearized system in canonical form can be thought of as the first step in a sequence of transformations, each of which simplifies the form of the terms of order i in (4),(5), for i=1,2,3,...

In the case in which F and G contain quadratic terms, we transform from x1,x2 to y1,y2 coordinates via a near-identity transformation with general quadratic terms:

$$(7) \qquad x1 = y1 + a20\ y1^2 + a11\ y1\ y2 + a02\ y2^2$$

$$(8) \qquad x2 = y2 + b20\ y1^2 + b11\ y1\ y2 + b02\ y2^2$$

where the aij and bij are to be determined.

We substitute (7),(8) into (4),(5) and neglect terms of order 3 (which do not influence the quadratic coefficients a20,...,b02). The resulting equations are linear in the derivatives yi'(t). We solve for the yi'(t) and again expand about y1=y2=0 and neglect terms of order 3. The result is of the form:

$$(9) \qquad y1' = a\ y1 + b\ y2 + c120\ y1^2 + c111\ y1\ y2 + c102\ y2^2$$

$$(10) \qquad y2' = c\ y1 + d\ y2 + c220\ y1^2 + c211\ y1\ y2 + c202\ y2^2$$

where cijk is the coefficient of $y1^j\ y2^k$ in the yi equation.

The cijk are known linear functions of the aij, bij. The linear terms in (9),(10) are identical to the linear terms in (4),(5) due to the near-identity nature of the transformation (7),(8).

The method consists of choosing the coefficients aij and bij so as to put equations (9) and (10) into a canonical form. The natural choice is to remove all the quadratic terms, c120=...=c202=0, although there are exceptional situations (involving resonances or repeated zero eigenvalues) in which this is not possible, see (3).

Once the coefficients in (7),(8) have been determined, we extend the transformation of xi to yi coordinates to include cubic terms. (Note that even if (4),(5) do not contain cubic terms, the transformation (7),(8) will generally introduce cubic terms.) Proceeding as before we compute the equations on the yi(t) neglecting terms of order 4, and choose the third order coefficients aij, bij in order to best simplify these equations. Note that the transformation up to and including cubic terms will not effect the already determined quadratic terms in the normal form.

Proceeding in this fashion we can (in principle) generate the desired transformation to any order of accuracy. Note, however, that the use of truncated power series can be expected to restrict the applicability of the method to a neighborhood of the origin.

CENTER MANIFOLDS

Center manifold theory is a related method which uses power series expansions of coordinates in order to reduce the dimension of a system of ordinary differential equations. The method involves restricting attention to an invariant subspace (called a center manifold) which contains all of the essential behavior of the system in the neighborhood of an equilibrium point, in the limit as time t approaches infinity.

This method is applicable to systems which, when linearized around an equilibrium point, have some eigenvalues which have zero real part, and others which have negative real part. (We assume that no eigenvalues have positive real part, since in such a case the center manifold is not attractive as t goes to infinity. Thus we assume that we are in the critical case of Lyapunov (see (4), p.150) in which the stability of the equilibrium point cannot be determined by the linearized equations.)

The idea of the method is that the components of the solution of the linearized equations which correspond to those eigenvalues with negative real part will decay as t goes to infinity, and hence the motion of the linearized

equations will asymptotically approach the space S1 spanned
by the eigenvectors corresponding to those eigenvalues with
zero real part. The center manifold theorem (see (5))
assures us that this picture (which is so far based on the
linearized equations) extends to the full nonlinear
equations, as follows:

There exists a (generally curved) subspace S2 (the
center manifold) which is tangent to the (flat) subspace S1
at the equilibrium point, and which is invariant under the
flow given by the nonlinear equations. All solutions which
start sufficiently close to the equilibrium point will tend
asymptotically to the center manifold. In particular, the
theorem states that the stability of the equilibrium point
in the full nonlinear equations is the same as its
stability in the flow on the center manifold (ref.(5),
p.4). Moreover, any additional equilibrium points or
periodic motions (limit cycles) which occur in a
neighborhood of the given equilibrium point on the center
manifold are guaranteed to exist in the full nonlinear
equations (ref.(5), p.29).

In order to illustrate this method, we consider the
following system of three differential equations:

(11) $\qquad x' = a\,x + b\,y + p(x,y,z)$

(12) $\qquad y' = c\,x + d\,y + q(x,y,z)$

(13) $\qquad z' = -z + r(x,y,z)$

where p,q and r are strictly nonlinear in x,y,z. Here the
linearized x and y equations are uncoupled from the
linearized z equation, and we assume that the coefficients
a,b,c,d are such that the linearized x,y equations have
eigenvalues with zero real part (i.e. either a pure
imaginary complex conjugate pair or a double zero.)

In this problem the center manifold is a surface which
is tangent to the xy plane at the origin. We may obtain an
approximate expression for it by writing

(14) $z = h(x,y) = K20\ x^2 + K11\ x\ y + K02\ y^2 + \ldots$

where constant and linear terms have been omitted in order that $z = h(x,y)$ be tangent to the xy plane.

The coefficients Kij in (14) are to be found by requiring $z = h(x,y)$ to be invariant under the flow (11)-(13). This may be accomplished by differentiating (14):

(15) $z' = 2\ K20\ x\ x' + K11\ (x\ y'+x'\ y) + 2\ K02\ y\ y' + \ldots$

and substituting expressions for x',y',z' given by (11)-(13). The resulting expression, which depends on x,y and z, can be made to depend on x and y only, by using (14). Finally we may collect terms and set the coefficients of x^2, xy and y^2 to zero in order to obtain K20, K11 and K02. By including higher order terms in (14), this process may be extended to arbitrary accuracy.

Note that once $z = h(x,y)$ is known, it may be substituted into (11),(12), thereby giving an abbreviated system of 2 differential equations representing the flow on the center manifold (or rather its projection onto the xy plane.)

We have just described the standard procedure for calculating center manifolds (5). In this work, however, we shall accomplish this computation in an equivalent but different manner. We shall consider the center manifold computation as a normal form problem, based on the following near-identity transformation from (x,y,z) to (u,v,w) coordinates:

(16) $x = u$

(17) $y = v$

(18) $z = w + h(u,v) = w + K20\ u^2 + K11\ u\ v + K02\ v^2$

where we have neglected terms of order 3.

When w=0 in (16)-(18), we obtain the previous expression (14) for the center manifold. Our procedure will be to substitute (16)-(18) into the differential equations (11)-(13), and to transform to new equations on

u,v,w (as described in the previous section on normal
forms).   These will be of the form:

(19)   u' = a u + b v + Q1(u,v,w)

(20)   v' = c u + d v + Q2(u,v,w)

(21)   w' = -w + Q3(u,v,w)

where $Qi(u,v,w)$ stands for quadratic terms in u,v,w.   We
then set w=0 in the w equation, (21), and obtain the
coefficients Kij by equating to zero the coefficients of
the remaining quadratic terms u^2, uv, v^2.

   We have adopted this scheme for performing center
manifold calculations using normal form theory in order to
use a single MACSYMA program to accomplish both kinds of
problems.   Although we have illustrated the procedure only
for the case of a system of 3 differential equations with a
2 dimensional center manifold (11)-(13), the same scheme of
embedding the center manifold calculation in a normal form
problem will work for any size system.

EXAMPLE

   We shall demonstrate our MACSYMA programs for
accomplishing normal form and center manifold calculations
by applying them to the following system of three
differential equations:

(22)   x' = y

(23)   y' = - x - x z

(24)   z' = - z + x^2

These equations represent a vibrating system with
parametric feedback control, and have been discussed in
(6).

   Eqs.(22)-(24) have an equilibrium point at the origin.
We shall be interested in the question of its stability.
This system is of the form (11)-(13) previously discussed,
and hence possesses a center manifold which is tangent to
the xy plane at the origin.

   In what follows we shall present the record of a
MACSYMA session in which we first obtain an approximation

to the center manifold, thereby reducing the dimension of the system from 3 to 2. Then we shall use normal forms to treat the resulting system, enabling us to determine the stability of the equilibrium point at the origin.

Before proceeding, we must consider the appropriate normal form for this problem. Takens (7) has shown that any system of the form

(25) $x' = y + f(x,y)$

(26) $y' = -x + g(x,y)$

where f and g are strictly nonlinear in x,y, can be put in the normal form

(27) $r' = a1\ r^3 + a2\ r^5 + \ldots$

(28) $\theta' = -1 + b1\ r^2 + b2\ r^4 + \ldots$

where r and θ are polar coordinates

(29) $u = r\cos\theta, \quad v = r\sin\theta$

and where u,v are related to x,y by a near-identity transformation. In rectangular coordinates, (27),(28) become

(30) $u' = v + a1\ (u^2 + v^2)\ u + b1\ (u^2 + v^2)\ v + O(5)$

(31) $v' = -u + a1\ (u^2 + v^2)\ v - b1\ (u^2 + v^2)\ u + O(5)$

Here, then, is the MACSYMA session. The reader is referred to (8) for an introduction to MACSYMA.

We begin by loading a file called NORMFORM7.MAC, and then displaying the user instructions which have been saved in a variable named GO (see (8), Chapter 2):

(C1) LOADFILE(NORMFORM7,MAC)$

(C2) GO;
(D2) THIS FILE CONTAINS NF( ), A NORMAL FORM FUNCTION.
WHEN ENTERING TRANSFORMATION, GEN(N) WILL GENERATE THE
GENERAL TERMS OF HOMOGENEOUS ORDER N IN 2 VARIABLES.
THE UTILITY DECOMPOSE( ) WILL ISOLATE THE COEFFS OF THE
NEW EQS., BUT WORKS ONLY FOR 2 EQS.

The MACSYMA program consists of the main function NF, and
of the two auxiliary functions GEN and DECOMPOSE. The use
of each of these functions will be illustrated in what
follows. We start the center manifold computation by
calling NF:

```
(C3) NF( )$
DO YOU WANT TO ENTER NEW VARIABLE NAMES (Y/N) ?
Y;
HOW MANY EQS
3;
SYMBOL FOR OLD X[ 1 ]
X;
SYMBOL FOR OLD X[ 2 ]
Y;
SYMBOL FOR OLD X[ 3 ]
Z;
SYMBOL FOR NEW X[ 1 ]
U;
SYMBOL FOR NEW X[ 2 ]
V;
SYMBOL FOR NEW X[ 3 ]
W;
DO YOU WANT TO ENTER NEW D.E.'S (Y/N) ?
Y;
ENTER RHS OF EQ. NO. 1 ,    D X /DT =
Y;
X  = Y
 T
ENTER RHS OF EQ. NO. 2 ,    D Y /DT =
-X-X*Z;
Y  = - X Z - X
 T
ENTER RHS OF EQ. NO. 3 ,    D Z /DT =
-Z+X**2;
        2
Z  = X  - Z
 T
```

Having entered the variable names and differential
equations (22)-(24), we next enter the transformation of
variables (16)-(18):

```
INPUT NEAR-IDENTITY TRANSFORMATION
(USE PREV FOR PREVIOUS TRANSFORMATION)
X = U + ?
0;
X = U
Y = V + ?
0;
Y = V
Z = W + ?
K20*U**2+K11*U*V+K02*V**2;
                2                    2
Z = W + K02 V   + K11 U V + K20 U
```

```
ENTER TRUNCATION ORDER (HIGHEST ORDER TERMS TO BE KEPT)
2;
```

The program now computes the transformed equations to order
2.   These correspond to eqs.(19)-(21) given previously:

```
U   +  .  .  .  = V +  .  .  .
 T
V   +  .  .  .  = - U - W U +  .  .  .
 T
                                          2
W   +  .  .  .  = - W + ((K11 - K20 + 1) U
 T
                                                    2
   + (- K11 - 2 K20 + 2 K02) V U + (- K11 - K02) V  ) +  .  .  .
DO YOU WANT TO ENTER ANOTHER TRANSFORMATION (Y/N) ?
N;
```

In order to find the coefficients Kij which specify the
center manifold, we set W=0 in the transformed equations.
Note the standard MACSYMA use of % to refer to the previous
result.

```
(C4)  %,W=0;
(D4)  [[U   = V,  V   = - U, 0 =
         T         T

                   2
(- K11 - K02) V   + (- 2 K20 - K11 + 2 K02) U V

                 2
  + (- K20 + K11 + 1) U  ]]
```

We use the MACSYMA function PICKAPART to obtain the
equations on the Kij:

```
(C5) PICKAPART(PART(%,1,3),3);
```

$$(E5) \qquad\qquad - K11 - K02$$

$$(E6) \qquad\qquad V^2$$

$$(E7) \qquad\qquad - 2\ K20 - K11 + 2\ K02$$

$$(E8) \qquad\qquad - K20 + K11 + 1$$

$$(E9) \qquad\qquad U^2$$

$$(D9) \qquad 0 = E5\ E6 + E7\ U\ V + E8\ E9$$

```
(C10) SOLVE([E5,E7,E8],[K20,K11,K02]);
```

$$(D10) \qquad [[K20 = \frac{3}{5},\ K11 = -\frac{2}{5},\ K02 = \frac{2}{5}]]$$

Having solved for the Kij, we plug them into the
transformation (called TRANS here), and substitute W=0 to
obtain an approximate expression for the center manifold:

```
(C11) TRANS,%;
```

$$(D11) \quad [X = U,\ Y = V,\ Z = W + \frac{2\ V^2}{5} - \frac{2\ U\ V}{5} + \frac{3\ U^2}{5}]$$

```
(C12) PART(%,3),W=0,U=X,V=Y;
```

$$(D12) \qquad Z = \frac{2\ Y^2}{5} - \frac{2\ X\ Y}{5} + \frac{3\ X^2}{5}$$

Our next step is to substitute the expression D12 for the
center manifold into the x and y equations, thereby
obtaining an approximation for the flow on the center
manifold. We once again call our function NF:

```
(C13) NF( )$
DO YOU WANT TO ENTER NEW VARIABLE NAMES (Y/N) ?
Y;
HOW MANY EQS
2;
SYMBOL FOR OLD X[ 1 ]
X;
SYMBOL FOR OLD X[ 2 ]
Y;
```

```
SYMBOL FOR NEW X[ 1 ]
U;
SYMBOL FOR NEW X[ 2 ]
V;
DO YOU WANT TO ENTER NEW D.E.'S (Y/N) ?
Y;
ENTER RHS OF EQ. NO. 1 ,    D X /DT =
Y;
X  = Y
 T
ENTER RHS OF EQ. NO. 2 ,    D Y /DT =
EV(-X-X*Z,D12);
```

$$X_T = Y$$

$$Y_T = -X\left(\frac{2Y^2}{5} - \frac{2XY}{5} + \frac{3X^2}{5}\right) - X$$

We wish to transform these equations via a near-identity
transformation with general coefficients. Note that since
all quadratic terms are absent from the differential
equations, our near-identity transformation begins with
cubic terms. To save typing, the function GEN is utilized:

```
INPUT NEAR-IDENTITY TRANSFORMATION
(USE PREV FOR PREVIOUS TRANSFORMATION)
X = U + ?
GEN(3);
```

$$X = A_{0,3}\,V^3 + A_{1,2}\,U V^2 + A_{2,1}\,U^2 V + A_{3,0}\,U^3 + U$$

```
Y = V + ?
GEN(3);
```

$$Y = B_{0,3}\,V^3 + B_{1,2}\,U V^2 + B_{2,1}\,U^2 V + V + B_{3,0}\,U^3$$

```
ENTER TRUNCATION ORDER (HIGHEST ORDER TERMS TO BE KEPT)
3;
```

The program now computes the transformed equations
representing the flow on the center manifold to order 3:

$$U_T + \ldots = V + ((A_{2,1} + B_{3,0})U^3$$

$$+ (B_{2,1} - 3A_{3,0} + 2A_{1,2})V U^2$$

$$+ (-2A_{2,1} + B_{1,2} + 3A_{0,3})V^2 U$$

$$+ (- A_{1, 2} + B_{0, 3}) V) + \ldots$$

$$V_{T} + \ldots = - U + ((5 B_{2, 1} - 5 A_{3, 0} - 3) U^{3}$$

$$+ (- 5 A_{2, 1} - 15 B_{3, 0} + 10 B_{1, 2} + 2) V U^{2}$$

$$+ (- 10 B_{2, 1} - 5 A_{1, 2} + 15 B_{0, 3} - 2) V^{2} U$$

$$+ (- 5 B_{1, 2} - 5 A_{0, 3}) V^{3})/5 + \ldots$$

DO YOU WANT TO ENTER ANOTHER TRANSFORMATION (Y/N) ?
N;

We now wish to isolate the coefficients of the terms in the transformed equations. As a convenient alternative to the MACSYMA function PICKAPART which we used previously (see C5), we use our function DECOMPOSE:

(C14) DECOMPOSE( )$
C[I,J,K] IS THE COEFF OF U **J V **K IN THE I-TH EQUATION

We aim for the Takens normal form of eqs.(30),(31). Using the coefficients C[I,J,K] assigned by DECOMPOSE, we produce the necessary equations:

(C15) C[1,3,0]=C[1,1,2];
(D15) $\quad B_{3, 0} + A_{2, 1} = - 2 A_{2, 1} + B_{1, 2} + 3 A_{0, 3}$

(C16) C[1,3,0]=C[2,2,1];
(D16) $\quad B_{3, 0} + A_{2, 1} = - 3 B_{3, 0} - A_{2, 1} + 2 B_{1, 2} + \frac{2}{5}$

(C17) C[1,3,0]=C[2,0,3];
(D17) $\quad B_{3, 0} + A_{2, 1} = - B_{1, 2} - A_{0, 3}$

(C18) C[1,0,3]=C[1,2,1];
(D18) $\quad B_{0, 3} - A_{1, 2} = - 3 A_{3, 0} + B_{2, 1} + 2 A_{1, 2}$

(C19) C[1,0,3]=-C[2,3,0];

$$(D19) \quad B_{0,3} - A_{1,2} = A_{3,0} - B_{2,1} + \frac{3}{5}$$

(C20) C[1,0,3]=-C[2,1,2];

$$(D20) \quad B_{0,3} - A_{1,2} = 2 B_{2,1} + A_{1,2} - 3 B_{0,3} + \frac{2}{5}$$

We now have 6 equations in 8 unknowns, A03,A12,A21,A30,B03, B12,B21,B30. We use the MACSYMA function SOLVE to automatically unravel this linear algebra problem. Expressions are returned for A03,A12,A21,A30,B03 and B12 in terms of B21 and B30 (called %R1 and %R2 respectively):

(C21) SOLVE([D15,D16,D17,D18,D19,D20],
[A[0,3],A[1,2],A[2,1],A[3,0],B[0,3],B[1,2],B[2,1],B[3,0]]);

$$(D21) \; [[A_{0,3} = -\frac{10 \, \%R1 - 1}{10}, \; A_{1,2} = \frac{20 \, \%R2 - 7}{20},$$

$$A_{2,1} = -\frac{20 \, \%R1 - 1}{20}, \; A_{3,0} = \frac{40 \, \%R2 - 13}{40},$$

$$B_{0,3} = \frac{40 \, \%R2 - 3}{40}, \; B_{1,2} = \frac{20 \, \%R1 - 3}{20}, \; B_{2,1} = \%R2,$$

$$B_{3,0} = \%R1 \,]]$$

We check the previous calculation by rerunning it with the values of the coefficients D21 substituted into the previous general near-identity transformation. Note the use of the variable PREV to refer to the previous transformation:

(C22) NF( )$
DO YOU WANT TO ENTER NEW VARIABLE NAMES (Y/N) ?
N;
DO YOU WANT TO ENTER NEW D.E.'S (Y/N) ?
N;

$$X_T = Y$$

$$Y_T = - X \left(\frac{2 Y^2}{5} - \frac{2 X Y}{5} + \frac{3 X^2}{5} \right) - X$$

```
INPUT NEAR-IDENTITY TRANSFORMATION

(USE PREV FOR PREVIOUS TRANSFORMATION)
X = U + ?
PREV,D21;
                       3                       2
        (10 %R1 - 1) V       (20 %R2 - 7) U V
X = - ---------------  +  ------------------
              10                    20


                            2               3
          (20 %R1 - 1) U  V   (40 %R2 - 13) U
        - ------------------ + ---------------- + U
                  20                  40
Y = V + ?
PREV,D21;
                    3                  2
       (40 %R2 - 3) V    (20 %R1 - 3) U V            2
Y = --------------- + ------------------ + %R2 U  V + V
            40                 20

                                                        3
                                            + %R1  U
ENTER TRUNCATION ORDER (HIGHEST ORDER TERMS TO BE KEPT)
3;
```

Now the program once again computes the transformed equations. Note that while the transformation involves the arbitrary quantities %R1 and %R2, the resulting equations are unique:

```
                3       2       2       3
              2 U  + 11 V U  + 2 V  U + 11 V
U  + . . . = V + ----------------------------- + . . .
 T                            40

                 3       2       2       3
              11 U  - 2 V U  + 11 V  U - 2 V
V  + . . . = - U - ----------------------------- + . . .
 T                            40
```

```
DO YOU WANT TO ENTER ANOTHER TRANSFORMATION (Y/N) ?
N;
```

Now we simplify the final equations, labeled D22, by transforming to polar coordinates:

```
(C23) DEPENDS([R,THETA],T);
(D23)                 [R(T),  THETA(T)]

(C24) [U=R*COS(THETA),V=R*SIN(THETA)];
(D24)      [U = R COS(THETA), V = R SIN(THETA)]
```

```
(C25)  D22,%,DIFF$
(C26)  SOLVE(PART(%,1),[DIFF(R,T),DIFF(THETA,T)]);
              3    2            3    2
          R  SIN (THETA) + R  COS (THETA)
(D26) [[R  = -------------------------------,
        T                  20


                  2    2             2    2
            11 R  SIN (THETA) + 11 R  COS (THETA) + 40
   THETA  = - -----------------------------------------]]
       T                        40

(C27)  TRIGSIMP(%);
                      3                        2
                      R                    11 R  + 40
(D27)        [[R   = --,   THETA   = - ----------]]
               T     20        T           40
```

The result of the computation follows from the derived
normal form D27:

$$(32) \qquad\qquad r' = r^3 / 20 + O(5)$$

$$(33) \qquad\qquad \theta' = -1 - 11\, r^2 /40 + O(4)$$

Eq.(32), used in conjunction with the center manifold
theorem, implies that the equilibrium point r=0 is
unstable.

THE MACSYMA PROGRAM

Before describing in detail the MACSYMA functions
which we used in the previous example, we continue the
previous run with FUNCTIONS and DISPFUN(ALL) commands,
thereby displaying all user-defined functions:

```
(C28) FUNCTIONS;
(D28) [NF( ), INP1( ), INP2( ), INP3( ), INP4( ), SETUP( ),
STEP1( ), STEP2( ), STEP3( ), GEN(NN), AUX(II, JJ, ROW),
DECOMPOSE( )]

(C29) DISPFUN(ALL);
(E29) NF( ) := BLOCK(TEST :
READ("DO YOU WANT TO ENTER NEW VARIABLE NAMES (Y/N) ?"),
IF TEST = N THEN GO(JUMP), INP1( ), SETUP( ), JUMP, INP2( ),
LOOP, INP3( ), INP4( ), STEP1( ), STEP2( ), STEP3( ),
BRANCH : READ("DO YOU WANT TO ENTER ANOTHER TRANSFORMATION
(Y/N) ?"), IF BRANCH = Y THEN GO(LOOP), TEMP4)
```

```
(E30) INP1( ) := (N : READ("HOW MANY EQS"),
FOR I THRU N DO X  : READ("SYMBOL FOR OLD X[", I, "]"),
                I

FOR I THRU N DO Y  : READ("SYMBOL FOR NEW X[", I, "]"))
                I

(E31) INP2( ) :=
(PRINT("DO YOU WANT TO ENTER NEW D.E.'S (Y/N) ?"),
TEST : READ( ), FOR I THRU N DO ( IF TEST = Y THEN
                                          I

RHS  : READ("ENTER RHS OF EQ. NO.", I, ",  D", X ,"/DT ="),
                                                 I

EQ  : DIFF(X , T) = RHS , PRINT(EQ ),
  I        I          I          I

EQS : MAKELIST(EQ , I, 1, N)))
                 I

(E32) INP3( ) := (PRINT("INPUT NEAR-IDENTITY TRANSFORMATION
(USE PREV FOR PREVIOUS TRANSFORMATION)"),
FOR I THRU N DO (ROW : I, PREV : TR ,
                                   I

TR  : READ(X , "=", Y , "+ ?"), PRINT(X , "=", Y  + TR )),
  I        I         I                 I         I     I

TRANS : MAKELIST(X  = Y  + TR , I, 1, N))
                  I    I     I

(E33) INP4( ) := M : READ("ENTER TRUNCATION ORDER (HIGHEST
ORDER TERMS TO BE KEPT)")

(E34) SETUP( ) := FOR I THRU N DO DEPENDS([X , Y ], T)
                                           I    I

(E35) STEP1( ) := TEMP2 : TAYLOR(EV(EQS, TRANS, DIFF),
MAKELIST(Y , I, 1, N), O, M)
          I

(E36) STEP2( ) := (FOR I THRU N
DO TEMP2 : SUBST(DUMMY , DIFF(Y , T), TEMP2),
                      I        I

TEMP3 : SOLVE(TEMP2, MAKELIST(DUMMY , I, 1, N)),
                                   I

FOR I THRU N DO TEMP3 : SUBST(DIFF(Y , T), DUMMY , TEMP3))
                                   I            I
```

```
(E37) STEP3( ) := (TEMP4 : TAYLOR(TEMP3,
MAKELIST(Y , I, 1, N), 0, M),
         I

FOR I THRU N DO PRINT(PART(TEMP4, 1, I)))

(E38) GEN(NN) := BLOCK( IF N > 2
 THEN (PRINT("GEN ONLY WORKS WITH 2 VARIABLES!"),
RETURN(0)) ELSE TEMPGEN : 0,
FOR II FROM 0 THRU NN DO (FOR JJ FROM 0 THRU NN
DO ( IF II + JJ = NN THEN (AUX(II, JJ, ROW),
```

$$\text{TEMPGEN} : \text{TEMPGEN} + Z_{II, JJ, ROW} \ Y_1^{II} \ Y_2^{JJ} \ ))), \text{TEMPGEN})$$

```
(E39) AUX(II, JJ, ROW) := IF ROW = 1
```

$$\text{THEN } Z_{II, JJ, ROW} : A_{II, JJ} \quad \text{ELSE } Z_{II, JJ, ROW} : B_{II, JJ}$$

```
(E40) DECOMPOSE( ) := IF N = 2
THEN (FOR I THRU 2 DO NEWRHS   :
                           I

EXPAND(RHS(PART(TEMP4, 1, I))),
FOR I THRU 2 DO (FOR J FROM 0 THRU M
DO (FOR K FROM 0 THRU M DO ( IF J + K = M
```

$$\text{THEN } C_{I, J, K} : \text{COEFF(COEFF(NEWRHS}_I , Y_1 , J), Y_2 , K)))),$$

$$\text{PRINT("C[I,J,K] IS THE COEFF OF", } Y_1 , \text{"**J", } Y_2 ,$$

```
"**K IN THE I-TH EQUATION"), " ")

 ELSE PRINT("DECOMPOSE ONLY WORKS FOR 2 EQUATIONS")

(D40)                          DONE
```

The main function NF first asks the user if new variable names are to be entered. If the answer is affirmative, the function INP1 is called which reads the number of equations, N, and the variable names from the keyboard. Note that the old and new variables are respectively referred to as X[I] and Y[I] throughout the program, although they are displayed in the MACSYMA output by their user-defined symbols. Note also that in MACSYMA,

X[I] is the same as X , the difference being that the
                     I
former is the user-input version, while the latter is the
MACSYMA-output version.

      Next the function SETUP declares X[I] and Y[I] to be
functions of T. The function INP2 then asks the user if
new differential equations are to be entered. In the case
of an affirmative reply, the right hand side of the ith
equation, RHS[I], is read from the keyboard. The N
differential equations are stored as a list called EQS.
The transformation is then input via the function INP3 and
is stored as a list called TRANS. Finally INP4 reads the
truncation order M.

      The input portion of the program having been
completed, NF then calls STEP1, STEP2 and STEP3, which
compute the transformed equations. STEP1 plugs the
TRANSformation into the EQuationS, Taylor expands them, and
calls the result TEMP2. STEP2 solves TEMP2 for the
derivatives of the new variables, and stores the result as
TEMP3. (In order to conveniently use the MACSYMA function
SOLVE, the derivatives Y[I]' were replaced in TEMP2 by
dummy variables.) STEP3 then Taylor expands TEMP3, stores
the final result as TEMP4 and displays it. NF finishes up
by asking the user if another transformation is to be
entered, and if not, returns as its value the transformed
equations TEMP4.

      The auxiliary function GEN builds up a general
homogeneous polynomial of arbitrary degree by use of nested
FOR-DO loops. The auxiliary function DECOMPOSE is included
as a convenience for isolating the coefficients of the
transformed equations, TEMP4. As written, both GEN and
DECOMPOSE only work for N=2 equations, but can obviously be
extended to a larger number if desired.

328

CONCLUSION

Normal forms and center manifolds have received considerable attention lately in the applied mathematics literature (3,5,9). In addition to providing a method for investigating the stability of equilibrium (as illustrated in this work), these methods have also been used to investigate the bifurcation of equilibria and limit cycles, as well as to approximate rates of decay. Their use as a practical tool has been discouraged, however, by the great quantity of algebra involved. It is hoped that the availability of MACSYMA software to perform these calculations will increase their utility and popularity.

ACKNOWLEDGEMENT

This work was partially supported by Air Force grant # AFOSR-84-0051.

REFERENCES
1. Nayfeh, A., Perturbation Methods, John Wiley & Sons (1973)
2. Cesari, L., Asymptotic Behavior and Stability Problems in Ordinary Differential Equations, Third Edition, Springer-Verlag (1971)
3. Guckenheimer, J. and Holmes, P., Nonlinear Oscillations, Dynamical Systems, and Bifurcations of Vector Fields, Springer-Verlag (1983)
4. Minorsky, N., Nonlinear Oscillations, D. Van Nostrand Co. (1962)
5. Carr, J., Applications of Centre Manifold Theory, Springer-Verlag (1981)
6. Moon, F.C. and Rand, R.H., Parametric Stiffness Control of Flexible Structures, NASA Conference Proceedings: Workshop on Identification and Control of Flexible Space Structures, San Diego, May 1984
7. Takens, F., Singularities of Vector Fields, Publ. Math. Inst. Hautes Etudes Sci. 43, 47-100 (1974)
8. Rand, R.H., Computer Algebra in Applied Mathematics: An Introduction to MACSYMA, Pitman Publishing (1984)
9. Arnold, V.I., Geometrical Methods in the Theory of Ordinary Differential Equations, Springer-Verlag (1983)

16

SYMBOLIC COMPUTATION OF THE STOKES WAVE

W.H. Hui and G. Tenti

Department of Applied Mathematics

University of Waterloo

Waterloo, Ontario, Canada

N2L 3G1

ABSTRACT

Despite their familiarity, our understanding of the dynamics of surface water waves is far from complete, mainly because of the nonlinearity of the basic equations. G.G. Stokes was the first to find a particular solution in the form of a perturbation series. His result has been improved upon only recently, when fast electronic computers allowed researchers to carry out Stokes' program numerically to very high order, although numerical noise precludes drawing definitive conclusions for large amplitude waves. However, the development of modern symbolic computation systems has made it possible to obtain exact results, and we report in this paper on the use of two such systems (MAPLE and MACSYMA) for this problem. Central to the success of the approach is a new mathematical formulation, particularly suitable for symbolic computation.

INTRODUCTION

A classical problem in fluid dynamics, whose solution is still far from complete, is the study of the behavior of surface waves over large bodies of water. Problems in coastal engineering, ship hydrodynamics, offshore structures, and air-sea interaction are but a few examples of the many applications where such studies are of great value.

The literature on the subject of water waves is enormous, and it is impossible to give here a fair review of the field. The classical formulation goes back to the pioneering work of Stokes [1,2], and argues convincingly that, in a first approach to the problem, the water in an ocean may be assumed to be an inviscid, incompressible and irrotational fluid. It then follows that there exists a velocity potential $\phi(x,y,z,t)$ which satisfies the Laplace equation

$$\nabla^2 \phi = 0, \tag{1}$$

where the coordinate system is chosen with the (x,z)-plane horizontally oriented and the y-axis vertically upwards. Assuming further that the ocean has infinite lateral extent and infinite depth (although the finite depth case can be handled in a similar fashion), and that the restoring force is due to gravity alone, Eq. 1 becomes supplemented by one bottom and two surface boundary conditions:

$$\phi_y = 0, \qquad y \to -\infty, \tag{2}$$

$$\eta_t + \phi_x \eta_x + \phi_z \eta_z - \phi_y = 0, \qquad \text{at} \quad y = \eta, \tag{3}$$

$$\phi_t + \frac{1}{2}(\nabla \phi)^2 + g\eta = \text{constant}, \qquad \text{at} \quad y = \eta. \tag{4}$$

Here the subscripts denote partial derivatives, the function $\eta(x,z,t)$ denotes the free surface - to be found as part of the solution - and g is the acceleration of gravity. It should be noted that it is the nonlinearity of Eqs. 3 and 4, and the free boundary nature of the problem which make its solution exceedingly difficult to obtain.

Stokes [2] looked for a periodic solution representing a wave of permanent shape, uniform in the z-direction and traveling at speed c in the x-direction. Seen from a frame solidly moving with the wave, the problem is then effectively reduced to one of steady flow in two dimensions. Thus the theory of complex variables may be brought into play, and in the end the solution emerges in parametric form as

$$x = -\frac{\phi}{c} - \sum_{n=1}^{\infty} \frac{a_n}{n} e^{-n(\psi/c)} \sin(n \frac{\phi}{c}), \tag{5}$$

$$y = -\frac{\psi}{c} + \sum_{n=1}^{\infty} \frac{a_n}{n} e^{-n(\psi/c)} \cos(n \frac{\phi}{c}), \tag{6}$$

where ψ is the stream function, $\psi = 0$ corresponding to the free surface. Of course, the surface boundary conditions impose severe constraints on the Fourier coefficients, requiring them to satisfy an infinite system of nonlinear algebraic equations. However, use of a perturbation method allows a systematic approximate calculation of all the a_n coefficients in terms of a_1 - the amplitude of the fundamental - by solving, to order n, a nonlinear system of n equations. Stokes himself did it by hand to the fifth order for the infinite depth case, and to the third order for finite depths. His program was carried further over the next century by many investigators, notably by Wilton [3], De [4], Schwartz [5] and Cokelet [6]. The last two works, in particular, take advantage of modern digital computers to evaluate the Stokes wave numerically to very high order, thus enabling one to address questions about the status of Stokes' theory which were inaccessible before. The salient points of these recent results are discussed in a recent review [7], to which the reader is referred.

The coming of age of powerful symbolic computation systems over the past few years makes it natural to attempt an exact calculation of the Stokes wave. Indeed this would appear to be an ideal arena where symbolic computation can show its worth, for the mathematical formulation is relatively well understood and only the complication and the sheer tediousness of the algebra involved make an evelution by hand impossible. The payoff would be considerable, especially for the case of large amplitude waves, where the numerical computations mentioned above run into the problem of computer-generated numerical noise. It is the purpose of this paper to report on some preliminary results from the first use of two such systems in the exact calculation of the Stokes wave to high order.

REFORMULATION OF THE MATHEMATICAL PROBLEM

As outlined in the previous section, the Stokes formulation of the problem - followed by most researchers in this area - leads to a nonlinear system of equations for the approximate calculation of the Fourier coefficients, with the result that a direct application of symbolic computation would quickly run into serious difficulties. Consequently, we first concentrate on reformulating the mathematical problem in a way that involves solving a linear problem at each order of the perturbation scheme. Surprisingly, it seems to have escaped everybody's attention that a way to attain this goal is simply to reverse the order of the Stokes procedure, that is to say to apply perturbation theory first. Thus we seek a solution in the form

$$x = -\phi - \sum_{n=1}^{\infty}\epsilon^n x^{(n)},\qquad(7)$$

$$y = -\psi + \sum_{n=1}^{\infty}\epsilon^n y^{(n)},\qquad(8)$$

where

$$x^{(n)} = \sum_{k=1}^{n}\beta_k^{(n)}e^{-k\psi}\sin k\phi,\qquad(9)$$

$$y^{(n)} = \sum_{k=1}^{n}\beta_k^{(n)}e^{-k\psi}\cos k\phi,\qquad(10)$$

and where ϕ and ψ stand for $\dfrac{\phi}{c}$ and $\dfrac{\psi}{c}$. The expansion parameter ϵ can be given the significance of the amplitide of the fundamental, as in Stokes' procedure, or of the wave height. This can be seen from Eq. 11 below, where for $k = 1$ the coefficients $\beta_1^{(n)}$ are free. It is then straightforward to show that the surface boundary condition requires the coefficients $\beta_k^{(n)}$ to satisfy a set of equations of the following type

$$\sum_{k=1}^{n}(k-1)\beta_k^{(n)}\cos k\phi = \sum_i \ldots \sum_j B_{i\ldots j}\cos(i+\ldots-j)\phi,\quad (n = 1,2,\ldots)\quad(11)$$

where the quantities $B_{i\ldots j}$ involve the β's only up to order $n-1$, and are therefore all known. As a matter of fact, one can use this formulation to extract from Eq. 11 a *recursion formula* for the Fourier coefficients; however, as this calculation involves a great deal of algebra, the details will be presented elsewhere. Suffice it to note, for the present purpose, that Eq. 11 can be used to ask the computer to pick like coefficients of $\cos k\phi$ on both sides, which most symbolic computation system can do in a straightforward manner. As an example, we reproduce below the results for

$$y(\psi,\phi)\big|_{\psi=0} = \sum_n B_n(\epsilon)\cos n\phi$$

obtained by using MAPLE running on a Vax at the University of Waterloo:

$$B_1(\epsilon) = \epsilon, \tag{12}$$

$$
\begin{aligned}
B_2 = {}& \epsilon^2 + \frac{1}{2}\epsilon^4 + \frac{29}{12}\epsilon^6 + \frac{1\,123}{72}\epsilon^8 + \frac{502\,247}{4\,320}\epsilon^{10} \\
&+ \frac{244\,787\,899}{259\,200}\epsilon^{12} + \frac{884\,130\,455\,111}{108\,864\,000}\epsilon^{14} \\
&+ \frac{3\,325\,337\,418\,580\,279}{45\,722\,880\,000}\epsilon^{16} \\
&+ \frac{12\,891\,044\,455\,831\,800\,281}{19\,203\,609\,600\,000}\epsilon^{18} \\
&+ \frac{25\,578\,862\,562\,531\,003\,535\,667}{4\,032\,758\,016\,000\,000}\epsilon^{20} + ...,
\end{aligned}
\tag{13}
$$

$$
\begin{aligned}
B_3(\epsilon) = {}& \frac{3}{2}\epsilon^3 + \frac{19}{12}\epsilon^5 + \frac{1183}{144}\epsilon^7 + \frac{475\,367}{8\,640}\epsilon^9 \\
&+ \frac{217\,414\,759}{518\,400}\epsilon^{11} + \frac{752\,904\,791\,921}{217\,728\,000}\epsilon^{13} \\
&+ \frac{2\,748\,410\,290\,014\,649}{91\,445\,760\,000}\epsilon^{15} + \frac{5\,209\,775\,835\,626\,994\,403}{19\,203\,609\,600\,000}\epsilon^{17} \\
&+ \frac{81\,283\,540\,354\,179\,077\,444\,303}{32\,262\,064\,128\,000\,000}\epsilon^{19} \\
&+ \frac{7\,131\,437\,999\,626\,246\,314\,138\,468\,179}{298\,101\,472\,542\,720\,000\,000}\epsilon^{21} + ...,
\end{aligned}
\tag{14}
$$

$$B_4(\epsilon) = \frac{8}{3}\epsilon^4 + \frac{313}{72}\epsilon^6 + \frac{103\,727}{4\,320}\epsilon^8 + \frac{43\,100\,119}{259\,200}\epsilon^{10}$$

$$+ \frac{140\ 875\ 978\ 961}{108\ 864\ 000} \epsilon^{12} + \frac{494\ 749\ 243\ 738\ 759}{45\ 722\ 880\ 000} \epsilon^{14}$$

$$+ \frac{1\ 824\ 399\ 982\ 578\ 318\ 671}{19\ 203\ 609\ 600\ 000} \epsilon^{16}$$

$$+ \frac{3\ 484\ 724\ 988\ 978\ 875\ 037\ 937}{4\ 032\ 758\ 016\ 000\ 000} \epsilon^{18}$$

$$+ \frac{1\ 804\ 877\ 430\ 941\ 033\ 590\ 618\ 501\ 121}{223\ 576\ 104\ 407\ 040\ 000\ 000} \epsilon^{20} + ..., \tag{15}$$

$$B_5(\epsilon) = \frac{125}{24} \epsilon^5 + \frac{16\ 603}{1\ 440} \epsilon^7 + \frac{5\ 824\ 751}{86\ 400} \epsilon^9$$

$$+ \frac{17\ 479\ 557\ 769}{36\ 288\ 000} \epsilon^{11} + \frac{58\ 271\ 163\ 593\ 861}{15\ 240\ 960\ 000} \epsilon^{13}$$

$$+ \frac{51\ 859\ 859\ 913\ 675\ 871}{1\ 600\ 300\ 800\ 000} \epsilon^{15}$$

$$+ \frac{515\ 023\ 640\ 529\ 872\ 558\ 839}{1\ 792\ 336\ 896\ 000\ 000} \epsilon^{17}$$

$$+ \frac{392\ 494\ 810\ 711\ 059\ 314\ 693\ 469\ 943}{149\ 050\ 736\ 271\ 360\ 000\ 000} \epsilon^{19}$$

$$+ \frac{102\ 245\ 704\ 741\ 743\ 069\ 448\ 142\ 772\ 474\ 647}{4\ 131\ 686\ 409\ 442\ 099\ 200\ 000\ 000} \epsilon^{21} + ..., \tag{16}$$

$$B_{19}(\epsilon) = \frac{5\ 480\ 386\ 857\ 784\ 802\ 185\ 939}{6\ 402\ 373\ 705\ 728\ 000} \epsilon^{19}$$

$$+ \frac{1\ 923\ 743\ 214\ 162\ 202\ 588\ 144\ 337\ 717\ 669}{212\ 917\ 852\ 147\ 586\ 826\ 240\ 000} \epsilon^{21} + ..., \tag{17}$$

$$B_{20}(\epsilon) = \frac{32\ 000\ 000\ 000\ 000\ 000}{14\ 849\ 255\ 421} \epsilon^{20} + ..., \tag{18}$$

$$B_{21}(\epsilon) = \frac{41\ 209\ 797\ 661\ 291\ 758\ 429}{7\ 567\ 605\ 760\ 000} \epsilon^{21} + ... \tag{19}$$

The dispersion relation is given by

$$\frac{c^2 k}{g} = 1 + \epsilon^2 + \frac{7}{2}\epsilon^4 + \frac{229}{12}\epsilon^6 + \frac{6175}{48}\epsilon^8$$

$$+ \frac{8\,451\,493}{8\,640}\epsilon^{10} + \frac{4\,162\,161\,883}{518\,400}\epsilon^{12}$$

$$+ \frac{13\,441\,768\,667}{193\,536}\epsilon^{14} + \frac{57\,077\,417\,875\,339\,637}{91\,445\,760\,000}\epsilon^{16}$$

$$+ \frac{110\,875\,985\,690\,364\,678\,853}{19\,203\,609\,600\,000}\epsilon^{18}$$

$$+ \frac{83\,926\,522\,731\,752\,447\,156\,327}{1\,536\,288\,768\,000\,000}\epsilon^{20} + \dots \tag{20}$$

As shown by Eq. 12, the parameter ϵ is taken to be the amplitude of the fundamental, and only minor changes are required if ϵ is identified with the wave height. Of course, similar results are obtained for $x(\psi,\phi)|_{\psi=0}$. It should be noted, in particular, that the dispersion relation (Eq. 20) is given exactly to order 21, and its first five (nonzero) terms agree with the values of Schwartz [5], who obtained them by recognizing repeating patterns in the numerical results.

EXPLICIT CALCULATION OF THE WAVE SHAPE

When the objective is the calculation of the Stokes wave profile, the previous formulation is not the ideal one. Indeed, the shape of the wave is only given in parametric form, $x = f(\phi)$ and $y = g(\phi)$, and the elimination of ϕ is not an easy task by hand or by symbolic computation. Once again it pays to first reformulate the problem in an appropriate setting. Thus, rather than working with the stream function $\psi = \psi(x,y)$ as in the traditional approach, we invert this relation and reformulate the basic equations in terms of $y = y(x,\psi)$. Then the mathematical formulation of the Stokes wave problem takes on the form

$$y_\psi^2 y_{xx} - 2y_x y_\psi y_{x\psi} + (1+y_x^2)y_{\psi\psi} = 0, \tag{21}$$

$$\frac{1}{2}(1+y_x^2)y_\psi^{-2} + gy = C, \quad \text{at } \psi = 0, \tag{22}$$

$$y \rightarrow 0 \quad \text{as} \quad \psi \rightarrow -\infty, \tag{23}$$

where the bottom condition, Eq. 23, refers to the case of infinite depth for simplicity. We then look for a solution in the form of a perturbation series of the form

$$y(x,\psi) = \psi + \sum_{n=1}^{\infty} \epsilon^n y^{(n)}(x,\psi), \tag{24}$$

which, after substitution in Eq. 21, leads to the linear equation at order n

$$y_{xx}^{(n)} + y_{\psi\psi}^{(n)} = F(\psi,y^{(1)},...,y^{(n-1)}), \quad n = 1,2,3,... \tag{25}$$

where the right-hand side contains all known quantities. It is not difficult to prove that the solution has the form

$$y^{(n)}(x,\psi) = \sum_{l=1}^{n} \sum_{m=0}^{l} A_{lm}^{(n)} e^{l\psi} \cos mx, \tag{26}$$

where the coefficients $A_{lm}^{(n)}$ are determined by Eq. 25 for $l \neq m$, and by the surface boundary condition for $l = m$. The latter, when expressed at order n, is analogous to Eq. 11 and can thus be used to obtain a machine recursion formula, much in the same spirit as before. In this case we used the symbolic computation system MACSYMA running on the MC machine (PDP-10) at the Massachusetts Institute of Technology. The results are already in the literature [8], and we refer the reader to that paper for the details.

CONCLUDING COMMENTS

We have outlined in this paper two methods for the exact computation of the Stokes wave via the use of symbolic computation systems such as MAPLE and MACSYMA. Rather than focussing on the details of the computations, we have stressed the fact that these powerful systems should be looked upon as an ideal took for the applied mathematician engaged in the study of complicated (real-life) problems. Their great advantage over numerical procedures is not only due to the absence of computer-generated noise, but, more importantly, to the interactive nature of symbolic systems, whereby the user can test ideas and methods of solutions with an immediate grasp of what is exactly going on, much in the same way as if paper and pencil were used. This, in particualr, stimulates the search for creative approaches to the solution of a given problem, as the Stokes wave calculations above show. It is natural, in a numerical computation, to simply follow the basic procedure of Stokes [2] as done, for instance, by Schwartz [5] and Cokelet [6]. However, since the traditional approach leads to a sequence of nonlinear problems which can only be solved correctly via a numerical iteration procedure, it is clear that the present formulation, leading to a sequence of linear problems, is superior even from a numerical point of view. From a symbolic computation standpoint, we have the further advantage of reducing the computing time and, above all, an assurance of the correctness of the results.

ACKNOWLEDGEMENTS

The work of both authors is supported by the Natural Science and Engineering Research Council of Canada. We thank Dr. G. Fee of the Maple Lab of the University of Waterloo for assistance with the computation.

REFERENCES

[1]   Stokes, G.G., Trans. Camb. Phil. Soc. *8:* 441-455, 1847.

[2]   Stokes, G.G., Math. and Phys. Papers, Vol. 1: 314-326, 1880.

[3]   Wilton, J.R., Phil. Mag. (6) *27:* 385-394, 1914.

[4]   De, S.C., Proc. Camb. Phil. Soc. *51:* 713-736, 1955.

[5]   Schwartz, L.W., J. Fluid Mech. *62:* 553-578, 1974.

[6]   Cokelet, E.D., Phil. Trans. Roy. Soc. A *286:* 183-230, 1977.

[7]   Schwartz, L.W. and Fenton, J.D., Ann. Rev. Fluid Mech. *14:* 39-60, 1982.

[8]   Hui, W.H. and Tenti, G., J. Appl. Math. Phys. (ZAMP) *33:* 569-589, 1982.

17

Simplifying Large Algebraic Expressions by Computer

Richard L. Brenner

Symbolics, Inc.
Cambridge Research Center

ABSTRACT

Computer simplification of very large algebraic expressions by direct methods is often impractical because of the exhaustion of available resources. Some of the causes of this difficulty are discussed, and a method of circumventing it for certain types of problems is presented. The method has been implemented for the Computer Algebra system MACSYMA and is being used in the calculation of one-loop corrections to the hadronic decay rate of quarkonium in Quantum Chromodynamics.

1. Introduction

Anyone who tries to simplify by computer a large algebraic expression learns quickly how easy it is to exhaust the available resources of computer memory, disk memory or computer time. Computer Algebra systems are capable of simplifying small problems much more rapidly and accurately than people can, but often their margin of superiority declines as the size of the problem increases. Although the cause of this difficulty varies from problem to problem and from program to program, their is a procedure, described in this paper, that enables the computer to reduce a large class of expressions that might otherwise prove difficult. This method enhances the capabilities of Computer Algebra systems by exploiting the ability of the machine to recall the results of intermediate calculations, and the ability of the user to invent techniques specialized to particular simplification problems. Although it is general, the method is flexible enough to accomodate specific features of the problem at hand because it is merely a

338

framework for constructing a program specifically designed to accomplish the required simplification. This method is being used in the calculation of one-loop corrections to the hadronic decay rate of quarkonium in the framework of Quantum Chromodynamics.

In Section 2 we describe the particular class of problems for which the procedure is useful, and the procedure itself is briefly described. This description of the procedure is meant to be independent of implementation. In Section 3, an implementation of the method for the Computer Algebra system MACSYMA [1] is described in some detail, and general instructions for its use in this implementation are given. Although some minimal familiarity with a Computer Algebra system is helpful, expertise in the use of MACSYMA itself is not necessary for understanding this section. An example of the use of the MACSYMA implementation of the method is given in Section 4. Finally in Section 5, we suggest some possible generalizations of this procedure that may make it more convenient to use.

2. Reducing Porphyrytic Expressions

Borrowing a term from Geology, we shall say that an expression is *porphyrytic* if it is a rational expression or a list [2] of rational expressions involving two classes of variables. The first class consists of a small number of symbolic quantities and numbers that do not require much reduction. The second class of variables consists of those subexpressions that do require extensive reduction or simplification. This second class might include, for example, integrals to be performed, special functions to be simplified, or products of Dirac matrices to be reduced. In addition, full evaluation of the variables of the second class must result in rational expressions that contain only a few additional variables, so that once the original expression has been fully expanded, it contains only a few variables. It is important to note that an expression that contains no elements from the second class is not porphyritic.

The reduction of a porphyrytic expression typically requires extensive processing of the quantities in the second class, followed by rational simplification of the quantities from the first class. Since one usually focusses upon the second class of quantities first, they are called the *foreground* kernels[3]. The rest of the expression is called *background*.

A simple example may clarify this terminology, and is useful in the discussions in the remainder of this section. Consider the set of functions defined by [4]:

$$F(0, x) = - \log(\frac{x - 1}{x}) \tag{2.1a}$$

$$F(n, x) = -x^n \log(\frac{x - 1}{x}) - \sum_{k=1}^{n} \frac{x^{n-k}}{k} \qquad n = 1, 2, 3 \cdots \tag{2.1b}$$

These functions arise when performing loop momentum integrals in perturbative treatments of certain quantum field theories in the context of dimensional regularization of ultraviolet divergences. They have a number of interesting properties, including:

$$F(n, x) = x F(n - 1, x) - \frac{1}{n} \tag{2.2}$$

In the discussion below, we shall use this recursive definition of $F(n, x)$ in preference to (2.2) because the recursive form better illustrates the important concepts. For the purposes of illustration, we shall also need a slightly more complicated structure:

$$Y(m, n, x, y) = \frac{y^m F(n, x) - x^n F(m, y)}{x y} \tag{2.3}$$

In our terminology, the right hand side of the equation

$$Y(3, 5, x, y) = \frac{y^3 F(5, x) - x^5 F(3, y)}{x y} \tag{2.4}$$

is porphyritic. Its foreground kernels are $F(5,x)$ and $F(3,y)$, and the background variables are x and y. We shall return to this example later to illustrate several possible reduction strategies.

A diagrammatic representation of these algebraic structures is also useful for discussing reduction strategies. Let us represent an algebraic expression as an n-ary tree. We can represent the rational operators *plus*, *times*, and *exponentiation* as nodes of this tree, and kernels as the leaves of the tree. We label a node p for *plus*, t for *times*, and e for *exponentiation*. We also assume that a complete set of rules determine the order of the branches attached to these nodes, so that we are assured that the mapping from the set of rational expressions to the set of trees is one-to-one. Thus we can represent the expression

$$x + y + Y(3,5,x,y) + \frac{1}{x} \tag{2.5}$$

as shown in Figure 1. If we apply (2.3) to the above expression, we obtain

$$x + y + \frac{y^3 F(5,x) - x^5 F(3,y)}{x\,y} + \frac{1}{x} \tag{2.6}$$

which is represented as shown in Figure 2.

Although these diagrams are useful for modeling the expressions themselves, they are not intended to represent the actual contents of any part of a computer memory. The contents of memory cells that are used to represent these expressions depend strongly on the computer algebra system being used. Nevertheless, this diagrammatic representation of the algebraic structure does provide a valuable conceptual framework, and can be useful for judging the complexity of the structures in question.

Now that we have established a suitable language, we turn to a discussion of strategies for reducing porphyritic algebraic expressions. The most significant characteristic of such reduction efforts is the practical difficulty of actually performing such a computation. When confronted

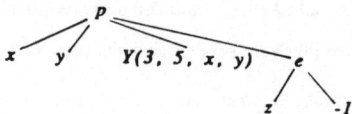

Figure 1.

The tree structure of the expression: $x \ + y \ + Y\,(3,\,5,\,x\,,\,y\,) + \dfrac{1}{z}$.

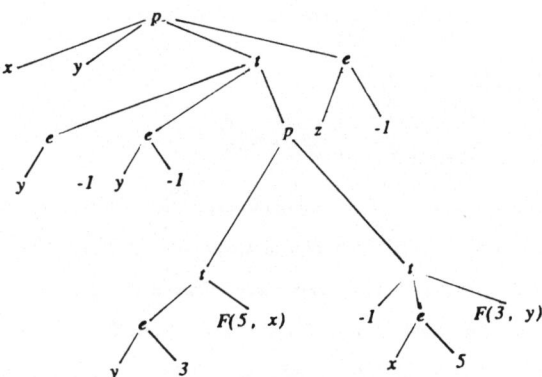

Figure 2.

The tree structure of the expression:

$$x \ + y \ + \frac{y^{3}\,F\,(5,\,x\,) - x^{5}\,F\,(3,\,y\,)}{x \ \ y} + \frac{1}{z\,.}$$

with a porphyrytic expression to simplify, one's first impulse is to apply the entire reduction procedure to the entire expression. If the reduction procedure is relatively simple, and the expression is relatively compact, this direct approach is perhaps the best. But when a large expression is being simplified, and the simplification procedure is relatively complicated, it may be wise to segment the calculation. This segmentation may be advisable for a variety of reasons:

[1] Non-fatal errors of conception or execution of a calculation may be discovered while the calculation is in progress. If the calculation has been segmented, then there is the possibility that only certain portions of it require correction. If the calculation has not been segmented, recovery of correct partial results may be difficult.

[2] The procedure may take so long to execute that much effort could be lost if the operation of the computer were interrupted due to either a programming error, hardware malfunction, or scheduled maintenance of equipment. This is possible, since it is not unusual for the time scale of a simplification effort to approach the scale of the mean time between failures for the equipment being used for that effort.

[3] Space may be at a premium, because of the size of the expression itself or the size of the programs that are required to operate on the expression, or both. For machines with limited address space, the size of the calculation may force segmentation.

[4] Most important, it may be possible to use intermediate results for parts of the calculation in subsequent parts of the calculation. If intermediate results have been saved, they can easily be extracted for later use. This is difficult unless the segmentation has been done systematically. This capability can greatly increase the efficiency of any procedure.

For these reasons, most large reduction efforts are eventually segmented, often by necessity if not by choice. In the remainder of this section, we shall describe an approach to this segmentation that systematically addresses the problems indicated abovew, while providing significant increases in efficiency. We shall carefully characterize the types of problems to which this approach is best suited, and compare the efficiency of this method to that of alternative methods. A family of MACSYMA programs called LTAB provides one example of a possible implementation of this segmentation, and it is the subject of Section 3.

To understand the need for, and the advantages of segmenting a reduction problem, it is necessary to examine the sourcess of the difficulties of reducing porphyritic expressions. Problems associated with reducing porphyritic expressions are due in large part to the well-known difficulty of simplifying rationally any large expression that contains many distinct kernels. The difficulty usually appears as a choice between rational simplification of a very large expression in only a few variables, and rational simplification of several smaller expressions involving large numbers of variables. In a typical situation, one begins with a relatively compact algebraic expression that is to be reduced according to some well-defined procedure to another relatively compact form. Unfortunately, the intermediate forms that are generated during this reduction procedure can be quitre large. If in addition, the intermediate results are rational functions of many variables, then intermediate rational simplification may be very costly, or even impossible in practical terms. For this reason, one might postpone rational simplification until the number of distinct kernels has been reduced. But this usually occurs late in the reduction, when the expression is larger still. Thus the size of the set of kernels has been traded for bulk of the rational expression, and one difficulty has been replaced by another. This is the unfortunate dilemma that one often faces when attempting to reduce a large porphyritic expression.

As a practical example of such a reduction problem, consider a typical Feynman diagram evaluation, a procedure that requires the application of several operations in sequence to reduce the original expression to its final form. For example, one might have to perform tensor contractions, then carry out traces of Dirac matrix prodects, then perform a Taylor expansion , and finally an integration. The end result might be a function of only a few kinematic invariants, but the intermediate results, viewed as rational expressions, might depend on several tens or even hundreds of kernels. Thus, any attempt to rationally simplify the intermediate results could lead to disaster.

This effect is easily demonstrated in terms of our simple example (2.3). Let:

$$Z(n) = \sum_{k=1}^{n} Y(n-k,n,x,y) \tag{2.7}$$

Consider the problem of reducing the expression:

$$\sum_{k=1}^{10} Y(10-k,10,x,y) \tag{2.8}$$

The final result is a rational function of only four distinct kernels: x, y, $\log(\frac{x-1}{x})$ and $\log(\frac{y-1}{y})$. On the other hand, the number of distinct kernels present at intermediate stages of the expansion can be much larger. After applying (2.3), the expression contains the kernels x, y, $F(n,y)$ and $F(10,x)$, for values of n from 1 through 10, so that there are 13 kernels in all, or roughly three times as many kernels as occur in the final result. If we try to avoid the difficulty of rationally simplifying such intermediate forms and instead choose to apply (2.2) to the result we have obtained so far, we reduce the number of distinct kernels to 4, but the resulting expression is very large. In this case, there are 10 occurrences of $F(10,y)$ alone, each with 11 terms. Although the scale of this particular example is well within the reach of modern Computer Algebra systems, one can easily imagine reduction problems that display the same expansion characteristics and present real difficulties to any existing system.

Explicit examples of the ideas described above should provide the reader with additional insight into these difficulties. The fundamental question is the timing of rational simplification. We shall discuss this issue in terms of two possible strategies, which we call Postponement and Interspersion.

As shown above, algebraic expressions can be represented diagramatically as tree structures. In this language, rational transformations can be represented as mappings from one diagram to another, and we can now construct a diagrammatic representation of the two most straightforward reduction streategies. For example, postponement of rational simplification until after all foreground kernels have been expanded is equivalent to scanning the tree for those leaves that represent foreground kernels, and then replacing them by subtrees that represent the expanded forms. Finally the whole structure is rationally simplified. We shall call this approach the Postponement Method. The other method that we shall consider is equivalent to first replacing some of the leaves that represent foreground kernels by partially expanded subtrees, simplifying the entire expression rationally, then alternately repeating replacement and simplification of the entire expression until the desired form is obtained. We shall call this approach the Interspersion Method.

We begin our discussion of these two approaches by comparing their advantages. Each can be useful for specific problems. In particular, Interspersion offers the possibility of avoiding duplication of effort in reduction problems that produce multiple copies of foreground kernels. In the Interspersion Method, it is possible to collect together many terms that involve a particular foreground kernel, and then to evaluate that kernel once, or perhaps only a few times. By comparison, in the Postponement Method, this evaluation may occur many more times, but of course, the actual amount of duplicated effort depends on the reduction procedure and on the Computer Algebra system itself. For some problems, duplication may not arise at all, in which case the Postponement Method becomes somewhat more attractive. For example, consider the expression $Y(3,5,x,y) + Y(2,5,x,y)$. Applying (2.4):

$$Y(3,5,x,y) + Y(2,5,x,y)$$

$$= \frac{y^3 F(5,x) - x^5 F(3,y)}{x\,y} + \frac{y^2 F(5,x) - x^5 F(2,y)}{x\,y} \qquad (2.9)$$

Proceeding by Postponement, we see that the evaluation of $F(5,x)$ is duplicated, whereas in Interspersion, judicious rational simplification of (2.9) before applying (2.2) can eliminate duplication:

$$= \frac{(y^3 + y^2) F(5,x) - x^5 (F(3,y) + F(2,y))}{x\,y} \qquad (2.10)$$

Such savings are conveniently obtained by the Interspersion method. Interspersion can also recognize cancellations, as shown below:

$$Y(3,5,x,y) - y\,Y(2,5,x,y)$$

$$= \frac{y^3 F(5,x) - x^5 F(3,y)}{x\,y} - \frac{y^2 F(5,x) - x^5 F(2,y)}{x}$$

$$= \frac{x^4}{y} (y\,F(2,y) - F(3,y)) \qquad (2.11)$$

However, Interspersion cannot recognize all such duplication. Duplications that occur within the same rational expression can be recognized, but there are many examples of duplications that appear in other ways. In terms of our example (2.4), the expression below illustrates this effect:

$$Y(3,5,x,y) + Y(2,4,x,y)$$

$$= \frac{y^3 F(5,x) - x^5 F(3,y)}{xy} + \frac{y^2 F(4,x) - x^5 F(2,y)}{xy} \tag{2.12}$$

Applying (2.2) to (2.12), we see that $F(4,x)$ again appears in the result, duplicating its appearance in (2.12). Thus unless the reduction is designed so that previously evaluated occurrences of $F(n,x)$ are stored, duplicate evaluations are necessary. Although many Computer Algebra systems do provide facilities for such storage, it is important to note that Interspersion alpne cannt eliminate all duplicated effort in all reduction problems.

We now consider examples of the failure of these two approaches. To see most clearly how the Postponement method can fail, let us expand the right hand side of (2.4). In Figure 3a, we illustrate the Postponement method, which in this case might be the sequential expansion of (2.4) by applying (2.2) throughout the expression repeatedly until all occurrences of F kernels have been eliminated. It is clear that the Postponement method leads to a large result similar to that shown in Figure 3b. Since rational simplification is to some extent a matter of taste, the form shown in Figure 3b is offered only as an example of what might be desirable for certain applications. In this example we see that rational simplification achieves considerable reduction in the complexity of the result. Because the expansion by (2.2) never increases the number of unique kernels in the expression, the Interspersion method does a bit better in this case. Since the complexity of the intermediate expressions is always comparable to the complexity of the final result, Interspersion is effective for this example, and superior to Postponement.

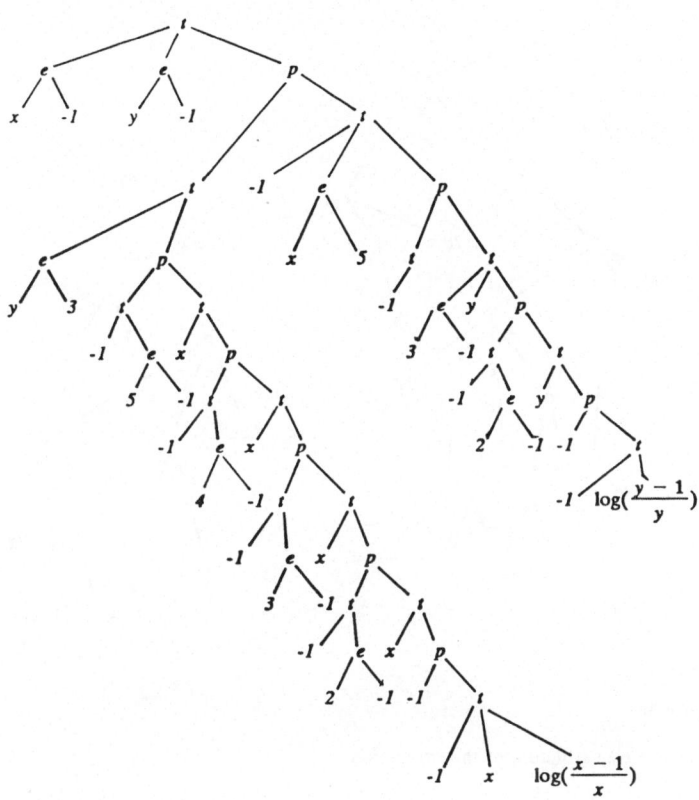

Figure 3a.

The tree structure of the expression:

$$((x\ (x\ (x\ (\ -\ x\ \log(\frac{x-1}{x}) - 1) - \frac{1}{2}) - \frac{1}{3}) - \frac{1}{4}) - \frac{1}{5})y^3$$

$$-\ x^5(y\ (y\ (\ -y\log(\frac{y-1}{y}) - 1) - \frac{1}{2}) - \frac{1}{3})/xy$$

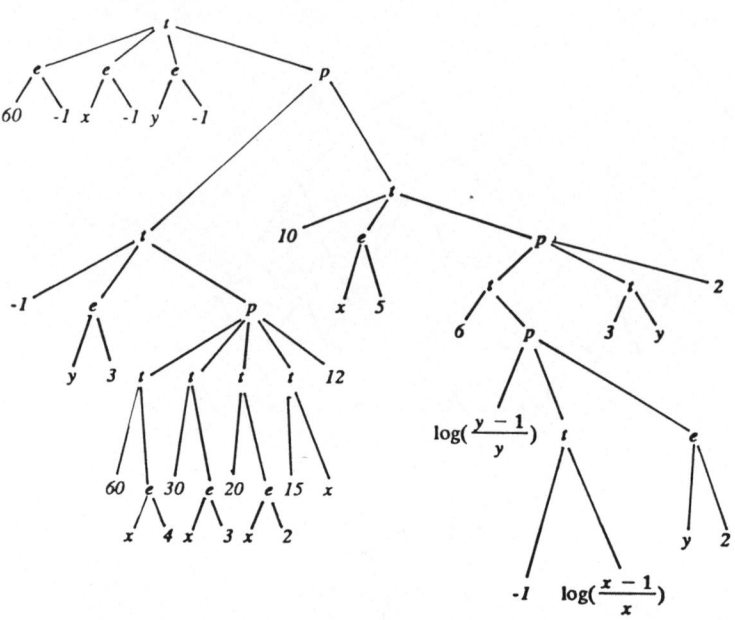

Figure 3b.

The tree structure of the expression:

$$(10x^{5}(6(\log(\frac{y-1}{y}) - \log(\frac{x-1}{x}) + y^{2}) + 3y + 2)$$

$$- (60x^{4} + 30x^{3} + 20x^{2} + 15x + 12)y^{3})/60xy$$

Figures 4a-4g illustrate a problem for which Interspersion is ineffective. We have chosen a problem similar to (2.8), but in the interest of brevity we have set $n = 5$. Referring to Figure 4, we see that although the intermediate expressions are only a bit more complex than the final result, they involve as many as twice the number of distinct kernels found in the final result. In this case, Interspersion requires the rational simplification of 6 intermediate expressions, each one more complex than the final result. Although duplicate evaluations of many F kernels are required, Interspersion itself does not prevent any duplication. The display of the tree structures of the intermediate expressions also clearly demonstrates the magnitude of the intermediate calculations, which casts some doubt on the wisdom of employing the Interspersion method. Moreover, Postponement offers no improvement, since this problem is actually composed of parts similar to the problem illustrated in Figure 3.

We now summarize the advantages and disadvantages of Postponement and Interspersion. Postponement avoids rational simplification of large expressions that contain many distinct kernels, but may lead to rational simplification of enormous expressions. Postponement does not allow for the possibility of duplicate foreground kernels unless intermediate evaluations of those kernels are stored. Interspersion can avoid rational simplification of very large expressions, but may require rational simplification of expressions that contain many distinct kernels. In addition, Interspersion may avoid duplicate evaluation of foreground kernels, but unless those kernels are stored, duplicate kernels are recognized only when they are present in the same rational expression. Thus we have shown that the two most straightforward reduction procedures are insufficient.

The method used by LTAB provides a desirable alternative to these two approaches. In this method, rational simplification is performed only on expressions that contain a small number of kernels, and since simplification is performed on intermediate subexpressions, the size of each expression that is subjected to rational simplification is limited. In this way, we

352

combine the virtues of both methods and reduce the difficulties associated with each. The procedure that is implemented in LTAB accomplishes this by postponing rational simplification, and maintaining tables of intermediate results. To avoid the problem of simplifying expressions that contain many distinct kernels, rational simplification is in fact postponed until expansion is complete, but instead of simplifying the expression as a whole, the simplification is applied to subtrees that represent the expanded forms of the foreground kernels. These expressions are then combined into progressively more inclusive subtrees and rationally simplified together. To avoid the duplication of effort that results from expanding multiple copies of the same foreground kernel, tables of intermediate results are maintained, so that previously obtained results can be used again whenever possible. Because rational simplification is performed on parts of the intermediate results, we shall call this method Dissection.

More specifically, Dissection begins by extracting all of the foreground kernels from the original expression. These kernels are compared for duplication, and then they are formed into a list. The first step of the reduction procedure is then applied to the elements of this list, which results in a new list, each entry of which is a porphyrytic expression. If a particular entry in this evaluated list is not a porphyritic expression, no further reduction is necessary, but some other entries may require more work. This part of the procedure is now complete for this level of the reduction, but it has resulted in a similar reduction problem, one step further along in the reduction procedure. We continue in this way, creating more lists of foreground kernels, one for each level, and their associated lists of further porphyrytic expressions until finally the reduction of a list of foreground kernels for some level leads to a list of expressions that consists entirely of background. Now we work back through the levels of lists it has generated, rationally simplifying at each level. Specifically, beginning at the penultimate level, the foreground kernels that were evaluated in the last level are replaced by their equivalent background expressions wherever they occur in the porhyrytic expressions of the penultimate level. The results are then rationally simplified. Next, this procedure is repeated, with the penultimate level now in the role of the last level. We continue in this way until the foreground kernels of the original expression have been replaced by their equivalent background expressions and the entire expression has been rationally simplified.

The advantages of this procedure result from three important features. First, rational simplification is never applied to expressions large expressions that contain more than a few distinct kernels. The problems inherent in the Interspersion Method, namely the rational simplification of expressions that involve many unique kernels, are avoided completely. Second, the extraction of the foreground kernels is done a way that avoids duplications, so some of the advantages of the Interspersion Method are recovered. Finally, the availability of intermediate results makes it possible to reclaim them for all parts of the calculation for which they are valid, even if those separate parts of the reduction have been carried out at different times. Thus, if the intermediate results are saved on disk or tape, it is unnecessary to duplicate any portions of the reduction that may later require those results, which can greatly improve the efficiency of the procedure.

One shortcoming of this method is that cancellations cannot be recognized automatically, because the intermediate rational expressions that involve the intermediate foreground kernels are never assembled. So if one expects numerous such cancellations, the Dissection method is less convenient than a segmented reduction interspersed with rational simplification, although in LTAB one can always reassemble the expression at any point where such cancellations are expected. Unless one expects frequent cancellations or small numbers of foreground kernels during the intermediate rational simplification, either Postponement or Interspersion will probably be considerably less efficient than Dissection.

The method implemented in LTAB is therefore best suited to problems that involve complicated reduction of large porphyritic intermediate expressions with porphyritic intermediate

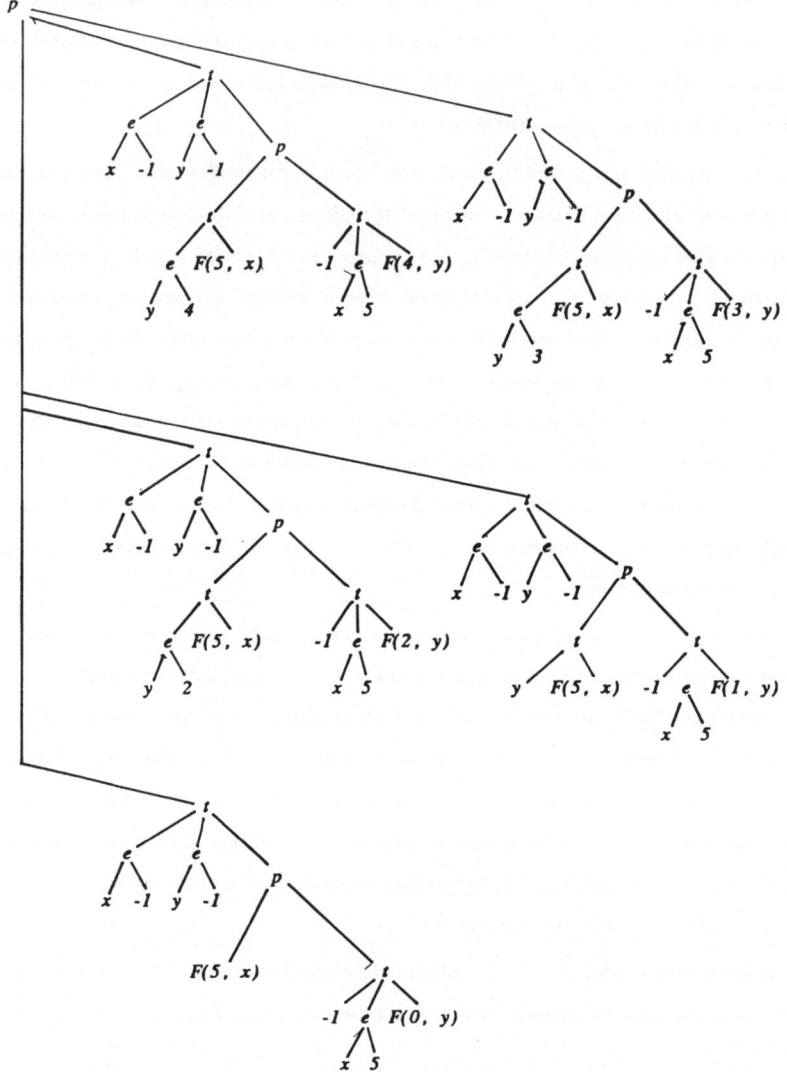

Figure 4a.

The tree structure of the expression:

$$\frac{F(5, x)y^4 - F(4, y)x^5}{xy} + \frac{F(5, x)y^3 - F(3, y)x^5}{xy} + \frac{F(5, x)y^2 - F(2, y)x^5}{xy}$$

$$\frac{F(5, x)y\,1 - F(1, y)x^5}{xy} + \frac{F(5, x) - F(0, y)x^5}{xy} +$$

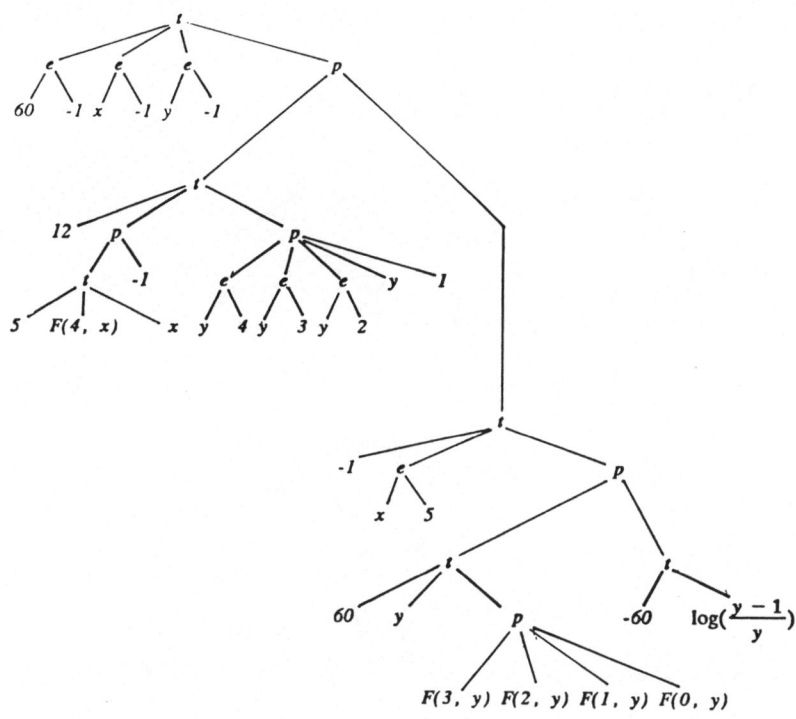

Figure 4b.

After applying (2.2) to the expression in Figure 4a, and after some simplification we obtain:

$$(12(5xF(4, x) - 1)(y^4 + y^3 + y^2 + y + 1)$$

$$- x^5(60y(F(3, y) + F(2, y) + F(1, y) + F(0, y)) - 60\log(\frac{y-1}{y}) - 125)/60xy$$

This figure illustrates the tree structure of this expression.

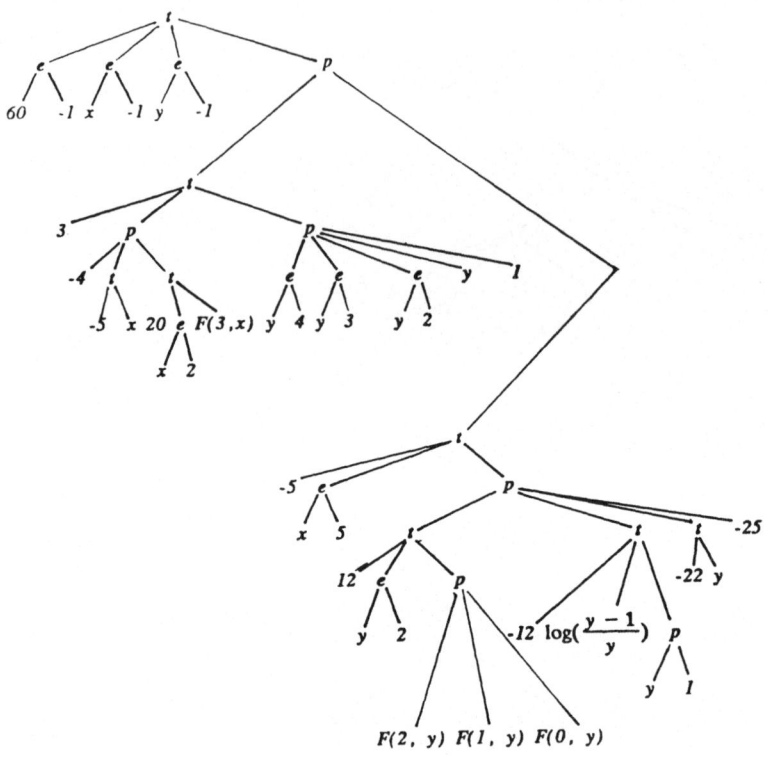

Figure 4c.

After applying (2.2) to the expression in Figure 4b, and after some simplification we obtain:

$$(3(20x\ ^2F\ (3,\ x\ )\ -\ 5x\ -\ 4)(y\ ^4 + y\ ^3 + y\ ^2 + y\ + 1)$$

$$-\ 5x\ ^5(12y\ ^2(F\ (2,\ y\ )\ +\ F\ (1,\ y\ )\ +\ F\ (0,\ y\ ))$$

$$-\ 12\log(\frac{y\ -\ 1}{y})(y\ +\ 1)\ -\ 22y\ -\ 25)/60xy$$

This figure illustrates the tree structure of this expression.

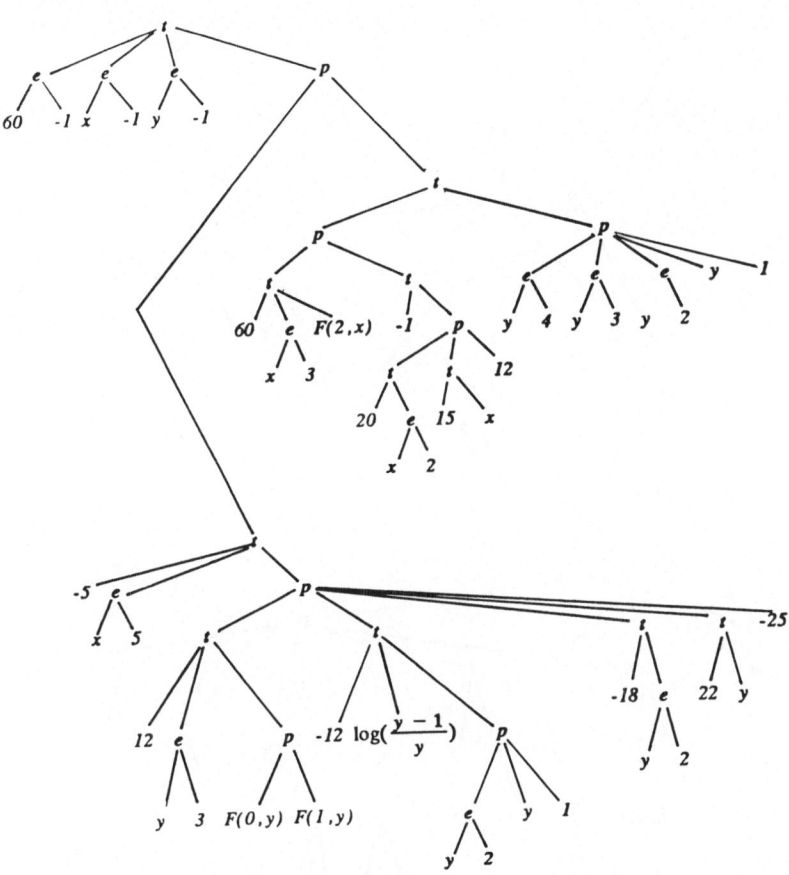

Figure 4d.

After applying (2.2) to the expression in Figure 4c, and after some simplification we obtain:

$$(60x^3F(2, x) - (20x^2 + 15x + 12)(y^4 + y^3 + y^2 + y + 1)$$

$$- 5x^5(12y^3(F(1, y) + F(0, y)))$$

$$- 12\log(\frac{y-1}{y})(y^2 + y + 1) - 18y^2 - 22y - 25)/60xy$$

This figure illustrates the tree structure of this expression.

358

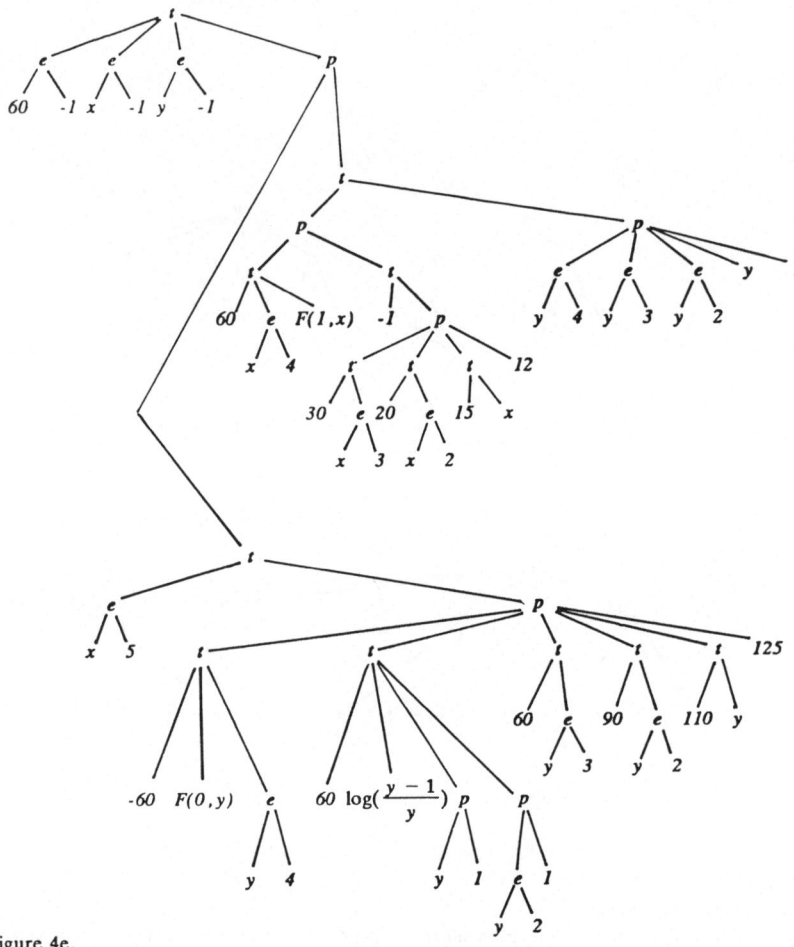

Figure 4e.

After applying (2.2) to the expression in Figure 4c, and after some simplification we obtain:

$$(60x\,^4F\,(1,\,x\,) - (30x\,^3 + 20x\,^2 + 15x\, + 12)(y\,^4 + y\,^3 + y\,^2 + y\, + 1)$$

$$x\,^5(\,-\,60F\,(0,\,y\,)y\,^4 + 60\log(\tfrac{y-1}{y})(y\, + 1)(y\,^2 + 1)$$

$$60y\,^3 + 90y\,^2 + 110y\, + 125))/\,60xy$$

This figure illustrates the tree structure of this expression.

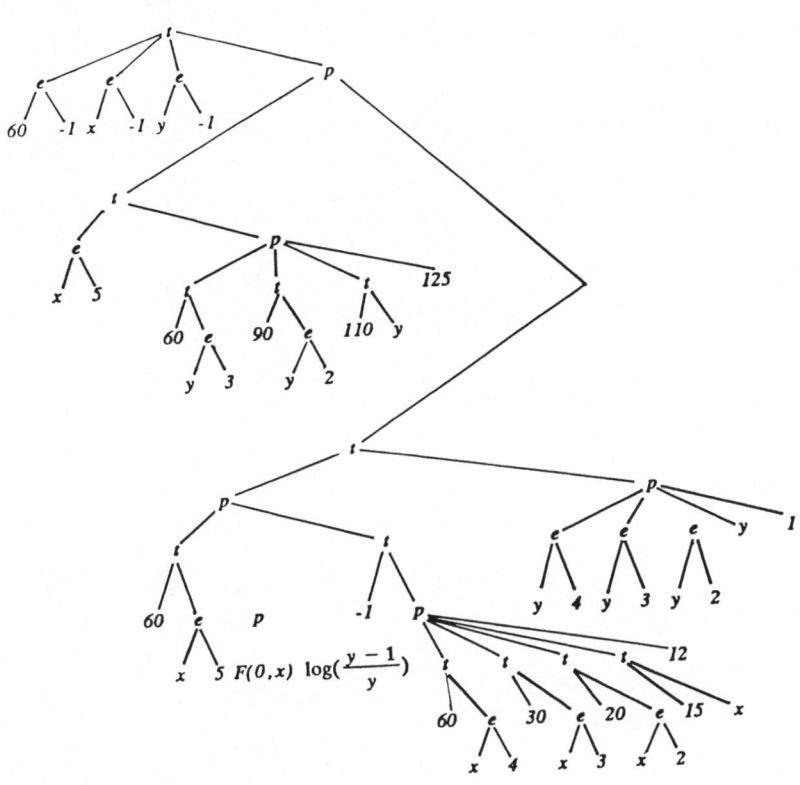

Figure 4f.

After applying (2.2) to the expression in Figure 4c, and after some simplification we obtain:

$$((60x\ ^5F\ (0,\ x\ )\ +\ \log(\frac{y-1}{y})))$$

$$-\ (60x\ ^4\ +\ 30x\ ^3\ +\ 20x\ ^2\ +\ 15x\ \ +\ 12)(y\ ^4\ +\ y\ ^3\ +\ y\ ^2\ +\ y\ \ +\ 1)$$

$$x\ ^5(60y\ ^3\ +\ 90y\ ^2\ +\ 110y\ \ +\ 125))/60xy$$

This figure illustrates the tree structure of this expression.

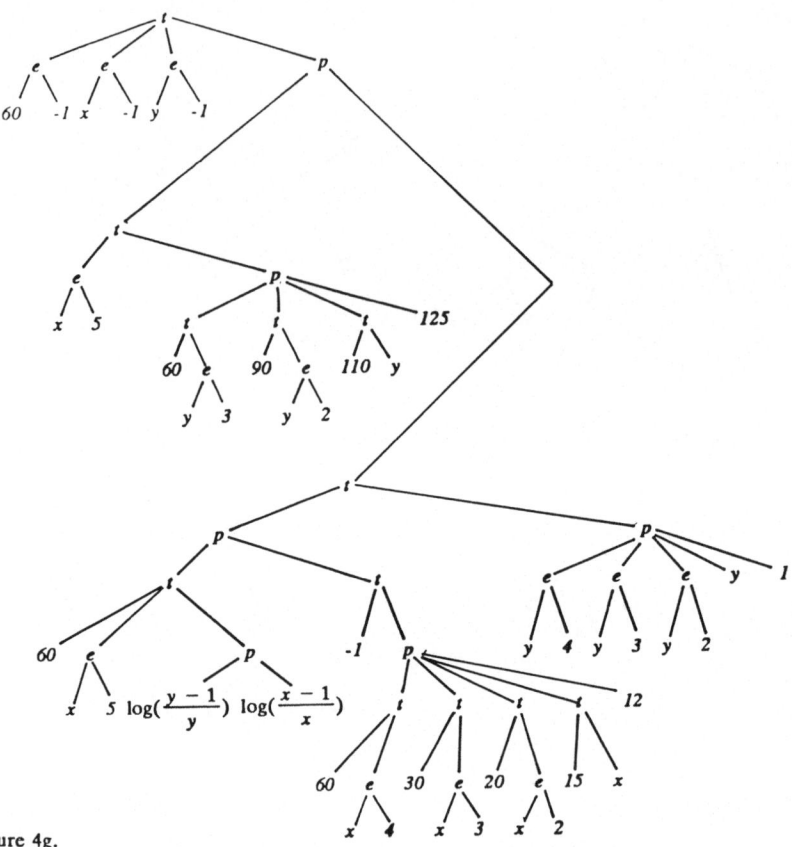

Figure 4g.

After applying (2.2) to the expression in Figure 4f, and after some simplification we obtain:

$$((60x^5(\log(\frac{x-1}{x}) + \log(\frac{y-1}{y}))$$

$$- (60x^4 + 30x^3 + 20x^2 + 15x + 12))$$

$$(y^4 + y^3 + y^2 + y + 1)$$

$$x^5(60y^3 + 90y^2 + 110y + 125))/60xy$$

This figure illustrates the tree structure of this expression.

expressions involving many intermediate foreground kernels that cannot be expected to cancel at intermediate stages.

To demonstrate the behavior of this alternative reduction procedure, consider the example used previously to examine the efficiency of Interspersion:

$$\sum_{k=1}^{5} Y(5-k,5,x,y) \tag{2.13}$$

LTAB is flexible. This flexibility can be exploited to implement the particular reduction procedure that is best suited to this problem. In this case one might rdeuce this expression as follows. It is usually wise to reorganize the expression while it is still compact, so that multiple copies of foreground kernels can be eliminated. Thus we obtain

$$\frac{F(5,x)(y^4+y^3+y^2+y+1)-(F(4,y)+F(3,y)+F(2,y)+F(1,y)+F(0,y))\,x^5}{x\,y}. \tag{2.14}$$

Next we extract from (2.14) all F kernels and enter them in separate lists. For this problem, it is probably best to organize them according to their first argument, so we obtain in this way six lists, as shown in Figure 5a. We have numbered these lists according to their level in this hierarchy. This organization of the problem is suggested by the recursion relation (2.2), which, in the course of expanding foreground kernels of a given level, generates foreground kernels that belong to the levels beneath it. The Dissection method as currently implemented in LTAB cannot deduce the reduction strategy that is best for a given problem. The strategy to be employed must be supplied by the user.

We now apply (2.2) to the highest level in the hierarchy, Level 1, to obtain list of partially evaluated forms, as shown in Figure 5b. In this case, there is only one entry in the list of foreground kernels of Level 1, but in general tfilere may be any number. These forms contain new foreground kernels, whose values are as yet unknown. They are then entered into the lists lower down in the hierarchy, for later evaluation. The result of this step is shown in Figure 5c. The first step of the reduction procedure is now complete.

For the next step of this procedure, we direct our effort at the second list, labeled Level 2, and procede as above. The result is shown in Figure 5d. Repeating this process for each level, we finally reach the state depicted in Figure 5e, in which all the foreground kernels that have been assigned to Level 6 have been expressed in terms of the background variables.

Next we substitute these values from Level 6 into the expressions entered in Level 5, as shown in Figure 5f. This upward substitution and simplification is then repeated for each level until we reach the state shown in Figure 5g. Finally all these results are inserted into the original expression, and the entire result is rationally simplified to obtain:

$$((60\,x^5\,(\log(\frac{y-1}{y})-\log(\frac{x-1}{x}))$$

$$-(60\,x^4+30\,x^3+20\,x^2+15\,x+12))(y^4+y^3+y^2+y+1)$$

$$+x^5(60\,y^3+90\,y^2+110\,y+125))/60\,x\,y \tag{2.15}$$

The effect of this procedure is exactly equivalent to applying (2.1b) to (2.13). Consequently, one might wonder whether anything has been gained by such efforts. For example, in light of the existence of (2.1b), one might object that the above rather intricate procedure is wasteful. However, in more practical problems than this one, closed form results analogous to (2.1b) do not necessarily exist. Furthermore, the steps of the reduction procedure of a practical problem are generally far more complex than those illustrated here, which unfortunately leads to the difficulties discussed above. Briefly, the reply to such objections is that the purpose of this illustration has been to compare the Dissection method to other methods that might be employed if (2.1b) were unavailable.

363

Level 1	
Kernel	Value
$F(5, x)$	

Level 1	
Kernel	Value
$F(5, x)$	$x \ F(4, x) - \frac{1}{5}$

Level 2	
Kernel	Value
$F(4, y)$	

Level 2	
Kernel	Value
$F(4, y)$	

Level 3	
Kernel	Value
$F(3, y)$	

Level 3	
Kernel	Value
$F(3, y)$	

Level 4	
Kernel	Value
$F(2, y)$	

Level 4	
Kernel	Value
$F(2, y)$	

Level 5	
Kernel	Value
$F(1, y)$	

Level 5	
Kernel	Value
$F(1, y)$	

Level 6	
Kernel	Value
$F(0, y)$	

Level 6	
Kernel	Value
$F(0, y)$	

Figure 5a.

The six-level hierarchy obtained by extracting the foreground kernels from (2.11).

Figure 5b.

The result of applying (2.2) to the highest level in Figure 5a.

Level 1	
Kernel	Value
$F(5, x)$	$x \ F(4, x) - \frac{1}{5}$

Level 2	
Kernel	Value
$F(4, x)$	$x \ F(3, x) - \frac{1}{4}$
$F(4, y)$	$y \ F(3, y) - \frac{1}{4}$

Level 3	
Kernel	Value
$F(3, x)$	
$F(3, y)$	

Level 4	
Kernel	Value
$F(2, y)$	

Level 5	
Kernel	Value
$F(1, y)$	

Level 6	
Kernel	Value
$F(0, y)$	

Figure 5c.

New foreground kernels generated in Figure 5b have been entered in the table for later evaluation.

365

Level 1	
Kernel	Value
$F(5, x)$	$x \ F(4, x) - \frac{1}{5}$

Level 2	
Kernel	Value
$F(4, x)$	$x \ F(3, x) - \frac{1}{4}$
$F(4, y)$	$y \ F(3, y) - \frac{1}{4}$

Level 3	
Kernel	Value
$F(3, x)$	$x \ F(2, x) - \frac{1}{3}$
$F(3, y)$	$y \ F(2, y) - \frac{1}{3}$

Level 4	
Kernel	Value
$F(2, y)$	

Level 5	
Kernel	Value
$F(1, y)$	

Level 6	
Kernel	Value
$F(0, y)$	

Figure 5d.

All foreground kernels that were inserted in the table from the original expression have been expanded according to (2.2). This resulted in the discovery of several new foreground kernels which were in turn inserted into the table and expanded. The procedure that was used was an extension of the one used to generate Figures 5a-c, applied sequentially to each of the 6 levels.

Level 1	
Kernel	Value
$F(5, x)$	$x \ F(4, x) - \frac{1}{5}$

Level 2	
Kernel	Value
$F(4, x)$	$x \ F(3, x) - \frac{1}{4}$
$F(4, y)$	$y \ F(3, y) - \frac{1}{4}$

Level 3	
Kernel	Value
$F(3, x)$	$x \ F(2, x) - \frac{1}{3}$
$F(3, y)$	$y \ F(2, y) - \frac{1}{3}$

Level 4	
Kernel	Value
$F(2, x)$	$x \ F(1, x) - \frac{1}{2}$
$F(2, y)$	$y \ F(1, y) - \frac{1}{2}$

Level 5	
Kernel	Value
$F(1, x)$	$x \ F(0, x) - 1$
$F(1, y)$	$y \ F(0, y) - 1$

Level 6	
Kernel	Value
$F(0, x)$	$-\log(\frac{x-1}{x})$
$F(0, y)$	$-\log(\frac{y-1}{y})$

Figure 5e.

New foreground kernels generated in Figure 5b have been entered in the table for later evaluation.

367

Level 1	
Kernel	Value
$F(5, x)$	$x\ F(4, x) - \frac{1}{5}$

Level 2	
Kernel	Value
$F(4, x)$	$x\ F(3, x) - \frac{1}{4}$
$F(4, y)$	$y\ F(3, y) - \frac{1}{4}$

Level 3	
Kernel	Value
$F(3, x)$	$x\ F(2, x) - \frac{1}{3}$
$F(3, y)$	$y\ F(2, y) - \frac{1}{3}$

Level 4	
Kernel	Value
$F(2, x)$	$x\ F(1, x) - \frac{1}{2}$
$F(2, y)$	$y\ F(1, y) - \frac{1}{2}$

Level 5	
Kernel	Value
$F(1, x)$	$-x\ \log(\frac{x-1}{x}) - 1$
$F(1, y)$	$-y\ \log(\frac{y-1}{y}) - 1$

Level 6	
Kernel	Value
$F(0, x)$	$-\log(\frac{x-1}{x})$
$F(0, y)$	$-\log(\frac{y-1}{y})$

Figure 5f.

The expansions of the kernels of Level 6 resulted in no new foreground kernels. These expressions were then substituted into the expansions of the foreground kernels that were being held in Level 5. The result is that now the expansions of the Level 5 kernels are expressed entirely in terms of background variables.

Level 1	
Kernel	Value
$F(5, x)$	$-\log\left(\frac{x-1}{x}\right) \quad x^5 - x^4 - \frac{1}{2}x^3 - \frac{1}{3}x^2 - \frac{1}{4}x \quad -\frac{1}{5}$

Level 2	
Kernel	Value
$F(4, x)$	$-\log\left(\frac{x-1}{x}\right) \quad x^4 - x^3 - \frac{1}{2}x^2 - \frac{1}{3}x \quad -\frac{1}{4}$
$F(4, y)$	$-\log\left(\frac{y-1}{y}\right) \quad y^4 - y^3 - \frac{1}{2}y^2 - \frac{1}{3}y \quad -\frac{1}{4}$

Level 3	
Kernel	Value
$F(3, x)$	$-\log\left(\frac{x-1}{x}\right) \quad x^3 - x^2 - \frac{1}{2}x \quad -\frac{1}{3}$
$F(3, y)$	$-\log\left(\frac{y-1}{y}\right) \quad y^3 - y^2 - \frac{1}{2}y \quad -\frac{1}{3}$

Level 4	
Kernel	Value
$F(2, x)$	$-\log\left(\frac{x-1}{x}\right) \quad x^2 - x \quad -\frac{1}{2}$
$F(2, y)$	$-\log\left(\frac{y-1}{y}\right) \quad y^2 - y \quad -\frac{1}{2}$

Level 5	
Kernel	Value
$F(1, x)$	$-\log\left(\frac{x-1}{x}\right) \quad x \quad -1$
$F(1, y)$	$-\log\left(\frac{y-1}{y}\right) \quad y \quad -1$

Level 6	
Kernel	Value
$F(0, x)$	$-\log\left(\frac{x-1}{x}\right)$
$F(0, y)$	$-\log\left(\frac{y-1}{y}\right)$

Figure 5g.

The procedure that was used to obtain Figure 5f was repeated for each of the other 5 levels.

369

What then has been achieved by this method? First, this segmented reduction procedure
has provided a derivation of (2.1b) for those kernels that appeared in our problem. In practical
problems for which closed forms are not known in advance, Dissection provides a method of
obtaining these necessary forms. Second, this result has been achieved by a method that
involved manipulating only those particular intermediate expressions that were directly related
to the goal of deriving the rquired closed forms. By ignoring the original expression (2.13)
throughout most of the procedure, expensive rational simplification of largely irrelevant structures was completely avoided. This principle was actually applied at all 6 levels of the segmented reduction procedure. Finally storage of intermediate results allowed us to avoid wasteful recalculation of the forms $F(n,y)$ for $n = 0, 1, 2,$ and 3, as would have been required in
either the Postponement or Interspersion methods described above.

These ideas are illustrated in Figure 6. The tree structure of (2.14), which is the starting
point of the reduction, is shown in Figure 6a. Since the entries in the lists of Figure 5 are all
relatively small, the tree structures of the intermediate structures that are stored there are not
shown. The largest structure, which is obtained when the results shown in Figure 5g are substituted into (2.14), is illustrated in Figure 6b. Since (2.14) was simplified by Interspersion as
shown in Figure 4, the result of simplifying the expression of Figure 6b is shown in Figure 4g.

Although the expression of Figure 6b is a bit larger than the expressions obtained when
the Interspersion method is applied to this problem, note that it contains only 4 distinct kernels.
By comparison, the intermediate forms that were simplified in the Interspersion treatment of
this problem contained as many as eight distinct kernels. The expression of Figure 6b is also
much smaller than the intermediate expression that would have been simplified if Postponement had been applied to this problem. Moreover, in the Dissection method, only one such
large expression was rationally simplified, whereas in the Interspersion method, five such intermediate forms were simplified. This shows how Dissection can reduce the number of kernels

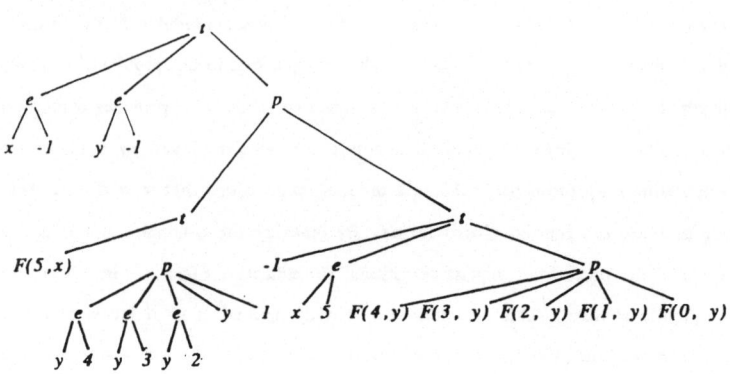

Figure 6a.

The tree structure of the expression of Eq. (2.12).

$$(F(5, x))(y^4 + y^3 + y^2 + y + 1)$$

$$- (F(4, y) + F(3, y) + F(2, y) + F(1, y) + F(0, y))x \, 5/xy.$$

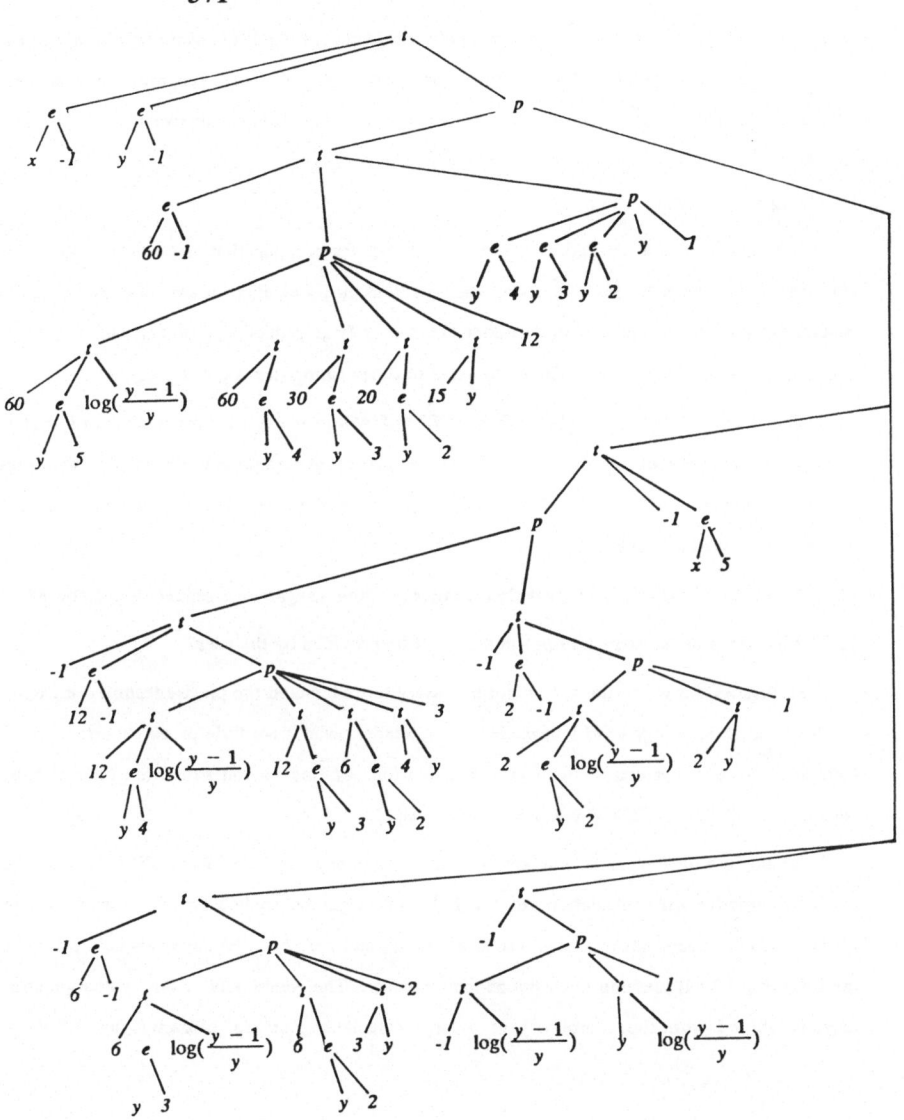

Figure 6b.

The tree structure of the expression obtained by substituting the results shown in Figure 5g into (2.12).

contained in expressions that must be rationally simplified, while at the same time reducing the amount of effort expended during the intermediate rational transformation. In this way, Dissection provides a useful compromise between Postponement and Interspersion.

3. Using LTAB in MACSYMA

We now turn to a detailed discussion of a computer program that makes use of the ideas preesented in Section 2. This discussion has a twofold purpose. First, it can serve as a guide to MACSYMA users who may wish to undertake a very large reduction problem, and second, it may suggest a useful reduction scheme to users of other Computer Algebra programs.

The reader who is unfamiliar with MACSYMA will probably be most interested in determining just what is required to achieve the efficiency improvements that are possible with segmented reduction. With this goal in mind, we have tried to organize this section to clearly answer two important questions.

[1]   What facilities does LTAB provide for implementing a segmented reduction procedure?

[2]   What steps of the reduction procedure must be provided by the user?

To discuss these questions, it is not necessary to understand the precise technical meaning of all of the terminology used below. However, readers who have little or no experience with Computer Algebra systems or who may find portions of this section somewhat obscure may wish to consult the MACSYMA Reference Manual[1].

We begin with a recommendation of caution. Although much of MACSYMA is oriented toward interactive use, calculations to which LTAB might be applied are best carried out in BATCH mode. Some advance planning is desirable, and may even be necessary as a glance at the following LTAB function descriptions may indicate. The complexity of this evaluation strategy, combined with the complexity of a large calculation, makes interactive use of these functions inadvisable. A BATCH mode approach is preferable for several reasons. First, the batch file provides an accurate record of the specific operations that were performed. This record is very useful for locating and understanding errors. Second, if an error is discovered, then it is often possible to recover from that error by simply running some of the batch files again, after making minor corrections. Finally, several attempts may be necessary before a particular step in the reduction can be achieved to satisfaction, and the availability of the batch file eliminates duplication ofd the simpler.portions of that effort.

Given this *caveat*, this section is divided into several parts. We begin with a general description (3.1) of the preparations that are necessary before one can begin to reduce an algebraic expression by means of the LTAB programs. Next we describe in detail (3.2) the structure of the data tables that are used by these programs. Finally, we describe (3.3) the functions that one actually uses to carry out the reduction.

3.1. Preparation for Segmented Reduction

To use LTAB in MACSYMA, one must first decide where to segment the reduction procedure. With experience, one will probably develop a feeling for choosing the segmentation points, but we offer the following suggestions. The choice of segmentation depends somewhat upon the computer facilities that are available. For example, if the reduction is expected to procede quickly, one might divide the reduction into only a few parts. On the other hand, a shortage of workspace may necessitate a more segmented approach. Often it is impossible to determine in advance the precise division points for a segmented reduction, and in the end, experimentation usually provides this information. The choice of segmentation is also directly influenced by the nature of the reduction problem. For example, the reduction may lead through some point where many identical foreground kernels are produced. Although it may not always be possible to anticipate that this will happen, it is usually advantageous to segment the reduction at such a point to avoid duplication of that part of the reduction procedure. Of course, the efficiency gains that result from the segmented depend upon the cost of reducing each kernel, but in general it is advantageous to segment a reduction procedure at those points where one expects multiple copies of identical foreground kernels to be produced.

Once one has determined a segmentation point, the next required preparatory step is to write a function that performs the step of the reduction procedure that carries the reduction from that segmentation point to the next one. In the discussion below, this function is called the PROCESSOR function. This function is specific to the problem at hand, so it cannot be provided by LTAB.

In the final preparatory step, one creates a blank table for each step in the segmented reduction. As the evaluation procedes, LTAB stores all necessary information in these tables. Although the structure of these tables is uniform independent of their level in the procedure, it is not possible for LTAB to generate these tables automatically, since the content of the tables is dependent on the procedure itself. The function LTAB_INITIALIZE has been provided for creating such blank tables.

3.2. Description of the Data Tables

In principle, preparation for a segmentation using LTAB consists of three steps: choosing the segmentation points, writing PROCESSOR functions for the corresponding reduction steps, and creating the requisite blank tables. In practice, however, things are more complicated for three reasons. First, choosing the segmentation points can be a difficult task. In many instances, the optimum choice of segmentation is not at all evident at the outset, and in the extreme case, it may not be possible to determine the next segmentation point until one has proceded part way into the reduction effort. Second, even the most careful advance planning of

375

the reduction procedure may fail to take account of certain special characteristics of the problem at hand. This failure can be expressed as a conceptual error, but more likely it appears as clumsiness or inefficiency in the reduction procedure. Finally, there is always the possibility of programming errors either in the user's programs or in system programs. For these reasons one almost inevitably needs to retrace some step or sequence of steps in reduction procedure. In such cases it is necessary to have a thorough understanding of the internal structure of LTAB and its data tables, so that recovery from errors can be as painless as possible. Therefor, detailed descriptions of the structure and content of the LTAB data tables ar provided below. The description of each component includes suggestions for its use whenever appropriate.

For the purpose of these descriptions, we think of a reduction procedure as if it consisted of a simple vertical chain, with the original porphyritic expression at the top and the final LTAB table at the bottom. Thus *down*, *below*, and *subsequent* refer to the direction of further reduction, while *up, above* and *previous* refer to the other direction. For convenience, we shall describe the contents of one such table named A. The table just above A is called PARENT, and A's successor is called CHILD.

In actual use however, there is no such restriction on the structure of the interrelations between the various LTAB tables. It is possible for any table to have several parents or children, and there are many instances when such structures are desirable. As an example of multiple children, suppose that at some stage in the reduction procedure two different types of foreground kernels are generated, and that further reduction of these kernels requires distinctly different methods. In such a case one might wish to process the two types of foreground kernels in separate tables, which would require that table A have two children. Multiple PARENT tables are also useful. Suppose that a given problem has been divided into two or more parts according to the validity of a certain approximation procedure or other special method particular to the problem. Algebraic expressions are then set up in each domain and must be reduced according to distinct procedures appropriate to each domain. However, suppose further that as the reduction procedes, it becomes possible to merge the domains and process the expressions according to a single procedure. This could happen if, for example, one has proceded past the point at which the special technique was applied. One might then save much effort by designating a particular table as CHILD of as many parents as possible, because of the sharing of effort that would then become automatic. Another example of a multiple parent structure is given in Section 4.

Each LTAB table is an array with 11 elements. Four of these elements refer to lists of raw or processed data: _VALLIST, _LABLIST, _NEXLIST, and _FULL_EV_LIST. Of the remaining seven elements, four are used by various LTAB functions when these data are being processed or stored. These elements are _PROCESSOR, _FULL_EV_FCN, _NEXT_TABLE and _MISCLIST. The remaining elements are used by the PARENT table when it is adding data to A. These elements are _COUNTR, _PREFIX and _OPERATORS. The formation and applications of each of these quantities are described below.

A[_VALLIST] This array element refers to a list of foreground kernels that were generated in the PARENT step of the evaluation procedure. During the PARENT step, whenever LTAB encountered one of these kernels, A[_VALLIST] was checked to see if it had already been entered there. If not, then an entry was made and a label generated for that kernel. All of these kernels were then replaced by the corresponding labels that were being held in A[_LABLIST]. These kernels are the input for the A step in the reduction procedure. After executing the PARENT step of the reduction procedure, one may examine the foreground kernels extraced in that step by asking for the value of A[_VALLIST]. One may also test the PROCESSOR of A by applying it to one of the elements of this list.

A[_LABLIST] This array element refers to a list of atomic symbols that were generated during the PARENT step in the reduction procedure. These are the symbols that were used in place of the foreground kernels that were discovered in that PARENT step, and which are being held in A[_VALLIST]. They were generated as labels for their corresponding foreground kernels by concatenating A[_PREFIX] with successive values of A[_COUNTER].

A[_PROCESSOR] This array element holds the name of the function that is used during the A step of the reduction procedure to process the foreground kernels that are being held in A[_VALLIST]. These kernels are the ones that were discovered during the step in the evaluation procedure that is just above A. If one wishes to change the name of the PROCESSOR function after the A table has been created, one merely sets A[_PROCESSOR] to that name. The processor function itself must be a function of one argument.

A[_NEXLIST] This array element refers to a list of expanded forms corresponding to the quantities held in A[_VALLIST]. The elements in this list are the results of applying the function A[_PROCESSOR] to the elements of A[_VALLIST]. The form of the elements in the list held in A[_NEXLIST] is as follows. Each entry consists of background kernels supporting the foreground kernels generated in the A step of the evaluation procedure. However, the newly generated foreground kernels have been replaced by new labels, using the alphabetic label prefix provided by CHILD, and stored in CHILD[_PREFIX]. If A is the last step in the procedure, then there are no such labels and the elements consists purely of background kernels.

This component of the A table is very useful as a safety mechanism. If one discovers an error in the reduction procedure somewhere below A, and if one has already back-substituted into the A table or perhaps even above it, one need correct only the rseults beginning at the point of the error, but not above it, if one uses the _NEXLIST of the tables above the error. Since back-substitution has been carried out above the error, one might think at first that all downward calculations would also be lost, but they can be recovered from the _NEXLIST, since it holds the forms calculated during the downward phase of the reduction. One simply executes A[_FULL_EV_LIST]:A[_NEXLIST] or equivalently, LTAB_FULL_EV_ERASE(A) to restore the A table to the state it was in prior to the back substitution of the incorrect forms. If only certain portions of A[_FULL_EV_LIST] have been affected, a correct structure can always be composed of pieces of A[_FULL_EV_LIST] and A[_NEXLIST], using the MACSYMA functions PART and SUBSTPART. In this way, any incorrect results can be carefully removed, without the necessity of duplicating results that are known to be correct.

A[_FULL_EV_LIST] This array element refers to a list of expanded forms corresponding to the quantities held in A[_VALLIST]. The form of the elements in this list, unlike the list held in A[_NEXLIST], can vary. Immediately after the processing of the elements in A[_VALLIST] with the function A[_PROCESSOR], the elements are in the form of the elements in A[_NEXLIST]. However, there are several circumstances that can result in changes in A[_FULL_EV_LIST]. Suppose for example, that one has evaluated all the tables beneath A, and then back-evaluated up to A, or possibly above A. The elements in A[_FULL_EV_LIST] then consist entirely of background. That is, the elements are "fully evaluated" (hence the name _FULL_EV_LIST). In this situation, they differ from the elements in A[_NEXLIST], which are always in the form of structures composed of background kernels and labels from the table beneath A. Yet another possible form of the elements in A[_FULL_EV_LIST] can result if one subsequently adds more elements to

A[_VALLIST] using the function LTAB_LABEL_UPDATE. This often happens, especially in a large calculation that has been segmented. The resulting form of the elements of A[_FULL_EV_LIST] is then mixed: some elements are full evaluated, and some are identical to their corresponding elements in A[_NEXLIST], awaiting further processing. Finally, the elements in A[_FULL_EV_LIST] can be pointers to other elements in that same list. This form results only when the function LTAB_OPT has been used to reduce the size of the A table.

A[_MISCLIST] This array element holds a list of the names of miscellaneous items that are to be saved along with A when the function LTAB_SAVE is used. For example, one might wish to store a predefined quantity for use with A's PROCESSOR function, or perhaps the names of auxiliary user-defined functions that PROCESSOR calls to carry out its task. Including the names of these quantities or functions in A[_MISCLIST] forces the function LTAB_SAVE to save these items whenever it save A. Elements can be added to this list at any time by the user, but they must evaluate to valid arguments to SAVE or FAS-SAVE.

A[_COUNTER] This array element holds an integer that is the number of the highest label generated so far by PARENT. Thus it is the number of foreground kernels stored in A.

A[_NEXT_TABLE] This array element holds the name of the table CHILD. If this array element holds a list, then the list is interpreted as a list of the CHILDREN of A. This feature allows multiple branching downward. If A has no children, A[_NEXT_TABLE] has the value END.

A[_FULL_EV_FCN] This array element holds the name of the function that is used by BACKEV when values from CHILD are inserted for CHILD's foreground kernels in A[_FULL_EV_LIST]. After BACKEV performs the substitutions of the values obtained from CHILD, A[_FULL_EV_FCN] is applied to the resulting expression. Examples of functions useful for this purpose are RATSIMP, FACTOR and SQFR, but the user may also provide the name of a more specialized function if desired. The only restriction is that the function must accept a single argument.

A[_OPERATORS] This array element holds a list of the leading operators of foreground kernels that are stored in A[_VALLIST], and is used by PARENT when it is searching for the foreground kernels to be inserted in A[_VALLIST]. Normally this is a list of one element, but it may take several other possible forms. For rg, there may be several different operators that can be the leading operators of foreground kernels stored in A. In this case, the list A[_OPERATORS] may contain several elements, including one for each operator. Also, one or more of the elements of the list A[_OPERATORS] can be of the form PREDICATE($predicate-name_1, predicate-name_2, \cdots$). In this case any kernel whose leading operator satisfies any of the predicates named in the argument list of the pseudo-function PREDICATE are also stored in A. Note that since this list is used by

PARENT, it is essential that the table A exist, at least in blank form, prior to the execution of the PARENT step of the reduction procedure.

A[_PREFIX; This array element holds the alphabetic string that PARENT uses when it generates labels for the foreground kernels of the A step. Bote that since this list is used by PARENT, it is essential that the table A exist, at least in blank form, prior to the execution of the PARENT satep of the reduction procedure.

3.3. Functions for the Reduction Process

Once the preparations are complete, one executes the reduction procedure by means of the functions provided. For example, to add the first foreground kernels to the first table in the reduction procedure, one uses the function LTAB_LABEL_UPDATE. To execute one step of the reduction procedure, one uses the function LTAB_VALUE_UPDATE. To substitute the final background kernels from one level into the table just above it, one uses the function BACKEV. To substitute the final values of foreground kernels back into the original expression, one uses the function LTAB_FINAL_SUBST. To store one or more LTAB tables in a disk file, one uses the function LTAB_SAVE. These and other useful functions are described below.

3.3.1. Creating a Table

A necessary first step for carrying out a segmented evaluation using LTAB is the creation of the desired evaluation tables. The following function is provided for this purpose.

LTAB_INITIALIZE(*name*, *processor*, *full_ev_fcn*, *next_table*, *string*, *ops*, *optional–args*) creates a table named *name* for a segmented evaluation. *processor* is the name of the function that is used to process the elements in the _VALLIST of the table *name*. *full_ev_fcn* is the name of the function that is applied by BACKEV to the elements of the _FULL_EV_LIST after substitutions have been made from lower level results. *next_table* is the name of table that is immediately below NAME in the segmented evaluation. If NEXT_TABLE is the atom END, then LTAB deduces that this table is the last of a chain.

string is the string of alphabetic characters that is to be used for generating labels to stand for the foreground kernels discovered either by the PROCESSOR of the table above *name* or by means of the function LTAB_VALUE_UPDATE. *Ops* is a list of the possible leading operators that the foreground kernels stored in *name* [_VALLIST] may have. It is also possible for *ops* to include elements of the form PREDICATE($< predicate -name_1, predicate -name_2, \cdots >$). If such elements are included, then any kernel that satisfies any of the predicates is also added to *name* [_VALLIST] when encountered by the *processor* of the table above *name*. *Optional —args* refers to any number of additional optional arguments that are the names of any objects that one might wish to save along with *name* when the function LTAB_SAVE is used, or when the switch PERIODIC_SAVE_FILENAME is not FALSE. For example, one might wish to have the *processor* function or the *full_ev_fcn* function stored in the same file as the table. If so, one would include the names of these functions among the arguments of LTAB_INITIALIZE, in the position indicated by *optional —args*.

3.3.2. Downward Evaluation

There are two different types of downward evaluation. In the beginning of an evaluation, one wishes to make entries in a table by processing an expression that is not itself a table. For this case one uses the function LTAB_LABEL_UPDATE. To make entries in a table A from a parent table of A, one uses the function LTAB_VALUE_UPDATE.

LTAB_VALUE_UPDATE(*name* , *n*) updates the values of any recent additions to the table named *name*. If no new additions are found, no changes are made in the table. If new additions are found, then the function whose name is held in *name* [_PROCESSOR] is used to process these new elements. The new foreground expressions that are generated in the course of this evaluation are automatically added to the table whose name is held in *name* [_NEXT_TABLE]. They appear in the table *name* only in an ISOLATEd form in *name* [_FULL_EV_LIST]. That is, only labels that stand for these newly generated foreground kernels appear in *name* [_FULL_EV_LIST]. The LINECHAR that is used for the generation of these labels is held in *name* [_NEXT_LABEL].

The second argument, *n*, is optional. If it is given, it must be a positive integer. When such an argument is given, it represents the maximum number of elements that are evaluated before the partially updated table *name* is written out into a disk file. Specification of a file for this purpose is given by setting PERIODIC_SAVE_FILENAME to the name of the desired file. If PERIODIC_SAVE_FILENAME is FALSE, then no such intermediate storage occurs. The switch DGVALFASSAVE[TRUE] determines whether SAVE or FASSAVE is used for this purpose. If DGVALFASSAVE is TRUE then FASSAVE is used, otherwise SAVE is used. It is reccommended that periodic storage be employed whenever a long computation is anticipated. In this way, recovery from errors is required only for the calculations done since the last writing.

LTAB_LABEL_UPDATE(*name*, *processor*, exp) is used to add new foreground kernels to the table *name* using the function *processor* on the expression exp. LTAB_LABEL_UPDATE is most useful for generating or adding to the first table in a chain of tables, or the root table in a tree of tables. The LINECHAR that is used for isolation of any new foreground kernels that might arise in the course of this evaluation is held in *name*[_PREFIX], and can be any string of alphabetic characters.

One switch controls the behavior of both **LTAB_LABEL_UPDATE** and **LTAB_VALUE_UPDATE** when they encounter foreground kernels that have already been fully or partially processed in a previous evaluation.

LABEL_UPDATE_FULL_EV determines precisely what quantities are substituted for the foreground kernels encountered by LTAB_LABEL_UPDATE and LTAB_VALUE_UPDATE. If a kernel has never been encountered in a previous evaluation of LTAB_LABEL_UPDATE or LTAB_VALUE_UPDATE then of course, there is no choice but to return a newly generated label. But if one is adding to an old table, and that table has had some further processing before the new additions are made, then its _FULL_EV_LIST may contain some fully- or partially-processed values of foreground kernels already in the table. Normally, one would prefer that these values be inserted whenever these kernels are reencountered. This is indeed the behavior when LABEL_UPDATE_FULL_EV is TRUE, the default. Setting this switch to FALSE forces substitution of the labels themselves, which may occasionally be preferable, especially when one wishes to make comparisons to expressions calculated earlier. Finally, setting this switch to ZERO_ONLY produces behavior similar to the FALSE setting except that if any of the fully evaluated kernels are known from previous calculation to be 0, that 0 is inserted instead of a label.

LTAB_LABEL_FIND(*name*) returns a list of the newly added and unevaluated foreground kernels that are held in *name*[_LABLIST]. This includes only those kernels that have not been processed in any way, and excludes those elements that have been expanded in terms of the next forground kernels of higher levels. It is unlikely that the user would ever require access to this function.

3.3.3. Correcting Errors

As noted earlier, it is frequently necessary to alter the contents of an evaluation table to correct errors. Two functions are provided for this purpose.

LTAB_ERASE(*name*) erases all labels and values that are already entered in the array *name*. This function is useful for purging the array of previously calculated erroneous results without changing other parameters or functions that might be associated with the array. It is also useful for obtaining a copy of the initialized array from one calculation for use in another.

LTAB_FULL_EV_ERASE(*arrayname*) is similar to LTAB_ERASE, except that the labels already stored in the array are not removed. _NEXLIST and _FULL_EV_LIST are, however, reinitialized. This function is useful for purging the array of previously calculated erroneous results while retaining the foreground kernels that are stored in the array.

3.3.4. Upward Evaluation

There are two possible kinds of upward evaluation. The first type involves substitution of the fully-evaluated foreground kernels of one level into the tabel one level immediately above it. This is done by means of the function BACKEV. The second type involves substitution into some expression that is not a table. This capability is needed for those tables that do not have parents. If the substituted quantities are to be foreground kernels then LTAB_LABEL_CLEAR is used. If the evaluated, reduced forms of these kernels are required, then LTAB_FINAL_SUBST is used.

BACKEV(*higher_level*, *lower_level*) BACKEV lifts the results of a table lower in the chain up to a table higher in the chain. This is accomplished by modifying the entries in *higher_level*[_FULL_EV_LIST] to reflect the values contained in *lower_level*[_FULL_EV_LIST]. The function whose name is held in *higher_level*[_FULL_EV_FCN] is applied to the entries in *higher_level*[_FULL_EV_LIST] after the substitutions are made.

LTAB_LABEL_CLEAR(*exp*, *name***)** examines the array *name* to determine what labels are stored in it as the result of any previous segmented evaluations. Then it substitutes the values that these labels stand for into the expression exp. This operation is the inverse of LTAB_LABEL_UPDATE, at least as far as exp is concerned. No alterations are made in *name*.

LTAB_FINAL_SUBST(*exp*, *name***)** examines the array *name* to determine what labels are stored in it as the result of any previous segmented evaluations. Then it substitutes the entries of the _FULL_EV_LIST that correspond to these labels into the expression exp. No alterations are made in *name*.

3.3.5. Storing and Retrieving the Tables

LTAB tables can be stored with any of the standard storage functions, but if one has several things to store along with the array itself, then it may be more convenient to use the function LTAB_SAVE. For example, one might wish to store a few global variables with the table, or perhaps the _PROCESSOR and _FULL_EV_FCN definitions. If so, then inclusion of the names of these objects in the _MISCLIST entry of the table permits one to exploit the convenience of LTAB_SAVE.

LTAB_SAVE(*filename*, arg_1, arg_2, $\cdots$ **)** saves the quantities named arg_1, arg_2, $\cdots$ in the file named filename. The form of the arguments of LTAB_SAVE is identical to the form of the arguments of SAVE or FASSAVE, except that LTAB_SAVE evaluates its arguments. Thus if any of the arguments has a value, it is necessary to present it to LTAB_SAVE in a "single-quoted" form. Failure to do so leads to an error when LTAB_SAVE passes such arguments to SAVE or FASSAVE. The use of SAVE or FASSAVE by LTAB_SAVE is determined by the value of the switch LTABFASSAVE[TRUE]. If LTABFASSAVE is TRUE then FASSAVE is used; if FALSE, SAVE is used. If any of the arg_i are hashed arrays, and the value of the array indexed by _MISCLIST is a list, then the elements of that list is also saved.

Retrieval of LTAB tables that have been stored on disk can be accomplished with the standard MACSYMA forms LOAD or LOADFILE.

384

4. An Example

It is difficult to construct a specific example of the application of this method that is general enough to be widely understood without special knowledge. Therefore we offer a generalized example in the hope that the important principles are clearly illustrated without introducing the obscurity that may result from a highly specialized example. Consider the following problem. We are given a rational expression Y that depends on variables ω, x, and s and on kernels of the form

$$f(n, x, p(x), s, \omega)$$

where p is a rational function of its argument. The precise form of p is not fixed, and may be different for each occurrence of f. The possible values of n are 1, 2, and 3. the function f is defined as:

$$f(n, x, p(x), s, \omega) = g(x, p(x), s)f(n-1, x, \frac{dp(x)}{dx}, s, \omega)$$

$$+ h(x, p(x), \log s, \log(1-\omega)) \qquad (n \neq 1)$$

and

$$f(n, x, p(x), s, \omega) = j(\frac{d p(x)}{dx}, \log s, \log(1-\omega), \omega) \qquad (n = 1)$$

where g, h, and j are known rational functions of their arguments. For all n, the expansions of f are cumbersome. We require a power series expansion for Y about $\omega = 0$. To this end, we have already applied the MACSYMA function TAYLOR (which produces a Taylor series expansion of its argument) to the result of expanding all of the f kernels in Y, and have found that the expanded form is much too large for our computer to work with efficiently. Therefore we try a segmented reduction.

We begin by choosing as segmentation points the points determined by the values of n. Thus we attempt to build tables of foreground kernels of the form:

$$f(n, x, p(x), s, \omega),$$

one table for each of the three possible values of n.

We are now prepared to write a program for performing this reduction by means of LTAB. First, we need functions that can detect f kernels, which appear in the expression Y as nouns. So:

```
F3P(EXP)  :=  IS(PART(EXP,0)  =  NOUNIFY('F)  AND  PART(EXP,1)  =  3)$
F2P(EXP)  :=  IS(PART(EXP,0)  =  NOUNIFY('F)  AND  PART(EXP,1)  =  2)$
F1P(EXP)  :=  IS(PART(EXP,0)  =  NOUNIFY('F)  AND  PART(EXP,1)  =  1)$
```

Each of these predicate functions returns TRUE for a given EXP if the value of n is as required and the leading operator of EXP is f. These functions will be used by LTAB to locate the kernels that must be expanded. Since we have decided to break the expansion at points that correspond to the different values of n, we now construct a PROCESSOR for each of the three levels. For this example, the PROCESSOR for $n = 3$ can also serve for $n = 2$:

```
PROCESSOR32(EXP) :=

  SUBST(LAMBDA([NVALUE,XVAR,RATFUN_X,X,OMEGA],

        'F(NVALUE-1,XVAR,DIFF(RATFUN_X,XVAR),X,OMEGA)

        *G(XVAR,RATFUN_X,S)

        + H(XVAR,RATFUN_X,LOG(S),LOG(1-OMEGA))

     NOUNIFY('F),EXP)$




PROCESSOR1(EXP) :=

  RATSIMP(SUBST(LAMBDA([NVALUE,XVAR,RATFUN_X,X,OMEGA],

           J(DIFF(RATFUN_X,XVAR),LOG(S),

             LOG(1-OMEGA),OMEGA)),

        NOUNIFY('F),EXP),

     LOG(1-OMEGA),OMEGA)$
```

The FULL_EV_FCN for each table should organize the results around terms involving ω, since we require a series   expansion in ω. This is desirable for all levels, so we shall use the function RATSIMP_LOG_OMEGA, defined below, as the FULL_EV_FCN of every level. The FULL_EV_FCN of the lowest table is ignored, so this restructuring with respect to ω is done in PROCESSOR1. For LEVEL2 and LEVEL3 we need:

```
RATSIMP_LOG_OMEGA(EXP)  :=  RATSIMP(EXP,LOG(1-OMEGA),'OMEGA)$
```

We are now prepared to create the needed blank tables. We choose the names of the tables to correspond to the values of n, and choose the alphabetic strings I, II and III as the prefixes for generating the labels for the kernels of tables LEVEL1, LEVEL2 and LEVEL3, respectively. Thus LEVEL1 is the lowest, LEVEL3 the highest of the tables in this reduction. We have included all functions associated with a particular table in the _MISCLIST of that table so that they will be saved with the table itself when we use LTAB_SAVE.

```
LTAB_INITIALIZE('LEVEL3,'PROCESSOR32,'RATSIMP_LOG_OMEGA,'LEVEL2,'III,
        [PREDICATE('F3P)],
        ['F3P,'PROCESSOR32,'RATSIMP_LOG_OMEGA])$
```

```
LTAB_INITIALIZE('LEVEL2,'PROCESSOR32,'RATSIMP_LOG_OMEGA,'LEVEL1,'II,
        [PREDICATE('F2P)],
        ['F2P,'PROCESSOR32,'RATSIMP_LOG_OMEGA])$
```

```
LTAB_INITIALIZE('LEVEL1,'PROCESSOR1,'RATSIMP_LOG_OMEGA,'END,'I,
        [PREDICATE('F1P)],
        ['F1P,'PROCESSOR1,'RATSIMP_LOG_OMEGA])$
```

We begin processing the expression Y by extracting the f kernels from Y and inserting them in their designated tables.

```
Y_EXTRACT_3:LTAB_LABEL_UPDATE('LEVEL3,'PROCESSOR32,Y)$
```

Y_EXTRACT_23:LTAB_LABEL_UPDATE('LEVEL2,'PROCESSOR32,Y_EXTRACT_3)$
Y_EXTRACT_123:LTAB_LABEL_UPDATE('LEVEL1,'PROCESSOR32,Y_EXTRACT_23)$

In the above sequence, the quantity Y_EXTRACT_3 has only the kernels of Level 3 extracted, Y_EXTRACT_23 has kernels of of both Level 2 and Level 3 extracted, and Y_EXTRACT_123 has all f's extracted. Y_EXTRACT_123 is the quantity into which we will substitute the final values of the expanded kernels. Y_EXTRACT_3 and Y_EXTRACT_23 are no longer needed. Now thathave made the initial entries of the foreground kernels into the tables LEVEL1, LEVEL2 and LEVEL3, we procede to evaluate the kernels stored in those tables. This is done easily:

LTAB_VALUE_UPDATE(LEVEL3)$
LTAB_VALUE_UPDATE(LEVEL2)$
LTAB_VALUE_UPDATE(LEVEL1)$

We have now reached the lowest level, and all foreground kernels in LEVEL1 are reduced to background. The next step is to substitute these results into the tables above:

BACKEV(LEVEL2,LEVEL1)$
BACKEV(LEVEL3,LEVEL2)$

All tables now contain expanded forms of al kernels, each one reduced to background and restructured so that the quantities involving ω are in leading positions. The next step is to substitute these results into Y. Since the original expression may have contained kernels from all three levels, iwe must make substitutions from all three tables.

Y_FINAL_3:LTAB_FINAL_SUBST(Y_EXTRACT_123,LEVEL3)$
Y_FINAL_23:LTAB_FINAL_SUBST(Y_FINAL_3,LEVEL2)$
Y_FINAL_123:LTAB_FINAL_SUBST(Y_FINAL_23,LEVEL1)$

Again, Y_FINAL_3 and Y_FINAL_23 are intermediate forms that are no longer useful. The result we have sought is Y_FINAL_123. All that remains is to apply TAYLOR to this form. The efficiency of this part of the procedure is improved if we first ISOLATE with respect to ω. In this case ISOLATE simply replaces expressions that do not contain ω with newly-generated atomic quantities.

ISOLATE_WRT_TIMES:TRUE$
Y_FINAL_ISOLATED:ISOLATE(Y_FINAL_123,'OMEGA)$
Y_FINAL_ISOLATED_TAYLOR:TAYLOR(Y_FINAL_ISOLATED,'OMEGA,0,3)$
Y_FINAL:EV(Y_FINAL_ISOLATED_TAYLOR)$

The final call to EV removes the isolation variables inserted by ISOLATE. The result we seek is Y_FINAL.

This problem has a feature that deserves special emphasis. The original expression Y contains kernels of each of the three types. Therefore it is possible that some of the kernels that are found in the original expression are also generated as a result of expanding higher order kernels that are themselves found in the original expression. A straightforward expansion of all kernels can therefore result in duplication of effort. In the approach that we have taken here, each unique kernel is evaluated only once.

5. Outlook

In its current implementation, LTAB provides a systematic method for applying intricate reduction procedures to large expressions. However, it is possible to construct programs which automate this procedure even further. Currently, the user must generate blank tables before the reduction process can begin, and provide names for the prefixes used in those tables, but one can define functions or macros that will carry out these steps automatically if necessary. These evaluation functions would need as arguments both the downward and upward processor functions for each level, as well as the expression to be reduced. Alternatively, one might construct a macro equivalent to a function definition operator, except that instead of producing a simple function definition, it produces a function definition that uses Dissection to procede from one step to the next within the body of the definition. The user might be required to specify the downward processors for each level, as well as their corresponding upward processors, all as statements in the "function definition". The only evidence that a Dissection-oriented reduction scheme was actually being used would be this tripling of statements in the definition. Such a scheme could greatly improve performance of Computer Algebra systems on machines with large address spaces.

In this way one can automate much more of the Dissection method than has been done in LTAB. However, the general problem of choosing a particular dissection for a given problem is more difficult. Although it may now be possible to automate the entire dissection procedure, including choice of segmentation, for some classes of problems, it is likely that general dissector programs capable of treating many kinds of reduction problems will continue to require human intervention.

Notes and References

[1] MACSYMA Reference Manual (Version 9), Mathlab Group, Laboratory for Computer Science, Massachusetts Insitute of Technology, Cambridge, Massachusetts, 1977.

[2] In Section 2 we use the term *list* to denote an ordered set of algebraic expressions. For many Computer Algebra systems, *list* has a specific technical meaning. This is not the sense in which the word is used here.

[3] A kernel of an expression is a subexpression that is not rationally simplified as a result of rational simplification of the expression itself. That is, its leading operator is not rational.

[4] G. Passarino and M. Veltmann, *Nucl. Phys.* B160, 151, (1979).

Acknowledgements

This work was carried out using MACSYMA, a symbol manipulation program developed at the MIT Laboratory for Computer Science and supported by the National Aeronautics and Space Administration under grant 1323, by the Office of Naval Research under grant ET-78-C-02-4687, and by the U.S. Air Force under grant F49620-79-C-020. The MACSYMA community, and especially J.P. Golden, has been very helpful. The author is also grateful to M.P. Shatz and to G.C. Fox for many useful discussions during the preparation of the manuscript.

18

A PROPOSAL FOR THE SOLUTION OF QUANTUM FIELD THEORY PROBLEMS USING
A FINITE-ELEMENT APPROXIMATION

CARL M. BENDER

Department of Physics, Washington University,
St. Louis, Missouri 63130

ABSTRACT

We show that the method of finite elements reduces intractable quantum operator differential equations to completely solvable operator difference equations. Early work suggests that this approximation technique is extremely accurate and very well suited to algebraic manipulation on a computer.

INTRODUCTION

Many of the approaches to numerical quantum field theory which have been developed to date start from the Euclidean path integral formulation of the theory. This is because it is possible, using path integrals, to write closed form, albeit formal and difficult to evaluate, expressions for the quantities of physical interest, namely the Green's functions. Thus, in this approach, the focus is not on finding the direct solution to the equations of quantum field theory, but rather on developing an effective and reliable method for evaluating the functional integrals representing the Green's functions.

The novelty of the present work[1] is that we have been able to use the method of finite elements, a technique widely employed to solve numerically partial differential equations arising in classical continuum mechanics and fluid dynamics,[2] to solve operator quantum field equations in Minkowski space directly. An accurate evaluation of a functional integral requires a great many Monte Carlo passes through the lattice. In our approach, we do operator time-stepping, which requires only one pass through the lattice. We believe that the finite element method is advantageous in quantum field theory because, unlike other methods, it treats bosons and fermions on an equal footing. This is possible because in this method one does not introduce a finite difference approximation for derivatives.

DESCRIPTION OF THE FINITE ELEMENT METHOD

The finite element method was developed for the numerical solution of ordinary or partial differential equations. In this method one begins by partitioning the domain D on which the equation is to be solved into a collection of nonoverlapping patches, called finite elements. The finite elements are usually chosen to be polygons, whose size and shape can vary over the domain in accord with the requirements of a particular problem. On each finite element, the unknown function is represented by a low order polynomial, whose coefficients are determined by two types of conditions. First, one wants to ensure that the local approximations to the function on each finite element can be pieced together consistently to give a global representation of the function. To do this, one requires the function, and sometimes its derivatives, to be continuous across the boundary of contiguous finite elements. For those finite elements which have a boundary coinciding with the boundary of the domain D, one must also impose the boundary conditions associated with the differential operator. The second set of conditions results from imposing the differential equation; this is often done by rewriting it in variational form. These two conditions give a set of equations which determine the coefficients of the polynomials on each finite element and hence an approximate solution to the differential equation.

We illustrate the finite element method by a very elementary ordinary differential equation example.

EXAMPLE

Given $y'(x) = y(x)$ and $y(0) = 1$, find $y(1)$. The exact solution of this problem is $y(1) = e = 2.718281828....$ How well can we compute e using the method of finite elements? We decompose the interval $0 \leq x \leq 1$ into N intervals of length $1/N$. Here, we consider the case of linear finite elements. On the ith interval we represent $y(x)$ by $a_i + b_i x$.

Since the differential equation is first order we impose only the constraint of continuity at the boundaries of the intervals. We clearly cannot impose the differential equation throughout each interval, so we impose it at just one point on every interval which we choose to be the center of the interval.

Suppose there is just one finite element ($N = 1$). Then for $0 \leq x \leq 1$, $y(x) = a_1 + b_1 x$. We impose the boundary condition $y(0) = 1$ by taking $a_1 = 1$. We impose the differential equation at $x = 1/2$, the center of the interval. (As we will see later, the requirements of quantization force us to impose the differential equation at the center.) This gives $b_1 = 2$. Thus, we have

$$y(x) = 1 + 2x \quad , \quad 0 \leq x \leq 1 ,$$

and we predict $y(1) = 3$ ($N = 1$), with a remarkably small relative error of 10%!

Now suppose there are two elements. Imposing $y(0) = 1$, continuity at $x = 1/2$, and the differential equation at $x = 1/4$ and $x = 3/4$, the center of each interval, we obtain $y(1) = 25/9 = 2.7777.... (N = 2)$ which differs from the exact answer by a relative error of less than 2%!

This problem is sufficiently simple that we can obtain $y(1)$ in closed form for any number N of finite elements. The result is

$$y(1)(N \ elements) = \left[\frac{2N+1}{2N-1} \right]^N .$$

For large N this expression rapidly approaches the exact answer e:

$$y(1)(N \ elements) \sim e \left[1 + \frac{1}{12N^2} \right] \quad , \quad N \to \infty .$$

The relative error,

$$\frac{y(1)(N \ elements) - y(1)(exact)}{y(1)(exact)} = \frac{1}{12N^2} \quad ,$$

rapidly decays to zero as N increases. For example, when $N = 10$ the relative error is less than 1 part in a thousand.

Richardson extrapolation can be used to improve the relative error by many orders of magnitude.

THE FINITE ELEMENT APPROXIMATION IN QUANTUM MECHANICS

The conventional methods of finite elements have often been applied to differential equation problems in classical physics involving fluid mechanics and solid mechanics. However, now the coefficients of the polynomials on the finite elements are operators, whose properties are determined by the equal time commutation relations.

To illustrate our procedure we consider the problem of solving the Heisenberg equations of motion for a one-dimensional quantum system. If the Hamiltonian is

$$H = p^2/2 + V(q) \quad . \tag{1}$$

The Heisenberg equations are

$$dq(t)/dt = p(t) \quad , \quad dp(t)/dt = f[q(t)] \quad , \tag{2}$$

where $f(q) = -V'(q)$. The quantum mechanical problem consists of solving (2) for the operators $p(t)$ and $q(t)$ given the equal time commutation relation

$$[q(t),p(t)] = i \quad . \tag{3}$$

We solve this problem first on a single finite element. We approximate $q(t)$ and $p(t)$ by linear functions of t:

$$q(t) = \left[1 - \frac{t}{h}\right] q_0 + \frac{t}{h} q_1 \quad , \quad p(t) = \left[1 - \frac{t}{h}\right] p_0 + \frac{t}{h} p_1 \quad , \quad 0 < t < h \quad . \tag{4}$$

Substituting (4) into (2) and evaluating the result at the *center* of the time interval $t_o = h/2$ gives

$$\frac{q_1 - q_0}{h} = \frac{p_0 + p_1}{2} \quad , \tag{5}$$

$$\frac{p_1 - p_0}{h} = f\left[\frac{q_0 + q_1}{2}\right] \quad . \tag{6}$$

Equations (5) and (6) are a discrete form of the equations of quantum mechanics. But, are they consistent with the equal time commutation relation (2)? At $t = 0$, (3) reads $[q_0,p_0] = i$. The question is, if we solve (5) and (6) simultaneously for the operators p_1 and q_1, will the commutator $[q_1,p_1]$ also have the value i? We can evaluate $[q_1,p_1]$ for any function f : Commute (5) on the right with $p_0 + p_1$,

$$[q_1 - q_0,p_0 + p_1] = 0 \quad , \tag{7}$$

and commute (6) on the left with $q_0 + q_1$,

$$[q_0 + q_1,p_1 - p_0] = 0 \quad . \tag{8}$$

Adding (7) and (8) gives

$$[q_1,p_1] = [q_0,p_0] = i \quad . \tag{9}$$

Thus, we have proved that the method of finite elements *is* consistent with the equal time commutation relation on a single finite element. But clearly, for a collection of N elements this argument may be used iteratively to show that the difference equations in (5) and (6) are consistent with (3) at the endpoints of each of the finite elements.

We make two observations. First, the equal time commutation relation holds only at the endpoints of a finite element, but not at any interior point. Second, the result in (9) depends crucially upon our having imposed the Heisenberg equations in (2) at the *center* of the finite element, $t = t_o = h/2$. For any other value of t_o, (9) is false. Thus the operator properties of quantum mechanics uniquely determine the value of t_o. In a classical problem, there is no such constraint on the value of t_o.

SOLUTION OF THE OPERATOR EQUATIONS

Even though (5) and (6) are operator equations, we can solve them explicitly for p_1 and q_1, in terms of p_0 and q_0. First, we solve (5) for p_1:

$$p_1 = \frac{2}{h}\left(q_1 - q_0\right) - p_0 \quad .$$
(10)

Next we use this result to eliminate p_1 from (6):

$$-\frac{2}{h}p_0 - \frac{4}{h^2}q_0 = g\left(\frac{q_1 + q_0}{2}\right) \quad ,$$
(11)

where $g(x) = f(x) - 4x/h^2$ is the function which completely characterizes the dynamical content of this quantum theory. The solution to (11) is given in terms of g^{-1}, where $g^{-1}(y)$ is the solution to the classical equation $y = g(x)$. From (11) we have

$$q_1 = -q_0 + 2g^{-1}\left(-\frac{2}{h}p_0 - \frac{4}{h^2}q_0\right)$$
(12)

and from (10) we obtain

$$p_1 = -p_0 - \frac{4}{h}q_0 + \frac{4}{h}g^{-1}\left(-\frac{2}{h}p_0 - \frac{4}{h^2}q_0\right) \quad .$$
(13)

The result in (12) and (13) is the one time-step solution of the quantum-mechanical initial value problem in (2).[3] We generalize to N time steps (N finite elements) by iterating (12) and (13) N times to express p_N and q_N in terms of p_0 and q_0. Note that the solution takes the form of a continued (nested) function $A + g^{-1}(A + g^{-1}(A + ...)))$. If $[q_1, p_1]$ had not turned out to be i, but had differed from i by a (presumably) small c-number part that vanishes with the spacing h, then we would not be able to obtain the N-finite-element solution by iteration, and indeed we would not even be able to solve for p_2 nd q_2.

HIGHER-ORDER FINITE ELEMENTS

We may replace the linear finite elements used in our calculations by quadratic finite elements. If we do, then $q(t)$ in (4) is replaced by

$$q(t) = \left(1 - \frac{t}{h}\right)q_o + \frac{t}{h}q_1 + \left(1 - \frac{t}{h}\right)\left(\frac{t}{h}\right)Q \quad .$$

The best numerical approximation is obtained by imposing the differential equations (2) *twice* on the interval $0 \leq t \leq h$, and imposing continuity at the boundaries between successive intervals. Requiring that q_2 and p_2 satisfy the same commutation relation as q_1 and p_1 determines the points t_1 and t_2. Where the differential equations must be imposed:

$$t_{1,2} = \left[\frac{1}{2} \pm \frac{1}{\sqrt{12}} \right] h \quad .$$

These are just the Legendre points for Gaussian quadrature.

In general, if we use D degree polynomials as finite elements we find that quantum mechanics determines that we must impose the Heisenberg equations at the D Gaussian quadrature points.

If we return to the simple classical example discussed earlier, $y' = y$, $y(0) = 1$, we find that the one-quadratic-finite-element prediction for $y(1)$ is $19/7 = 2.714$ which is extremely close to e ($-1/8\%$ relative error). For N quadratic finite elements the relative error decays like N^{-4}. Also, if we use Dth degree finite elements the error decays like N^{-2D}, where N is the number of finite elements.

SOME NUMERICAL RESULTS FOR SIMPLE QUANTUM SYSTEMS

For the harmonic oscillator defined by $H = p^2/2 + m^2 q^2/2$ the operator equations (5) and (6) are linear and the solution in (12) and (13) can be written in matrix form:

$$\begin{pmatrix} p_1 \\ q_1 \end{pmatrix} = M \begin{pmatrix} p_0 \\ q_0 \end{pmatrix} \quad .$$

Although M is not symmetric

$$M = \frac{1}{1 + m^2 h^2/4} \begin{bmatrix} 1 - h^2 m^2/4 & -hm^2 \\ h & 1 - h^2 m^2/4 \end{bmatrix} \quad . \tag{14}$$

it can be written as a similarity transform of a diagonal matrix D, $M = QDQ^{-1}$, whose entries are

$$d_{11}, d_{22} = (1 - h^2 m^2/4 \pm ihm)/(1 + m^2 h^2/4) = \exp\{ \pm i \, \arcsin[mh/(1 + m^2 h^2/4)] \} \quad . \tag{15}$$

The solution for p_n and q_n is

$$p_n = A(d_{11})^n + B(d_{22})^n \tag{16}$$

and

$$q_n = C(d_{11})^n + D(d_{22}^n) \tag{17}$$

where $A = (p_0 + imq_0)/2$, $B = (p_0 - imq_0)/2$, $C = A/(im)$, $D = iB/m$. Observe that in the continuum limit $nh = T$, $mh \to 0$, the diagonal entries d_{11}, d_{22} become

$$d_{11}, d_{22} = e^{\pm imT} \quad .$$ (18)

Thus, the fields p and q have a time oscillation given by (18) and we can read off the exact answer for the energy gap ΔE (the excitation of the first energy level over the vacuum)

$$\Delta E = E_1 - E_0 = m \quad .$$ (19)

We can also compute the two-point function, which in the continuum is defined by

$$G(t,t') \equiv <0 \mid T[q(t)q(t')] \mid 0> \quad .$$ (20)

$G(t,t')$ satisfies the differential equation

$$\frac{\partial^2 G}{\partial t^2} + m^2 G = -i\delta(t-t') \quad ,$$

whose solution is

$$G(t,t') = \frac{1}{2m} \exp(-im \mid t-t' \mid) \quad .$$ (21)

To compute G on the lattice we substitute (17) into the formula

$$G_{NM} = <0 \mid \theta(N-M)q_N q_M + q_M q_N \theta(M-N) \mid 0>$$ (22)

$$= \frac{1}{2m}(d_{22}^N \, d_{11}^M \, \theta_{N-M} + d_{22}^M \, d_{11}^N \, \theta_{M-N})$$

$$= \frac{1}{2m} e^{-i \mid N-M \mid \arc \sin[mh/(l+m^2h^2/4)]} \quad .$$

In the continuum limit $T = t-t'$, $mh \to 0$, $t = Nh$, $t' = Mh$, G_{NM} in (22) becomes $G(t,t')$ in (21).

For the anharmonic oscillator, whose Hamiltonian is $H = p^2/2 + \lambda q^4/4$, the energy gap is known to be $\Delta E = E_1 - E_0 = (1.08845)\lambda^{1/3}$. Using just *one* linear finite element we find that $\Delta E = 1.1447...\lambda^{1/3}$ which has a relative error of 5%. By going from one to two time steps the error goes down by a factor of five, just as in the simple calculation of e in the Example. Using one quadratic finite element we find that $\Delta E = 1.0822...\lambda^{1/3}$ which has a relative error of $-6/10\%$.

HIGHER-DIMENSIONAL QUANTUM FIELD THEORY

Now we show how to generalize this method to quantum field theory. Consider a scalar field theory in two-dimensional Minkowski space. We write the operator field equations as a coupled first-order system so that we can continue to use linear approximations to the fields on finite elements

$$\pi = \phi_t \quad , \quad \gamma = \phi_x \quad , \quad \pi_t - \gamma_x + f(\phi) = 0 \quad . \tag{23}$$

We introduce rectangular finite elements whose length in the time direction is h and in the space direction is k. On the m, n-element, the field ϕ is approximated by the bilinear polynomial

$$\phi(x,t) = \left[1 - \frac{t}{h}\right]\left[1 - \frac{x}{k}\right]\phi_{m-1,n-1} + \left[1 - \frac{t}{h}\right]\left[\frac{x}{k}\right]\phi_{m,n-1} \tag{24}$$

$$\left[\frac{t}{h}\right]\left[1 - \frac{x}{k}\right]\phi_{m-1,n} + \left[\frac{t}{h}\right]\left[\frac{x}{k}\right]\phi_{m,n} \quad ,$$

where the coefficient $\phi_{m,n}$ is the value of the field operator at the site (m,n). The fields π and γ are represented in a similar way.

Our objective is to show how to advance one step in the time direction. We consider a single horizontal row of M finite elements and impose (23). The result is the following system of difference equations:

$$\frac{1}{4}(\pi_{m-1,0} + \pi_{m,0} + \pi_{m-1,1} + \pi_{m,1}) = \frac{1}{2h}(\phi_{m,1} + \phi_{m-1,1} - \phi_{m,0} - \phi_{m-1,0}) \; ,$$

$$\frac{1}{4}(\gamma_{m-1,0} + \gamma_{m,0} + \gamma_{m-1,1} + \gamma_{m,1}) = \frac{1}{2k}(\phi_{m,1} + \phi_{m,0} - \phi_{m-1,1} - \phi_{m-1,0}) \; ,$$

$$\frac{1}{2h}(\pi_{m,1} + \pi_{m-1,1} - \pi_{m,0} - \pi_{m-1,0}) - \frac{1}{2k}(\gamma_{m,1} + \gamma_{m,0} - \gamma_{m-1,1} - \gamma_{m-1,0}) \; ,$$

$$= f\left[\frac{1}{4}(\phi_{m-1,0} + \phi_{m,0} + \phi_{m-1,1} + \phi_{m,1})\right] \quad . \tag{25}$$

$m = 1,2,...M$. We take each finite element for ϕ and π to represent one degree of freedom and define the dynamical variables

$$\Phi_{m,n} = \frac{1}{2}(\phi_{m,n} + \phi_{m-1,n}) \quad , \quad \Pi_{mn} = \frac{1}{2}(\pi_{m,n} + \pi_{m-1,n}) \quad . \tag{26}$$

If we eliminate γ_{mn} from (25) and express the resulting equations in terms of the dynamical variables we obtain a system of 2M equations whose general structure is

$$\Pi_{m,0} + \Pi_{m,1} = \frac{2}{h}(\Phi_{m,1} - \Phi_{m,0}) \quad , \tag{27}$$

$$\Pi_{m,1} - \Pi_{m,0} = \sum_{g=1}^{M} S_{m,j}(\Phi_{j,1} + \Phi_{j,0}) + F(\Phi_{m,1} + \Phi_{m,0}) \tag{28}$$

$m = 1,2,3...M$. Here S is a symmetric matrix[4] and F is a nonlinear function simply related to f in (15).

When written in terms of $\Phi_{m,n}$ and $\Pi_{m,n}$, the ETCR's for the fields, $[\phi(x,t),\pi(y,t)] = i\,\delta(x-y)$, become

$$[\Phi_{j,n},\Phi_{l,n}] = 0 \quad ,[\Pi_{j,m},\Pi_{l,m}] = 0 \quad ,[\Phi_{j,m},\Pi_{l,m}] = \frac{i}{k}\delta_{j,l} \quad . \tag{29}$$

The consistency problem here is to show that if (29) holds for $n = 0$ then by virtue of (27) and (28) it also holds for $n = 1$.

The proof of consistency is not simple. There are three steps. First, we eliminate $\Pi_{m,1}$, $m = 1,2,3...M$, from (28) using (27). Thus (28) takes the form

$$\Pi_{m,0} + \frac{2}{h}\Phi_{m,0} = \sum_{j=1}^{M} S_{m,j}(\Phi_{j,1}+\Phi_{j,0}) + G(\Phi_{m,1}+\Phi_{m,0}) \quad , \quad j = 1,2,...M \tag{30}$$

where G is simply related to F. Because

$$\left[\Pi_{j,0} + \frac{2}{h}\Phi_{j,0}\,,\,\Pi_{l,0} + \frac{2}{h}\Phi_{l,0}\right] = 0 \quad ,$$

we can in principle solve (30) for $\Phi_{m,1}+\Phi_{m,0}$ in terms of $\Pi_{m,0}+\frac{2}{h}\Phi_{m,0}$. It follows from (27) at $n = 0$ that

$$\left[\Phi_{j,1} + \Phi_{j,0}\,,\,\Phi_{l,1} + \Phi_{l,0}\right] = 0 \quad . \tag{31}$$

Second, we replace G in (30) by ϵG, where ϵ is a small parameter. Then, assuming that G has a Taylor series, we solve for $\Phi_{m,1}$ as a perturbation series in powers of ϵ. We can show that to all orders in powers of ϵ,

$$[\Phi_{j,1},\Phi_{l,1}] = 0 \quad , \tag{32}$$

so long as S is a symmetric matrix. This is the difficult part of the proof.

Third, combining (31) and (32) gives

$$[\Phi_{j,1} - \Phi_{j,0},\Phi_{l,1} - \Phi_{l,0}] = 0 \quad . \tag{33}$$

We complete the proof by using the same procedure as that leading to (9): we commute (27) and (28) with $\Phi_{j,0} \pm \Phi_{j,1}$ and add the resulting commutators to establish (29) for $n = 1$. It is quite interesting that this proof depends critically on the symmetry of the matrix S; the result is not sensitive to the choice of the nonlinear function F except for the assumption that it has a Taylor series expansion. Having established the consistency for the first time step, by induction (29) holds for all values of n. This proof also shows how to solve the operator equations (27) and (28) algebraically.

THE FERMION PROBLEM

There is a well-known difficulty encountered in lattice fermion theories. To illustrate this problem, we consider the free Dirac equation

$$(i\,\partial + \mu)\psi = 0 \quad . \tag{34}$$

On a Kogut-Suskind lattice (discrete space and continuous time) (34) becomes

$$i\,\gamma^0 \partial_0 \psi_m(t) + \frac{i\,\gamma^1}{h}(\psi_{m+1} - \psi_m) + \mu\psi_m = 0 \quad (1 \le m \le M) \; , \tag{35}$$

where h is the lattice spacing and we take a forward difference for the space derivative.

Equation (34) can be obtained by varying the continuum action

$$S = \int\int dx\,dt \; \bar{\psi}(i\,\partial + \mu)\psi \tag{36}$$

with respect to $\bar{\psi}$. Similarly, (35) can be derived by varying the discrete action

$$S = h \sum_{m=1}^{M} \int dt \; \bar{\psi}_m \left[i\,\gamma^0 \partial_0 \psi_m + \frac{i\,\gamma^1}{h}(\psi_{m+1} - \psi_m) + \mu\psi_m \right] \tag{37}$$

with respect to $\bar{\psi}_m$.

There is a serious problem: the discrete action in (37) is *not* Hermitian. Under Hermitian conjugation the forward difference becomes a backward difference:

$$i\sum\int dt \; \bar{\psi}_m \, \gamma^1(\psi_{m+1} - \psi_m) \xrightarrow{H.C.} i\sum\int dt \; \bar{\psi}_m \, \gamma^1(\psi_m - \psi_{m-1}) \quad .$$

Thus, the lattice theory is not unitary.

An alternative way to view this problem is to find the dispersion relation corresponding to (35). We take the continuous Fourier transform of (35)

$$\tilde{A}(\omega) \equiv \int_{-\infty}^{\infty} dt \; A(t)\, e^{i\omega t}$$

followed by the lattice Fourier transform

$$\tilde{A}_p \equiv \sum_{m=1}^{M} A_m \; e^{2\pi imp/M}$$

and obtain the dispersion relation

$$\omega^2 = \mu^2 + \frac{4}{h^2}\sin^2(\pi P/M)e^{-2\pi iP/M} \quad . \tag{38}$$

Note that the energy ω is *complex*. (One would not expect a real energy spectrum from a non-Hermitian action.)

402

An alternative approach which voids the non-Hermiticity problem is to use a symmetric difference for the space derivative; (35) is replaced by

$$i\gamma^0\partial_0\psi_m(t) + \frac{i\gamma^1}{2h}(\psi_{m+1} - \psi_{m-1}) + \mu\psi_m = 0 \quad . \tag{39}$$

Now the action is Hermitian and the dispersion relation

$$\omega^2 = \mu^2 + \frac{1}{h^2}\sin^2(2\pi P/M) \tag{40}$$

is real. However, a new problem has surfaced: note that $\omega = \mu$ at $p = 0$ *and* at $p = M/2$. Thus, there are *two* species of fermions.

These problems apparently cannot be avoided. The no-go theorems concerning lattice fermions tell us that there is no local difference scheme which gives a real dispersion relation, does not exhibit species doubling, and is chirally invariant in the massless limit (the upper and lower components of ψ are treated equally at each lattice site).[5]

The finite-element method provides an interesting resolution of this problem. The finite-element method requires *averaged* forward differences in order to preserve the equal-time anticommutation relations. Thus, the finite-element discrete version of (25) is

$$\frac{i\gamma^0}{2h}(\psi_{m,n+1} + \psi_{m+1,n+1} - \psi_{m,n} - \psi_{m+1,n}) \tag{41}$$

$$+ \frac{i\gamma^1}{2h}(\psi_{m+1,n+1} + \psi_{m+1,n} - \psi_{m,n+1} - \psi_{m,n})$$

$$+ \frac{\mu}{4}(\psi_{m,n+l} + \psi_{m+1,n+1} + \psi_{m,n} + \psi_{m+1,n}) = 0 \quad .$$

This operator difference equation has the following properties:

(i) It is the only way to discretize the continuum Dirac equation (34) which exactly preserves the equal-time anticommutation relations.

(ii) The discrete action from which (41) follows is Hermitian. Moreover, the dispersion relation associated with (41) is real:

$$\omega^2 = \mu^2 + \frac{4}{h^2}\tan^2\left[\frac{p\pi}{M}\right] \quad . \tag{42}$$

(iii) There is no species doubling.

(iv) The difference scheme in (41) is local (that is, there are only a finite number of terms in the discrete derivatives) as opposed to nonlocal differencing schemes like the

SLAC derivative.

(v) Equation (41) is chirally symmetric in the massless limit ($\mu \to 0$).

(vi) Equation (41) is avoids the dire consequences of the no-go theorems because while the action is local the Hamiltonian is not.

It is interesting that the finite-element method leads uniquely and unambiguously to the difference equation (41) with the above nice properties. The avoidance of fermion doubling comes about through an effortless and natural procedure.[6]

GAUGE INVARIANCE

It is possible to incorporate Abelian gauge invariance into the finite-element framework. Consider the continuum equations of two-dimensional electrodynamics:

$$\partial_\mu A_\nu - \partial_\nu A_\mu = F_{\mu\nu} \quad , \tag{43}$$

$$\partial_\nu F^{\mu\nu} = J^\mu \quad , \tag{44}$$

$$(i \not\partial + e \not A + \mu)\psi = 0 \quad , \tag{45}$$

where $J^\mu = e\, \bar\psi \gamma^\mu \psi$. Note that the current J^μ must be carefully defined as a point-split limit, or else the theory will be chirally symmetric and will have a vanishing anomaly.

To show that the equations (43)-(45) are gauge invariant we perform an infinitesimal gauge transformation

$$\psi \to \psi + \delta\psi \quad , \quad \delta\psi = ie\,\delta\Lambda\,\psi \quad , \tag{46}$$

$$A_\mu \to A_\mu + \delta A_\mu \quad , \quad \delta A_\mu = \partial_\mu\,\delta\Lambda \quad , \tag{47}$$

and show that to first order in $\delta\Lambda$, (43)-(45) are invariant.

The finite-element transcription of (47) is

$$(\delta A_0)_{m,n} = \frac{1}{2h}\left[\delta\Lambda_{m+1,n+1} + \delta\Lambda_{m,n+1} - \delta\Lambda_{m+1,n} - \delta\Lambda_{m,n}\right] \quad , \tag{48}$$

$$(\delta A_1)_{m,n} = \frac{1}{2h}\left[\delta\Lambda_{m+1,n+1} + \delta\Lambda_{m+1,n} - \delta\Lambda_{m,n+1} - \delta\Lambda_{m,n}\right] \quad , \tag{49}$$

where we impose periodic boundary conditions in the space direction $(a_\mu)_{0,n} = (A_\mu)_{m,n}$, $(\delta A_\mu)_{0,n} = (\delta A_\mu)_{M,n}$, $(\delta\Lambda)_{0,n} = (\delta\Lambda)_{M,n}$. Equations (48) and (49) ensure that the lattice transcription of (43) and therefore the current J^μ in (44) is gauge invariant.

The direct lattice transcription of the Dirac equation (45) is not gauge invariant. One must seek an interaction term in the difference equation for ψ which exhibits gauge invariance and has (45) as its continuum limit. To accomplish this we write down the finite-element transcription of (46):

$$\delta\left(\psi_{m+1,n+1} + \psi_{m+1,n} + \psi_{m,n+1} + \psi_{m,n}\right)$$

$$= ie\,\delta\Omega_{m,n}\left(\psi_{m+1,n+1} + \psi_{m+1,n} + \psi_{m,n+1} + \psi_{m,n}\right) \quad, \tag{50}$$

where $\delta\Omega_{m,n} = \left(\delta\Lambda_{m+1,n+1} + \delta\Lambda_{m+1,n} + \delta\Lambda_{m,n+1} + \delta\Lambda_{m,n}\right)$.

The next step is to solve (41) for $\delta\psi_{m,n}$:

$$\delta\psi_{m,n} = (-1)^n\,\delta\psi_{m,0} + \frac{ie}{2}\sum_{n'=0}^{n-1}\left(\sum_{m'=1}^{m-1} - \sum_{m'=m}^{M}\right)(-1)^{m+m'+n+n'} \tag{51}$$

$$\times\left(\psi_{m'+1,n'+1} + \psi_{m'+1,n'} + \psi_{m',n'+1} + \psi_{m',n'}\right)\delta\Omega_{m'n'} \quad,$$

where we have imposed the boundary conditions $\psi_{M,n} = (-1)^{M+1}\psi_{0,n}$ and $\delta\psi_{M,n} = (-1)^{M+1}\delta\psi_{0,n}$.

We use the result in (51) to perform an infinitesimal gauge variation of the free Dirac equation. To vary the lattice transcription of $i\,\partial\psi$,

$$\frac{i\gamma^0}{h}\delta\left(\psi_{m+1,n+1} + \psi_{m,n+1} - \psi_{m+1,n} - \psi_{m,n}\right)$$

$$+ \frac{i\gamma^1}{2h}\delta\left(\psi_{m+1,n+1} + \psi_{m+1,n} - \psi_{m,n+1} - \psi_{m,n}\right) \quad,$$

We must use (51) eight times. Next, by inspection we write down an interaction term $(I_1)_{m,n}$ with the property that if we vary I_1 with respect to A and use the free Dirac equation, terms of order $\delta\Lambda_{m,n}$ cancel. However, varying I_1 with respect to ψ gives terms of order $\delta\Lambda$ which do not cancel. Thus, by inspection, we write down a second interaction term $(I_2)_{m,n}$ with the property that if it is varied with respect to A it cancels the terms that arise from varying I_1 with respect to ψ (if the Dirac equation with I_1 is used). However, varying I_2 with respect to ψ gives new terms of order $\delta\Lambda$ which do not cancel.

We continue this process *ad infinitum* and obtain a series of interaction terms $I_1 + I_2 + I_3 \cdots$. This series can be summed in closed form (it exponentiates) to produce a completely gauge invariant interaction term for the discrete Dirac equation:

$$I_{m,n} = \frac{i\gamma_0}{h}\sum_{n'=0}^{n}(-1)^{n+n'}\exp\left[ieh\sum_{n''=n'+1}^{n}B_{m,n''}\right]\left(e^{iehB}-1\right)\left(\psi_{m,n'}+\psi_{m+1,n'}\right)$$

$$- \frac{i\gamma^0}{h}(-1)^n\exp\left[ieh\sum_{n''=1}^{n}B_{m,n''}\right]\left(e^{iehB_{m,1}/2}-1\right)\left(\psi_{m,0}+\psi_{m+1,0}\right)$$

$$- \frac{i\gamma^1}{2h}\sec\left[eh\sum_{m'=1}^{M}C_{m',n}/2\right]\sum_{m''=1}^{M}\operatorname{sgn}(m''-m)(-1)^{m+m''}$$

$$\times \exp\left[\frac{ieh}{2}\sum_{m'''=1}^{M}\text{sgn}(m'''-m)\text{sgn}(m'''-m'')\text{sgn}(m''-m)C_{m''',n}\right]$$

$$\times\left(e^{iehC_{m'',n}}-1\right)\left(\psi_{m'',n}+\psi_{m'',n+1}\right) \quad , \tag{52}$$

where $B_{m,n}=[(A_0)_{m,n}+(A_0)_{m,n-1}]/2$ and $C_{m,n}=[(A_1)_{m,n}+(A_1)_{m-1,n}]/2$. Note that the point splitting in the interaction term is built in automatically. It is easy to verify that the discrete Dirac equation with the interaction term in (52) has (45) as its continuum limit.

THE ANOMALY

It is not hard to solve the two-dimensional discrete Dirac equation with the interaction term I in (43) (Schwinger model). We choose a gauge in which $A_0=0$. The exact answer for the anomaly (mass of the boson) in the continuum is

$$\omega^2=e^2/\pi \quad . \tag{53}$$

Using the method of finite elements we obtain

$$\frac{e^2}{N\sin(\pi/N)} \quad , \tag{54}$$

where M is the number of finite elements in the space direction. Thus, the relative error between (53) and (54) decays like $1/N^2$, just as in the simple example presented at the beginning of this talk.

CONCLUSIONS

Using the method of finite elements we have been able to formulate and solve operator difference equations on a lattice. These difference equations are consistent with the requirements of quantum mechanics. Moreover, they do not suffer from the problems of fermion doubling and can be made invariant under Abelian gauge transformations. Our next objective is to try to incorporate non-Abelian gauge invariance. We hope to bring to bear the enormous power of algebraic computational systems on the complicated but solvable operator equations that will arise in this research.

ACKNOWLEDGMENT

The author is grateful to the U.S. Department of Energy for financial support.

REFERENCES

1. This talk is based on published as well as unpublished work: C. M. Bender and D. H. Sharp, Phys. Rev. Lett. **50**, 1535 (1983); C. M. Bender, K. A. Milton, and D. H. Sharp, Phys. Rev. Lett. **51**, 1815 (1983); C. M. Bender, K. A. Milton, and D. H. Sharp, "Gauge Invariance and the Finite-Element Solution of the Schwinger Model," submitted to Phys. Rev.

2. Useful general references on the finite element method are G. Strang and G. J. Fix, *An Analysis of the Finite Element Method*, (Prentice-Hall, Inc., Englewood Cliffs, 1973) and T. J. Chung, *Finite Element Analysis in Fluid Dynamics* (McGraw-Hill, New York, 1978).

3. There are several interesting remarks to be made here. One intriguing question is whether (12) and (13) might be used in combination with $[q_0, p_0] = i$ to find a spectrum generating algebra. Second, one may ask what happens when the equation $y = g(x)$ has multiple roots; that is, what role is played by instantons in these lattice calculations?

4. The matrix S is a numerical matrix containing the lattice spacings h and k. It is symmetric because with properly chosen boundary conditions the operator ∇^2 in the continuum is symmetric.

5. For a detailed discussion see L. H. Karsten and J. Smit, Nucl. Phys. **B183**, 103 (1981); H. B. Nielsen and M. Ninomiya, Nucl. Phys. **B185**, 20 (1981); J. M. Rabin, Phys. Rev. **D24**, 3218 (1981).

6. By experimenting with various types of difference schemes, R. Stacey independently discovered the dispersion relation (42) and the Kogut-Suskind version of (41). See Phys. Rev. **D26**, 468 (1982).

19

EXACT SOLUTIONS FOR SUPERLATTICES AND HOW TO RECOGNIZE THEM WITH COMPUTER
ALGEBRA

GENE COOPERMAN, LIONEL FRIEDMAN, WALTER BLOSS
GTE Laboratories, Inc., Waltham, MA 02254

ABSTRACT

The study of superlattices has been motivated by the possibility of
"custom-engineering" new solid state materials. Using the Kronig-Penney equations for superlattices, we found a novel series expansion solution for
several of its physical properties. We derived the first two terms by hand,
providing an accurate estimate of the physical quantities of interest. With
the aid of MACSYMA, we were later encouraged to derive still higher order
terms. To our surprise, the higher order terms reduced to zero for a physically important special case. This motivated further analysis, in which we
were able to show our original two-term solution to be an exact, closed form
solution in this special case.

INTRODUCTION

We report on a discovery process, by which we found an approximate
analytical solution to a model for superlattices. After deriving and justif-
ying the equations by hand, we reviewed them using the computer algebra
package, MACSYMA. We discovered that our answer was much more accurate than
we had expected.

Superlattices are semiconductor structures consisting of alternating
layers of semiconductor materials. The model we solved is the Kronig-Penney
model, based on Schrodinger's equation. Each layer of a superlattice is on
the order of 50 angstroms wide. Since the materials of the alternating layers
are different, an electron in such materials will see periodic variations in
the bandgap energy which may vary from layer to layer by several tenths of an
electron volt. The properties of the material, such as group velocity, electron effective mass, electron mobility, optical nonlinearity, etc.. are determined by the quantum mechanical behavior of electrons in this material. By
varying the semiconductors, their doping, and the width of the materials, one

408

holds forth the possibility of custom-engineering these materials to specifications.

For an fuller discussion of superlattices, [1] is recommended. A more technical description of our own physical results is contained in [2,3,4]. Reference [3] discusses the mathematical derivation described in this paper.

DESCRIPTION OF PROBLEM

The theoretically interesting aspect of superlattices concerns the motion of electrons through the layers, since in moving along the layers the electron sees a constant potential energy. In travelling through the layers, the electron encounters a series of square quantum potential wells.

For review, we recall Schrodinger's equation. (see [5])

$$\frac{d^2\psi}{dx^2} + \frac{2\mu(E-V)\psi}{\hbar^2} = 0 \tag{1}$$

This is the time-independent, one-dimensional equation with constant potential, V. It must be solved for the wavefunction ψ, and the energy E, which mathematically acts as an eigenvalue. Its solution is the following.

$$\psi(x) = A \exp(ikx)+B \exp(-ikx) \tag{2}$$

$$k=(2\mu(E-V))^{1/2}/\hbar$$

k is the wavenumber, and is physically related to the momentum of the electron. (1) and (2) are the simplest form of the Schrodinger equation with an exact solution, commonly discussed.

The next simplest equation with exact solution, is that for the square quantum well. It has the same form as (1), but V is no longer constant. Instead, $V=-V_0$ between $x=-L/2$ and $x=L/2$, while $V=0$ elsewhere. Recall that that equation has multiple solutions for the "bound" electron with discrete eigenvalue, E, and solutions for the "free" electron which can take on a continuous spectrum of energy values, E.

Our problem consists of a series of such square quantum wells. Consider a periodic potential V of period d, such that $V=-V_0$ for $(2n)d < x < (2n+1)d$, and $V=0$ for $(2n+1)d < x < (2n+2)d$. Hence, we have a series of quantum wells of depth V_0 and width a, with barriers between the wells of width b, and period $d=a+b$. The solution to the Schrodinger equation for this potential is given below.

$$\cos kd = \cosh 2k_a a \cos 2k_b b + (\varepsilon/2) \sinh 2k_a a \sin 2k_b \tag{3}$$

$$k_a = (2m^*(V_0 - E))^{1/2}/\hbar, \quad k_b = (2m^*E)^{1/2}/\hbar, \quad \varepsilon = k_a/k_b - k_b/k_a$$

The solution shares many of the features of the solution for the single square quantum well problem. However, the "bound" solutions are no longer restricted to a single energy value, E. Instead, there are bands of allowed values of E, with a solution, and intermediate "forbidden" regions for which values of E, there exists no solution. We are primarily concerned with the value of E and its derivatives as a function of k. This can be interpreted as an energy-momentum relation, and is usually called the dispersion relation.

Equation (3) is known as the Kronig-Penney model. (see [5]) k_a is the wavenumber for the layer making up the well, and k_b is the wavenumber for the layer making up the barrier. k is the overall wavenumber of an electron. Unfortunately, this provides an explicit solution for the momentum (or wavenumber, k) as a function of energy (E). It is numerically feasible to solve the implicit equation to find E as a function of k, but we also require the fourth derivative of the energy, E, with respect to k, since this has implications for the optical nonlinearity of the material. To proceed numerically would require very high numerical precision (higher than quadruple precision) to avoid roundoff errors and other numerical errors associated with fourth differences.

If we assume E small compared to V_0, a and b small, and noting $\cosh 2k_a a$, $\cos 2k_b b \rightarrow 1$ in the limit, one can approximate (3).

$$E = [(1 - \cos kd)/2abV_0] [\hbar^2/2m^*] = t(1 - \cos kd) \tag{4}$$

This solution is well-known in the literature, as the tight-binding formula, and t is the tight-binding constant. However, while the approximation is useful for many cases, we were concerned whether the formula was sufficiently accurate to justify taking a fourth derivative. Accordingly we looked for a more accurate approximation. We were fortunate enough to find such a series solution for which the tight-binding formula was the first term.

MATHEMATICAL SOLUTION

In the spirit of mathematical abstraction, we replace the right hand side of equation (3) by a general function of E, and write

$$\cos kd = f(E). \tag{5}$$

Then expanding $f(E)$ about $E=E_0$ corresponding to (k=0), we have

$$\cos kd = c_0 + c_1(E-E_0) + (c_2/2!)(E-E_0)^2 + \ldots \tag{6}$$

where $c_0=1$, and $c_k = (\partial^k f/\partial E^k)|_{E=E_0}$. Neglecting second order corrections and higher ($c_k = 0$, k>2), the tight binding form may be recovered,

$$E(k) = t(1 - \cos kd), \tag{7}$$

with $t=-1/c_1$. Using the full Taylor series (6), we obtain

$$E = E_0 + (\cos kd - c_0)/(c_1 + (c_2/2!)(E-E_0) + \ldots). \tag{8}$$

This is solved by the method of successive substitutions. The entire right hand side of (8) is substituted for E in equation (8). To third order, our iterative solution becomes

$$
\begin{aligned}
E = E_0 &+ (1/c_1)(\cos kd - c_0) \\
&- (c_2/2c_1^3)(\cos kd - c_0)^2 \\
&+ ((c_3 c_1 - 3c_2^2)/6c_1^5)(\cos kd - c_0)^3 + \ldots.
\end{aligned} \tag{9}
$$

We apply this when E_0 corresponds to k=0 (bottom of miniband). In this case, $c_0=1$.

Finally, in order to know where the extra terms of our series solution would be required, we looked for a criterion under which the validity of the tight-binding, or first order approximation, equation (7), is justified. Equation (3) may be looked at as the scalar product,

$$\cos kd = \mathbf{x} \cdot \mathbf{y} \tag{10}$$

with $\mathbf{x} = (\cos 2k_b b, \ \sin 2k_b b)$,
$\mathbf{y} = (\cosh 2k_a a, \ (\varepsilon/2)\sinh 2k_a a)$.

Note that $|\mathbf{x}|=1$ and $|\mathbf{y}|>1$ in (10) for all possible values of a, b, V_0, and E. If $k_a a \gg 1$ (decay length in barrier $\ll$ barrier width), then in fact $|\mathbf{y}| \gg 1$. Since $|\cos kd| \leq 1$, $\mathbf{x}$ and $\mathbf{y}$ must be nearly orthogonal. Since $\mathbf{x}$ and $\mathbf{y}$ rotate in opposite directions with increasing E, it follows that equation (3) can be satisfied only for a small range of E, over the entire miniband. The situation is illustrated in Fig. 1. Physically, these ranges of E for which the equation is satisfied correspond to the energy bands of a semiconductor.

Since equation (3) is satisfied for such a small range of E, one would expect a small variation of f(E) over the entire band. So, f(E) can be replaced by the zeroth and first order terms of its Taylor series expansion. This means replacing the right hand side of (3) with the zeroth and first order terms of its Taylor series is justified when $k_a a \gg 1$. Under these conditions, $\mathbf{x}$ and $\mathbf{y}$ are nearly perpendicular.

Strictly, replacing f(E) by the first two terms of its Taylor series can be justified only by comparing the magnitudes of successive terms. Nevertheless, the preceding argument can be made more precise. While the mathematical details are left out, the more precise argument leads to the conclusion that for $\hbar \ll 2(m^* V_0)^{1/2} b$, the tight-binding criterion will be valid when

$$[b^2 m^* V_0 / (\hbar^2)] \ \sinh^2(a(8m^* V_0)^{1/2}/\hbar) \gg 1. \tag{11}$$

In what follows, we use the second order approximation corresponding to (9). All coefficients of $(\cos kd - c_0)^n$, for $n \geq 3$, are set to zero. To this order, we have:

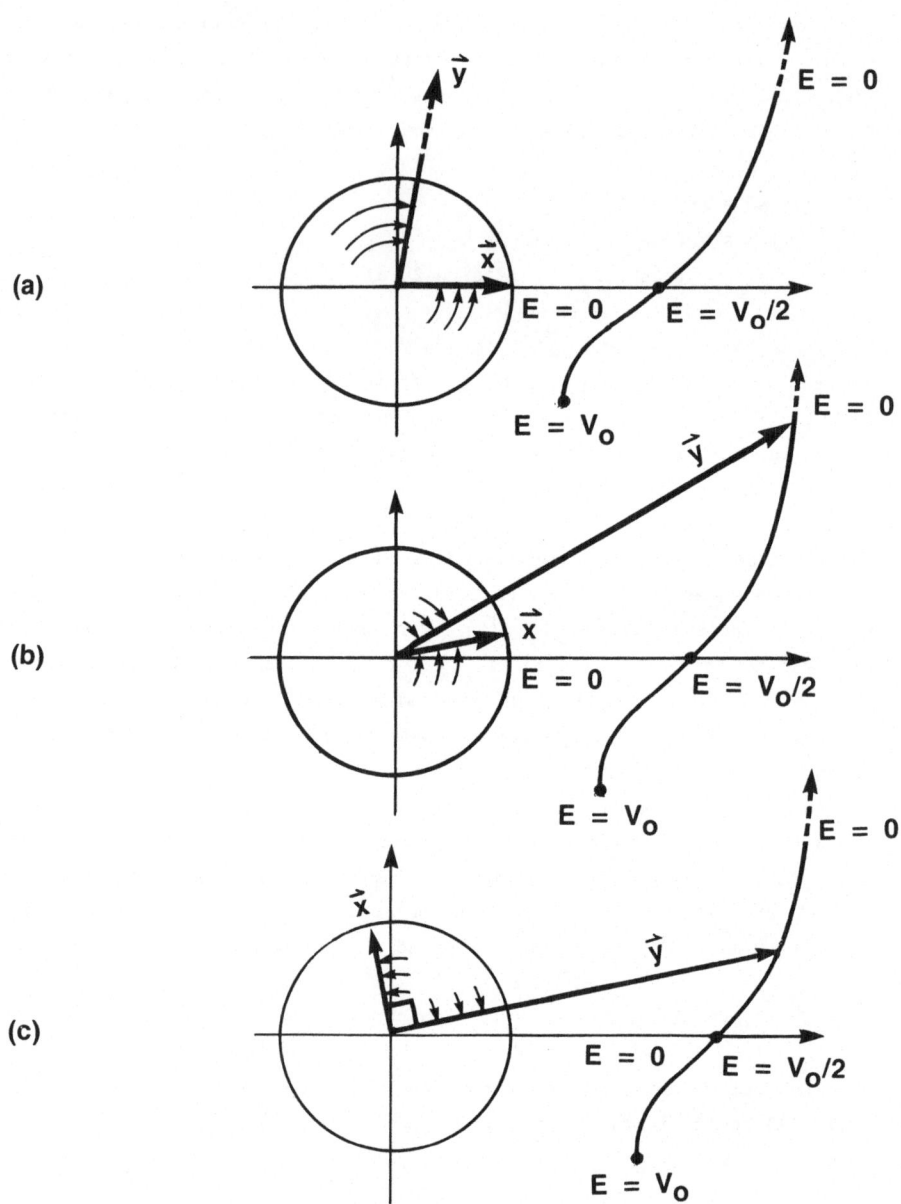

FIGURE 1:   GEOMETRIC INTERPRETATION OF KRONIG-PENNEY MODEL

$$E = E_0 + (1/c_1)(\cos kd - 1)[1 - (c_2/2c_1{}^2)(\cos kd - 1)] \tag{12a}$$

$$E' = -(1/c_1)\sin kd + (c_2/c_1{}^3)[-\sin kd + (1/2)\sin 2kd] \tag{12b}$$

$$E'' = -(1/c_1)\cos kd + (c_2/c_1{}^3)[-\cos kd + \cos 2kd] \tag{12c}$$

$$E''' = (1/c_1)\sin kd + (c_2/c_1{}^3)[\sin kd - 2\sin 2kd] \tag{12d}$$

$$E'''' = (1/c_1)\cos kd + (c_2/c_1{}^3)[\cos kd - 4\cos 2kd] \tag{12e}$$

where the primes on the left hand side denote differentiation with respect to kd.

Successive differentiations increase the magnitude of the correction term, as noted previously. Since we are interested in the fourth derivative, the correction terms are more important than for the dispersion (energy vs. wavenumber) relation. The relatively large size of the fourth derivative of the correction term also hints at the difficulties of numerical precision that one would encounter with a purely numerical approach in a traditional language such as FORTRAN.

USE OF COMPUTER ALGEBRA

We then used MACSYMA to check our derivation on the computer. While the original hand analysis had been carried out to order two, the power of computer algebra tempted us to "go on a fishing expedition." So, we not only checked our original derivation, but with a trivial modification, carried out our analysis to fourth order. Equation (13) was our not so elegant result.

$$E'''' = (1/c_1)\cos kd + (c_2/c_1{}^3)[\cos kd - 4\cos 2kd]$$
$$- ((c_1 c_3 - 3c_2{}^2)/8c_1{}^5)[5\cos kd - 32\cos 2kd + 27\cos 3kd]$$
$$+ ((c_1{}^2 c_4 - 10c_1 c_2 c_3 + 15c_2{}^3)/24c_1{}^7)[7\cos kd - 56\cos 2kd + 81\cos 3kd$$
$$- 32\cos 4kd] \tag{13}$$

Attempting to physically understand this solution, we evaluated the expression at the bottom of the band (kd=0). To our surprise, both the third and fourth terms reduced to zero.

Operating after the fact, we reexamined equation (9). Since all higher terms for E'''' correspond to fourth derivatives of $(\cos kd - c_0)^n$, $n \geq 3$, they clearly had to reduce to zero. Hence, our equation (12e) to second order,

414

which we had thought to be approximate, was exact for the physically important case, kd=0, and very accurate close to that value. Happily, the tight-binding approximation is given only to first order, and so this exact equation is a direct generalization of a previously known approximate solution.

CONCLUSION

We derived by hand the solution (12e) to equation (3) by approximate methods. We took advantage of the computer algebra package, MACSYMA, to do the same derivation to fourth order. With the insight provided, we discovered that the second order solution was exact for the physically important case, kd=0, and all higher terms were zero.

The problem illustrates an evolution in computer algebra which mirrors that of FORTRAN. Initially, the system is used to verify previous calculations, and do further calculations which would be tedious or impossible to do by hand. Such systems eventually become integral to the process of finding new insights. This finally gives rise to a qualitative rather than quantitative difference in the type of research which can be done.

REFERENCES

1. Dohler, Gottfried, "Solid-State Superlattices," Scientific American, 249(5), pp. 144-151, Nov., 1983.

2. W.L. Bloss and L. Friedman, Applied Physics Letters, Vol. 41(11), p. 1023, 1982.

3. G. Cooperman, L. Friedman, and W.L. Bloss, "Corrections to Enhanced Optical Nonlinearity of Superlattices," Applied Physics Letters, Vol. 44(10), pp. 977-979, May 15, 1984.

4. L. Friedman, W.L. Bloss, and G. Cooperman, "Enhanced Optical Nonlinearities of Superlattices within the Kronig-Penney Model Incorporating Inherent Bulk Nonlinearities," J. of Superlattices and Microstructures, to appear.

5. E. Merzbacher, Quantum Mechanics, Second Edition, John Wiley and Sons (1970), p. 100.

20

Computer Generation of Symbolic Generalized Inverses and Applications to Physics and Data Analysis

W.J. Frawley
Fundamental Research Laboratory,
GTE Laboratories, Waltham, MA 02254

Abstract. Problems in data analysis, electrical networks, and finite element methods often involve linear models having singular, square matrices or matrices which are not square. The concept of generalized inverse extends the ranges of application of the notion of matrix inverse to such matrices and provides powerful mathematical tools for handling them. The use of symbolic generalized inverses during the analyses of these problems is equally powerful but requires more manipulation than can be reasonably performed by a human analyst. In this paper, the symbolic calculation of the Moore-Penrose generalized inverse, based on Albert's limit formulation, will be expressed in Macsyma and examples of its use will be given. In particular, the computer derivation of a novel form of the generalized inverse of a covariance matrix typically used in statistical analysis will be shown.

Notation. For the purposes of this exposition, a matrix will be taken to be a rectangular array having n rows and s columns of real numbers. Matrices are made up of entries or elements indexed by their row-column positions, such as the $(2,1)$ element of the matrix A shown below.

$$A = \begin{bmatrix} 2 & 3 & -1 \\ -10 & -2 & 5 \end{bmatrix} \text{ has 2 rows and 3 columns.}$$

$$A_{2,1} = -10 = A[2,1].$$

Representing an n-vector by an n-by-1 (single column) matrix, allows a matrix to be viewed in terms of its column vectors, as in

$$A = [U_1|U_2|U_3] \text{ with}$$

$$U_1 = \begin{bmatrix} 2 \\ -10 \end{bmatrix}, U_2 = \begin{bmatrix} 3 \\ -2 \end{bmatrix}, \text{ and } U_3 = \begin{bmatrix} -1 \\ 5 \end{bmatrix}.$$

415

The transpose of a matrix is obtained by interchanging its rows and columns. Thus, A-tranpose is:

$$A^T = \begin{bmatrix} 2 & -10 \\ 3 & -2 \\ -1 & 5 \end{bmatrix} \ from \ which$$

$$A = \begin{bmatrix} V_1^T \\ \overline{V_2^T} \end{bmatrix} \ with \ V_1 = \begin{bmatrix} 2 \\ 3 \\ -1 \end{bmatrix} \ and \ V_2 = \begin{bmatrix} -10 \\ -2 \\ 5 \end{bmatrix}.$$

These vector-based forms are of value in treating matrices as linear transformations of vector spaces.

Matrix algebra consists of the usual operations of (1) scalar multiplication, (2) elementwise addition, and (3) row-by-column noncommutative multiplication. In terms of the scalar (real) x and the matrices $A = (a[i,j])$ and $B = (b[i,j])$, they are described by:

$$C = xB \ \ iff \ \ c[i,j] = x * b[i,j],$$
$$C = A + B \ \ iff \ \ c[i,j] = a[i,j] + b[i,j], and$$
$$C = A \cdot B \ \ iff \ \ c[i,j] = \sum_{k=1}^{s} a[i,k] * b[k,j].$$

In particular, the usual dot product of two (column) vectors, e.g.

$$DOT(A, B) = A_3 B_3 + A_2 B_2 + A_1 B_1 \ for \ 3-vectors,$$

is expressible as a matrix product, viz. $DOT(U,V) := TRANSPOSE(U) \cdot V$, and the product of two matrices can be expressed in terms of the dot product,

$$C = A \cdot B \ iff \ c[i,j] = dot(transpose(row(A,i)), column(B,j)).$$

Finally, the length of quadratic norm of a vector V is defined as the square root of the sum of the squares of its components:

$$\|V\|^2 = DOT(V,V).$$

Nonsingular Matrices and Their Look-Alikes. The identity matrix, I, of size n-by-n, having ones on the diagonal (that is, those elements whose row and column indices are identical) and zeroes elsewhere, is the only matrix, X, for which $A \cdot X = X$. $A = A$ for all square matrices A having side n. Since reducing a given matrix to an identity by multiplication by an appropriate factor can simplify a form under analysis, the computation of such appropriate multipliers, called matrix inverses, has received much attention. A square matrix A has the matrix B as its inverse, displayed as

$$B = A^{-1}, \; if \; B \cdot A = I = A \cdot B;$$

the matrix A is nonsingular if such a matrix inverse exists. Note that this is an algebraic, rather than a constructive, definition. The recognition of those matrices not having inverses and the hand calculation of the inverses of small symbolic matrices, such as

$$\begin{bmatrix} 0 & 1 \\ 0 & 0 \end{bmatrix}, \quad \begin{bmatrix} x & \dfrac{1}{x+1} \\ 1 & \dfrac{2}{x(x+1)} \end{bmatrix}, \; and \; \begin{bmatrix} sin(Y) & cos(Y) \\ tan(Y) & 1 \end{bmatrix},$$

is usually achieved with pencil and paper by determinantal and row-reduction methods, along with the use of simplifying identities and substitutions. The first of the above three matrices clearly has zero determinant, adduced "by inspection", and hence has no inverse. A user of mathematics would calculate quite quickly that that the second matrix has determinant $1/(X+1)$. But s/he will also have the "mathematical" realization that the matrix "blows up" at $X = 0$ and $X = -1$. (Facts which, by the way, numerical routines come upon only by trying and failing; there is no notion in Pascal, for example, that $1/(X+1)$ has a singularity at $X = -1$.) Finally, the first row in the third matrix is equal to the second row multiplied by $cos(Y)$, hence is "rank deficient"; its determinant, $sin(Y) - cos(Y)tan(Y)$, is identically zero.

For calculations of any scale, the user of mathematics turns to the digital computer with its packages of programs for solving known problems using efficient numerical techniques quite different from those of the hand solver. What happens when the user presents to numerical routines a problem the package cannot solve? Will it be recognized? As a concrete case, consider the third matrix above, which has no inverse whatever value of Y is chosen. Yet, after substituting 0.6 for Y, one finds that

$$\begin{bmatrix} 0.5646425 & 0.8253356 \\ 0.6841368 & 1 \end{bmatrix} has \begin{bmatrix} 3.2319382e7 & -2.6674336e7 \\ -2.211088e7 & 1.8248896e7 \end{bmatrix}$$

computed incorrectly as its inverse. The entries in the inverse are meaningless except for the fact that their exponents indicate (correctly) that the machine used for the computation employed between 7 and 8 places of accuracy in its floating point implementation. Running the same algorithm to greater accuracy will merely increase the exponents. When the two matrices above are multiplied together, they do not produce a 2-by-2 identity matrix. Rather, depending on the order of multiplication, the products are

$$\begin{bmatrix} 0.0 & 1.0 \\ -2.0 & 0.0 \end{bmatrix} and \begin{bmatrix} 0.0 & 2.0 \\ -1.0 & 0.0 \end{bmatrix}$$

These neat integral entries represent units of roundoff error in the least significant place.

The example above is due neither to a poor algorithm nor to insufficient accuracy. Rather, the user has asked a matrix inversion routine to invert what symbolically is a rank-deficient square matrix, one not having an inverse, whose numerical form, due to roundoff error, appears to be nonsingular. It is a "look-alike" of an invertible matrix. This is a simple case of what happens with regularity in scientific calculation: the scientist, with paper and pencil, generates a complex mathematical form, one that in its generality is too unwieldly for complete analysis, and turns prematurely to fixed computer routines to provide insight via numbers, charts, and graphs. The computer routines, thus abused, produce consistent but meaningless results, some of which are taken as true and others which lead the scientist to damn the software. What is needed is the use of powerful symbolic tools to explore the mathematical properties of the scientists' expressions before they are reduced to calculation.

Generalized Inverses of Rectangular Matrices. When a square matrix A has the inverse B, the question "given the vector b, what vector x satisfies $A \cdot x = b$?" has one and only one answer, namely, $x = B \cdot b$. Even when a matrix has no inverse, determining the input x which gives rise to output $b = H \cdot x$ is, in practice, an important activity which led to the discovery in 1920 [refs. 5 and 7] and the rediscovery in 1955 [ref. 6] of the generalized inverse of a matrix. Starting with the rectangular matrix H, a new matrix H^+ is sought to handle the three cases of the broadened question as to what input produces a given output: (1) if exactly one x_0 satisfies $b = H \cdot x$, then $H^+ \cdot b = x_0$; (2) if many (a linear manifold of) vectors, x, map into b under H, then $H^+ \cdot b$ is the "best" one of these in the least squares sense; and (3) if no x maps into b, $H^+ \cdot b$ is the shortest vector x_0 among those vectors x which minimize $b - H \cdot x$, again, in the least squares sense. The main theoretical result follows [ref. 1].

For any real, rectangular matrix, H, having n rows and s columns, there exists a Moore-Penrose generalized inverse matrix, H^+, of s rows and n columns, uniquely defined by three equivalent conditions:

1. For every Y, $X_0 = H^+ \cdot Y$ is the least norm minimizer of

$$\|Y - H \cdot X\|^2;$$

2. H^+ is the unique matrix which satisfies

    (*i*) $H \cdot H^+ \cdot H = H$,

    (*ii*) $H^+ \cdot H \cdot H^+ = H^+$, *and*

    (*iii*) *both* HH^+ *and* H^+H *are symmetric*;

3. H^+ is defined by the limit formula (attributed to ref. 2):

$$H^+ = \lim_{x \to 0} ((H^T \cdot H + x^2 I)^{-1} \cdot H^T)$$

Note that the first of these conditions specifies a minimization criterion and the second, algebraic criteria. In a sense, the "minimization" condition gives meaning to the notion of generalized inverse. If H is invertible, the "fitting error" is reduced to zero. When H has, for example, linearly independent columns, as can happen in overdetermined data analysis problems, the generalized inverse reduces to the solution of the "normal equations" familiar in least-squares fitting. That is,

$$H^+ = (H^T\,H)^{-1}H^T.$$

The algebraic conditions provide a test: if a matrix satisfies $(i), (ii)$, and (iii), then it must be the generalized inverse. Only the last condidtion above provides a computational definition, but one which is not numeric in nature, since the symbolic parameter x persists throughout a standard matrix inversion.

There is a hint of magic in this, getting a purely algebraic result from an analytic process. The theoretical roots of computing a generalized inverse using the limiting form of a matrix inverse lie deeper, and are simpler, than matrix theory. Surprisingly, the value of

$$\lim_{\delta \to 0} \frac{x}{x+\delta} \; is \; IF \; x = 0 \; THEN \; 0 \; ELSE \; 1.$$

Since, for real, symmetric matrices, the important projection theorems are based on arguments using their real eigenvalues, which are merely numbers, and since the generalized inverse can be described in terms of projections and singular values, these latter being numbers, too, the apparent mystery in the generalized inverse computation above is due solely to this simple limit.

Computing the Generalized Inverse. Numerical routines for computing the generalized inverse face the same problems, cast in terms of eigenvalues, as do the traditional matrix inverse routines, whose possibility of failure was illustrated above. When a rectangular matrix has one or more singular values which are zero, that is, when the square matrix $H^T \cdot H$ has one or more zero eigenvalues, the routine being used must, in effect, compute precise zeroes. Worse yet, the error in the calculated generalized inverse is inversely proportional to the nonzero value attributed to what is a zero eigenvalue.

Starting with H, be it symbolic or numeric, all the steps prescribed by the limit formula can be carried out by hand, though for matrices of size greater than three-by-three the process is both time consuming and subject to error. First, one must left multiply H by its transpose, add to that product a diagonal matrix with the square of X on the diagonal, invert the sum, multiply that inverse by the transpose, and take the elementwise limit of that matrix as the variable X approaches 0. In the MACSYMA symbolic mathematics program [ref. 4], this procedure is defined succinctly by the statement

```
geninv(h):=
        limit(ratsimp
                ((diagmatrix(length(transpose(h)),x^2)+
                    (transpose(h).h))^^(-1)  .  transpose(h)),
            x,0);
```

Only a few terms need explaining. RATSIMP causes a rational simplification of all of the terms in the matrix to be carried out. (In hand calculation, forms are constantly being re-expressed and simplified.) The LENGTH of a matrix is the number of its rows, so that the LENGTH of the TRANSPOSE of a matrix is the number of columns it has. (This is needed to say what size of diagonal matrix is to be used.) X^2 is the scalar X squared, $A \cdot B$ represents a product of matrices, and M^^N raises the matrix M to the Nth power.

Having defined the function GENINV in MACSYMA, it is instructive to examine the results for some simple cases. Applying GENINV to an undefined symbol, R, yields

$$\lim_{x \to 0}\left[\frac{1}{x^2 + transpose(r) \cdot r}\right] \cdot transpose(r),$$

which allows no further simplification. A bit of study of this formula shows that any number except 0 has its own reciprocal as its generalized inverse, while the generalized inverse of 0 is 0. (Note, though, that the number N as argument to GENINV must be expressed as a 1-by-1 matrix, that is, $[N]$.) In order to preserve precision, MACSYMA deals with rational numbers, just as a user of mathematics would use 1/3 in hand calculations rather than some typical but inaccurate approximation, such as .333333. For example, GENINV([56.78]) is exactly

$$\frac{36617965722441}{2079168093720200}$$

which converts to 0.017611835 in floating point. Another example of exact rational arithmetic is that the generalized inverse of the TEST matrix

$$\begin{bmatrix} 1 & 2 \\ 3 & 4 \\ 5 & 6 \end{bmatrix} \quad is \; GENINVTEST \; = \; \begin{bmatrix} -\dfrac{4}{3} & -\dfrac{1}{3} & \dfrac{2}{3} \\ \dfrac{13}{12} & \dfrac{1}{3} & -\dfrac{5}{12} \end{bmatrix}.$$

Finally, as a check on the correctness of these results, one can verify using MACSYMA that

$$TEST \cdot GENINVTEST \cdot TEST - TEST = \begin{bmatrix} 0 & 0 \\ 0 & 0 \\ 0 & 0 \end{bmatrix} \quad and$$

$$GENINVTEST \cdot TEST \cdot GENINVTEST - GENINVTEST = \begin{bmatrix} 0 & 0 & 0 \\ 0 & 0 & 0 \end{bmatrix}.$$

The Generalized Inverse of Cross Product Operation. Consider now an elementary example of the type that presents itself in rotational mechanics problems. Often in matrix differential equations of motion, the cross product of an angular frequency and a position vector is represented as a linear operator, that is, a matrix, acting on the $[x, y, z]$ column vector. Note that multiplying a vector $[x, y, z]$ by the antisymmetrical matrix

$$P = \begin{bmatrix} 0 & -c & b \\ c & 0 & -a \\ -b & a & 0 \end{bmatrix}.$$

is equivalent to taking the cross product of the column vectors $[a, b, c]$ and $[x, y, z]$. Since a vector crossed into itself yields the zero vector, the matrix P times the vector $[a, b, c]$ is zero, whence P has no inverse, even though it is square. (According to MACSYMA, RANK(P) is 2.)

What matrix GP "undoes" the rotation by $[a, b, c]$ represented by P? The GENINV function provides a clear-cut symbolic answer:

(c30) gp : geninv (p) ;

$$(d30) \quad \begin{bmatrix} 0 & \dfrac{c}{c^2 + b^2 + a^2} & -\dfrac{b}{c^2 + b^2 + a^2} \\ \dfrac{c}{c^2 + b^2 + a^2} & 0 & \dfrac{a}{c^2 + b^2 + a^2} \\ \dfrac{b}{c^2 + b^2 + a^2} & -\dfrac{a}{c^2 + b^2 + a^2} & 0 \end{bmatrix}.$$

GP is a scaled rotation "the other way"; it is proportional to $-P$ with the square of the length of $[a, b, c]$ as the factor.

This particular result can be obtained by careful, geometric reasoning and need not rely on such sophisticated matrix machinery. But for higher dimensions the problem increases in complexity. Suppose an antisymmetric n-by-n matrix is constructed from a vector having $n(n-1)/2$ components by distributing the vector components, much as was done in creating the 3-by-3 rotation matrix. Viewing this matrix as a geometric linear operator, what matrix reverses its operation? Does this generalized inverse retain the same form? For n greater than 3, mechanized symbolic tools are needed to handle even the simple cases whose answers light the path to the general answer.

Generalized Inverse of Multinomial Covariance Matrix. Any n-vector of positive entries which sum to 1,

$$p = column(p_i, \cdots, p_n),$$

can be regarded as a probability vector determining a multinomial probability distribution. The variance-covariance matrix for estimating the elements of p from an N-sample experiment is V given by

$$V_{ij} = N \cdot (p_i \cdot \delta_{ij} - p_i \cdot p_j),$$

which, ignoring the factor N, has the form

$$V = D(p) - p \cdot p*,$$

$D(p)$ being the diagonal matrix, $p_i \cdot \delta_{ij}$. The importance of V stems from the fact that covariance matrices and their inverses (or generalized inverses, as the case may be) provide the key links between least squares and best unbiased linear estimation. In particular, such an estimate of x from $y = T \cdot x + r$, where r is a zero-mean random vector whose covariance is Q, is

$$\hat{x} = [(T * Q^+ \cdot T)^+ \cdot T * Q^+] \cdot y.$$

Taking u to be an n-vector of ones, $p * u = 1 = u * p$ denotes that the entries of p sum to 1, from which it follows that V is singular with null space spanned by u. Hence, its generalized inverse is required for use in statistical analysis.

When $Q = V$, the multinomial case, various approximations are brought to bear, the simplest being replacing V with $D(p)$, whose (generalized) inverse is the diagonal matrix formed from the reciprocals of the p_i. To the author's knowledge, only recently has the exact generalized inverse been described [ref. 3] and that was done with the help of GENINV using MACSYMA. Starting with the formula for the entries in V,

$$f_{i,j} := p_i(-p_j + (IF \ i = j \ THEN \ 1 \ ELSE \ 0))$$

the following command generates an n-by-n matrix, with $n = 2$ chosen here in order to conserve space,

  v : genmatrix (f, 2, 2);

$$\begin{bmatrix} (1 - p_1)p_1 & -p_1 p_2 \\ -p_1 p_2 & (1 - p_2)p_2 \end{bmatrix}$$

whose determinant, invoked by "factor(determinant(v));",

$$-p_1 p_2 (p_2 + p_1 - 1),$$

tells an important part of the story, that, for 2-by-2, the determinant is zero in the precise case of interest, that is, when the components of the vector p sum to 1. Analysis verifies that this is the case for all sizes of V. Note, though, that unless explicitly constrained to be 1, this symbolic determinant,

$$\Pi p_i * (1 - \sum p_k),$$

does not reduce to zero, so that $V\hat{\ }\hat{\ }(-1)$, the apparent inverse of V, can be computed for various small sizes, leading to the provable guess that

$$V_{i,j}^{-1} = \frac{\delta_{i,j}}{p_i} + \frac{1}{1 - \sum p_k}$$

which shows clearly what would happen if an attempt were made to invert V numerically. Error having the order of the reciprocal of the inaccuracy in computing the difference between 1 and the sum of $P's$ elements would be spread uniformly over the matrix. Finally, the exact generalized inverse, given by GENINV(V), is examined.

$$V^+ = \begin{bmatrix} \dfrac{1}{4p_1 p_2} & -\dfrac{1}{4p_1 p_2} \\ -\dfrac{1}{4p_1 p_2} & \dfrac{1}{4p_1 p_2} \end{bmatrix}$$

Using various small orders of V, suspected "ingredients" of its generalized inverse are subtracted out and the resultant expressions subjected to further scrutiny. For example, the 3-by-3 generalized inverse has a 9 in the denominator of each entry; it is easier to deal with $n^2 V^+$. These probes guide the analyst's efforts before s/he gets down to testing a formula against the algebraic criteria. The final form, true for all values of n, is

$$(V^+)_{ij} = \frac{\sigma_{ij}}{p_i} - \frac{1}{n}\left(\frac{1}{p_i} + \frac{1}{p_j}\right) + \frac{1}{n^2} \cdot \sum_{i=1}^{n} \frac{1}{p_i}$$

Conclusion. What makes the generalized inverse a powerful tool is that it exists for all matrices; hence, it applies to the analysis of all linear systems. Unfortunately, the insight typically provided by hand calculation of highly illustrative special cases is unavailable since the algebraic criteria for generalized inverses are not constructive and since computer programs which do construct these matrices are strictly numerical. The limit formulation of the generalized inverse highlighted in this paper defines a straightforward procedure for computing the generalized inverse of matrices that retain the variables of the theory under study, but carries with it the drawback that the user must be able to invert certain symbolic matrices. To invert symbolic matrices reliably and without undue effort requires mechanized assistance. Along with its matrix computer algebra capabilities, the MACSYMA symbolic manipulation program can take limits and perform rational simplifications. Integration of these facilities leads to a simple function, GENINV, which computes the generalized inverse of its symbolic or numeric matrix arguments. Two applications of this function in probing the form taken by the generalized inverse of matrices have been discussed as a way of displaying the power inherent in computer-based symbolic manipulation.

Bibliography

1. Albert, A.E., Regression and the Moore-Penrose Psuedoinverse, Academic Press, New York, 1972.

2. den Broeder, G.G., and Charnes, A., Contributions to the Theory of Generalized Inverses for Matrices, ONR Research Memo No. 39, Northwestern University, 1962.

3. Frawley, W.J., "The Generalized Inverse of the Covariance Matrix of the Multinomial Distribution," GTE Laboratories Technical Note 85-050.1.

4. Mathlab Group, MACSYMA Reference Manual, MIT, 1977.

5. Moore, E.H., (Abstract only), Bull. Am. Math. Soc. **26**, 394-5, 1920.

6. Penrose, R., "A Generalized Inverse for Matrices," Proc. Cambridge Philos. Soc. **51** 406-413, 1955.

7. Reid, W.T., "Generalized Green's Matrices for Compatible Systems of Differential Equations," Am. Journ. Math. **53** 443-459, 1931.

INDEX

429